MAGNETISM, PLANETARY ROTATION, AND CONVECTION IN THE SOLAR SYSTEM: RETROSPECT AND PROSPECT

In Honour of Prof. S.K. Runcorn

Edited by W. O'REILLY

The University of Newcastle-upon-Tyne

Reprinted from Geophysical Surveys, Vol. 7, Nos. 1-3 (1984-5)

D. Reidel Publishing Company / Dordrecht / Boston

Library of Congress Cataloging in Publication Data

Main entry under title:

Magnetism, planetary rotation, and convection in the solar system.

"Reprinted from Geophysical surveys, vol. 7, nos. 1-2-3 (1985)."
1. Magnetism, Terrestrial—Congresses. 2. Planets, Theory of—Congresses.
3. Convection (Astrophysics)—Congresses. 4. Paleomagnetism—Congresses.
5. Runcorn, S. K. I. Runcorn, S. K. II. O'Reilly, W.
QC811.M34   1985   538'.7   85-18371

ISBN-13:978-94-010-8886-2   e-ISBN-13:978-94-009-5404-5
DOI:10.1007/978-94-009-5404-5

Published by D. Reidel Publishing Company,
P.O. Box 17, 3300 AA Dordrecht, Holland.

Sold and distributed in the U.S.A. and Canada
by Kluwer Academic Publishers,
190 Old Derby Street, Hingham, MA 02043, U.S.A.

In all other countries, sold and distributed
by Kluwer Academic Publishers Group,
P.O. Box 322, 3300 AH Dordrecht, Holland.

TABLE OF CONTENTS

MAGNETISM, PLANETARY ROTATION, AND CONVECTION IN THE SOLAR SYSTEM: RETROSPECT AND PROSPECT

In Honour of Prof. S. K. Runcorn

Edited by

W. O'REILLY

TABLE OF CONTENTS

PREFACE

On the 6th, 7th and 8th April 1983, a conference entitled "Magnetism, planetary rotation and convection in the Solar System" was held in the School of Physics at the University of Newcastle upon Tyne. The purpose of the meeting was to celebrate the 60th birthday of Prof. Stanley Keith Runcorn and his, and his students' and associates', several decades of scientific achievement. The social programme, which consisted of excursions in Northumberland and Durham with visits to ancient castles and churches, to Hexham Abbey and Durham Cathedral, and dinners in Newcastle and Durham, was greatly enjoyed by those attending the meeting and by their guests. The success of the scientific programme can be judged by this special edition of *Geophysical Surveys* which is derived mainly from the papers given at the meeting.

The story starts in the late 1940s when the question of the origin of the magnetic field of the Earth and such other heavenly bodies as had at that time been discovered as having a magnetic field, was exercising the minds of several scientists, notably P. M. S. Blackett at Manchester, W. M. Elsasser at the University of Pennsylvania and E. C. Bullard at Cambridge. Two alternative mechanisms were proposed. In one the magnetic field was in some way connected with the distributed angular momentum of a rotating body; in the other, electric currents in conducting parts within the body were proposed as the source of magnetic field. The attempts to resolve the problem led, in one way or another, to much fruitful scientific activity (Figure 1), contributing to discoveries made about our planet, from its core to its atmosphere, and beyond to our neighbouring Moon. The present brief account of developments in geophysics over the past few decades is *not* intended to be comprehensive, nor would it be possible to mention each of the numerous gifted scientists who have played their part in building up our picture of the Earth. The emphasis here, in this preface to a celebratory issue of the journal, is only on the work carried out within the ambit of Keith Runcorn.

Returning now to Manchester forty years or so ago, the need to test the angular momentum hypothesis, which Blackett had revived, resulted in the famous 'negative experiment' for which Blackett constructed an astatic magnetometer of unprecedented sensitivity to measure the weak fields expected from a metal sphere in the laboratory. The availability of this magnetometer was an important factor in the development of palaeomagnetic studies in Britain, and indeed the astatic magnetometer remained the bread-and-butter tool of British palaeomagnetism for two decades before being generally superceded by the improved spinner magnetometer. The coal mine measurements of Runcorn, D. H. Griffiths, A. F. Moore, and A. C. Benson provided an observational test of the angular momentum hypothesis. The negative results here, together with those of Blackett's laboratory experiment, effectively brought this line of enquiry to an end (Figure 1).

Meanwhile, the contribution of electric currents in the core of the Earth to the field was the subject of a farewell lecture given by Bullard who had resigned his post at Cambridge. This was to inspire the laboratory model work of F. J. Lowes at

W. O'REILLY

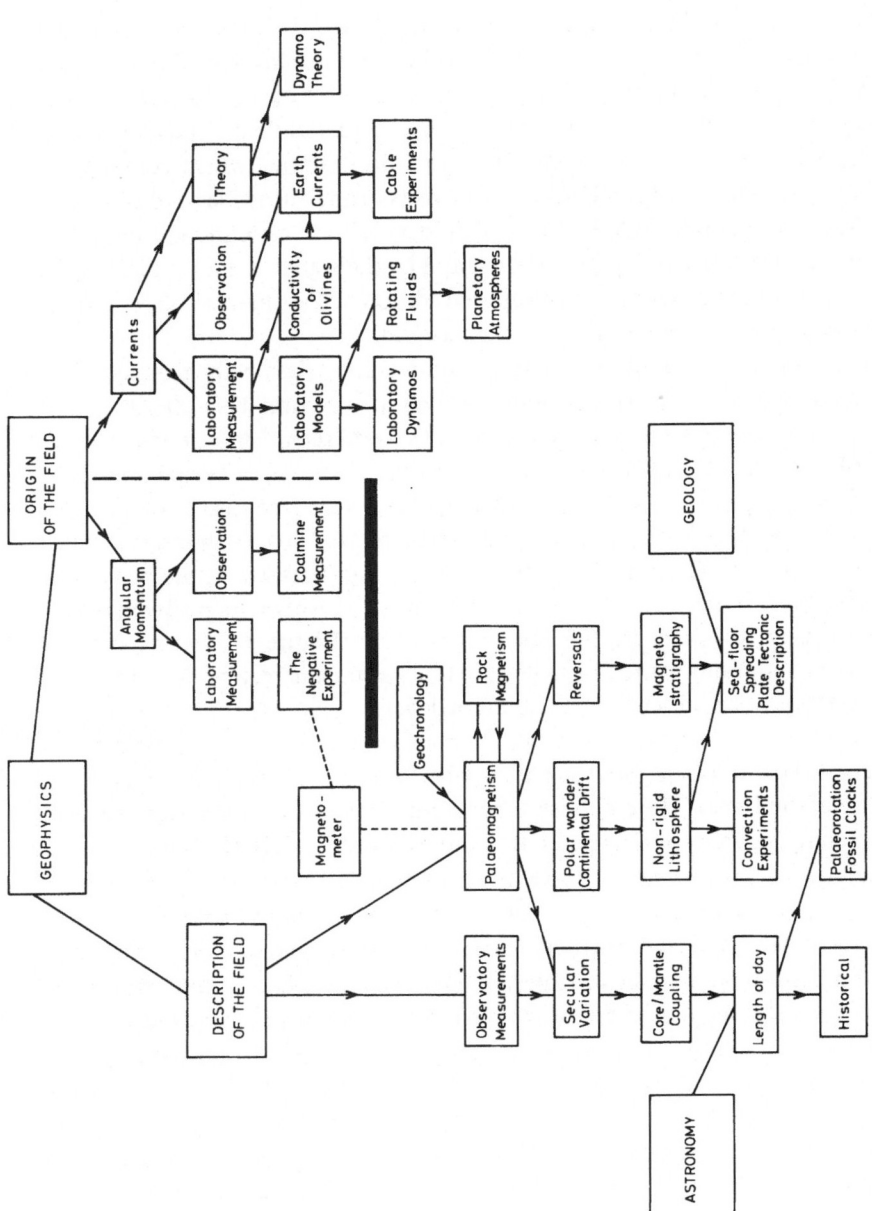

Fig. 1. Attempts to understand the source of the magnetic fields of the Earth and other heavenly bodies, and to obtain a description of the geomagnetic field over geological time, were responsible for the making of many discoveries in geophysics. Tests of the angular momentum hypothesis, despite producing only negative results in both observational and laboratory experiments, bore fruit in sparking off the study of palaeomagnetism by British scientists. The ideas of currents in the conducting core as a source of the field led to a wide variety of theoretical and experimental work, to laboratory models of the core, mantle and atmosphere and the construction of laboratory homogeneous dynamos.

Manchester and the theoretical work of A. Herzenberg which led, some years later, to the publication of his 'existence theorem' for a self-exciting homogeneous dynamo.

In 1951 Runcorn moved to Cambridge, appointed to the post vacated by Bullard. Palaeomagnetic work, aimed at providing a description of the geomagnetic field over geological time, expanded with the arrival of E. Irving, K. M. Creer, J. Hospers from the Netherlands and D. W. Collinson, later to be joined by J. C. Belshé and P. du Bois from North America and A. E. M. Nairn. Apart from the existence of reversals in the ancient geomagnetic field, palaeomagnetism was also to provide evidence for wander of the magnetic poles and the close association of magnetic and spin poles. This required the lithosphere to be deformable because of the non-spherical surface of the Earth.

The description of the field in historical time, in the form of observatory data, was analysed by A. F. Moore who found what might now be called 'jerks' in the field, and raised the possibility of differential core/mantle rotation rates as a source of the secular variation of the field. The quest for the source of the field continued with the search for currents leaking out of the conducting core and passing through the mantle and crust. P. H. Roberts and Lowes made theoretical calculations of the magnitude of such currents and Lowes attempted to observe them. The magnitude of the Earth 'currents' depends on the variation of mantle resistivity with radius. The question of to what extent the secular variation of the field originating in the core would be screened at the surface of the Earth by the intervening material also required a model for the resistivity field within the Earth which would, in turn, depend on the composition, temperature and pressure fields. These questions led to laboratory measurements of the resistivity of olivines by H. Hughes and of the band gap in olivines by Runcorn. D. C. Tozer carried out experimental studies of the effect of pressure on the band gap of semiconductors. An interest in the magnetic properties of the minerals and rocks of the crust grew out of the palaeomagnetic work, and J. H. Parry began the development of the laboratory instruments needed for such studies.

A further expansion in laboratory geophysics took place with R. Hide's experiments on rotating fluids. Perhaps there had been some intention of throwing light on the fluid dynamics of the core and contributing to the picture of the geodynamo but, in the event, the experiments were to become important in the understanding of planetary atmospheres. Magnetohydrodynamic experiments aimed more directly at the core of the Earth were at this time being carried out in the Engineering Laboratories by E. R. Niblett.

In 1955 Runcorn was appointed to the senior chair in the Department of Physics at King's College, Newcastle upon Tyne, then the Newcastle wing of the University of Durham. He was joined there by Lowes, Hide, Creer, Collinson, Roberts, Tozer, Parry, and Nairn forming the nucleus of what was to become, about ten years later, the Department of Geophysics and Planetary Physics of the University of Newcastle upon Tyne. Palaeomagnetism was a major activity, of course, with the construction of the polar wandering curves for North and South America and Europe providing quantitative evidence of continental drift. K. M. Creer was especially interested in the need to establish a firm physical foundation for the palaeomagnetic method, and this led to the growth of the study of rock magnetism at Newcastle.

The source of the geomagnetic field was studied via the laboratory realization by Lowes and I. Wilkinson of the Herzenberg dynamo, and in terms of dynamo theory by Roberts. Tozer extended the work on the electrical properties of olivines, growing synthetic single crystals in collaboration with J. Wilson and determining their resistivity at elevated temperatures. The need to know the temperature regime within the Earth was later to lead Tozer to carry out experiments modelling convection in the mantle, and to develop a wider interest in the thermal evolution of planets. R. J. G. Strens determined high pressure optical properties of mantle minerals.

The rotating fluids work of Hide, and C. W. Titman, continued with work on the formation of Taylor columns over obstacles and holes, and the interesting speculation emerged as to whether this mechanism might account for the Great Red Spot of Jupiter. In addition to rotating systems, modelling the atmosphere and oceans, the geophysical fluid dynamics research group also carried out convection experiments, e.g. those in a fluid with internal heat generation (D. J. Tritton and M. N. Zarraga), pertinent to the interiors of planets.

The interest in the rotation of the Earth, which had been inspired in this context by the secular variation in the geomagnetic field, was taken up by F. R. Stephenson who, in order to decipher ancient documents recording astronomical information, had first to learn Chinese. A further approach, and one which held the possibility of determining the variation in the rate of rotation over even longer times, was the connection between the Earth's rotation and the biological world. Thus began the attempt at Newcastle to extract geophysical information from the growth rhythms in the skeletal structures of invertebrates – the palaeontological clock. Papers by W. W. Hughes and G. D. Rosenberg appear in this special issue. Other participants in the research programme were D. J. Barnes, J. W. Dolman, and C. B. Jones.

The rate of rotation of the Earth is affected also by external influences such as the tidal loss in the gravity field of the Moon and Sun. This prompted an interest in the Moon as a neighbouring terrestrial-type planet (R. G. Hipkin and Runcorn) which was greatly increased when direct involvement in experiments on the Moon, and laboratory studies of lunar samples, were made possible by the Apollo programme. The data from magnetometers, seismometers and gravity determinations were discussed with great excitement at Newcastle. P. C. Sellers was involved in the analysis of Lunar seismic data. The results of magnetic studies of the Lunar samples (by Collinson and A. Stephenson) are contained in this present volume. Collinson was later a participant in the Mariner programme.

Terrestrial palaeomagnetism continued with studies of secular variation using lake sediments (Creer and R. Thompson). D. H. Tarling used palaeomagnetism to study the evolution of the crust and also in archaeological problems. J. G. Mitchell developed a K-Ar geochronology laboratory to date palaeomagnetic material and attack geochemical problems. The study of recent crustal processes, especially rifting, was pursued by R. W. Girdler using geophysical field measurements.

The search for steady Earth currents in the mantle was resumed when obsolescent telegraph cables in the Pacific became available as long leads by which to measure the potential difference between widely separated points on the surface of the Earth (L. Molyneux, M. L. Richards, and Mrs. M. R. Strens).

F. J. Lowes analysed the spatial and temporal variations of the geomagnetic field, and rock and mineral magnetism developed into a subject in its own right (Stephenson, Collinson, and W. O'Reilly).

The message which seems to emerge from this rather cursory catalogue of names (many of whom appear as faces in Figure 2) and activities, is that, although it may be possible in pure or applied laboratory science to specialize in a particular field with an indifference to other fields within the same broad area without decreasing either intellectual satisfaction or the progress of knowledge, this is not so in geophysics. The Earth is an interconnected whole and does not allow it. Although this present version of a geophysical history (Figure 1) emphasises the central role played by the quest for the origin and description of the geomagnetic field, other quests, if embraced at the beginning of this fruitful post-war period of scientific research, might have equally played such a role. Perhaps in the same history, written by other writers, they do.

W. O'REILLY

Fig. 2.   The participants.

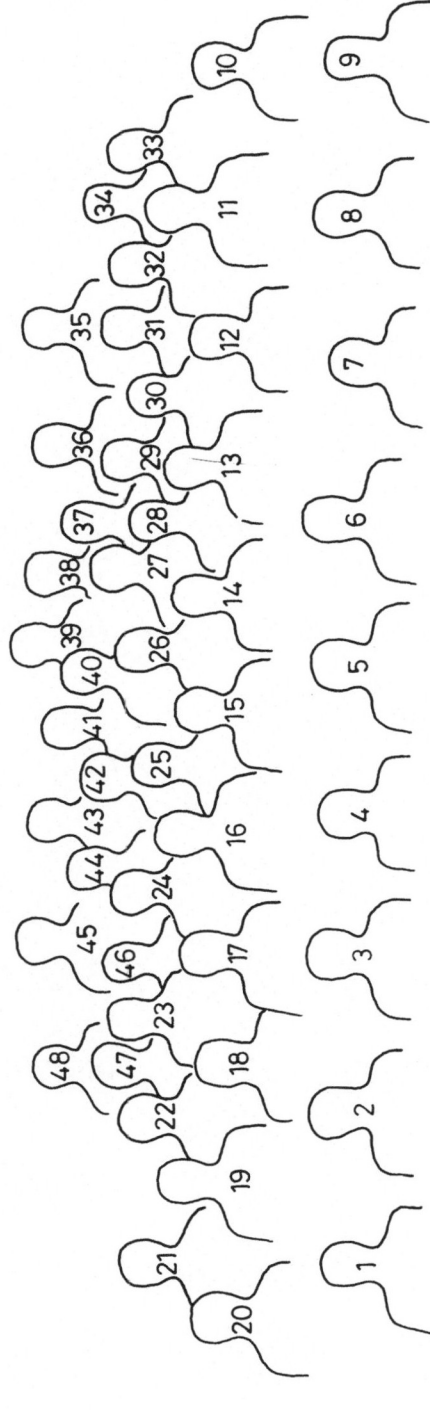

Key to photograph

(1) E. Irving. (2) R. Hide. (3) D. H. Griffiths. (4) K. M. Creer. (5) S. K. Runcorn. (6) J. Hospers. (7) F. J. Lowes. (8) A. F. Moore. (9) D. W. Collinson. (10) J. Palmer. (11) R. Thompson. (12) R. W. Girdler. (13) J. C. Belshé. (14) M. D. Fuller. (15) J. A. Jacobs. (16) F. R. Stephenson. (17) L. V. Morrison. (18) J. W. Dolman. (19) A. Stephenson. (20) I. Wilkinson. (21) Not at Conference. ??? (22) E. A. Hailwood. (23) W. Bainbridge. (24) L. R. Pattinson. (25) D. V. Crossley. (26) D. Facey. (27) S. Chowdhary. (28) J. H. Parry. (29) P. Johnson. (30) M. Gross. (31) C. Snape. (32) P. C. W. Davies. (33) L. Molyneux. (34) W. O'Reilly. (35) W. W. Hughes. (36) G. D. Rosenberg. (37) M. Maslanyi. (38) Not at Conference. ??? (39) A. Rice. (40) D. Potter. (41) D. J. Tritton. (42) K. Brown. (43) T. C. Xu. (44) M. Hopkinson. (45) P. Smith. (46) D. Kerridge. (47) B.A.-M. Al.-Azem. (48) D. Tamsett.

PROLOGUE

Sir Ronald Fisher, who as a Fellow of Gonville & Caius College became guide, philosopher and friend not only to me but to my research students in the Department of Geodesy and Geophysics at Cambridge, once said "Research (or graduate) students are the most important people in a University". I cannot quite recall where he said it but I remember those present being taken aback: after all, was it not self evident that professors, dons and tutors* were, or if not, undergraduates – Cambridge then had and may still have some memory of its role as a finishing school for young gentlemen? Fisher was always provocative and had there been administrators or politicians around he would have enjoyed their discomfiture. In 1984 Government policy has largely closed the door on research students from abroad and Fisher, though somewhat to the right of Mrs. Thatcher, would, I am sure, have strongly disapproved.

My work with my research students has been, of course, the great experience of my life. In the early days at Manchester University and Cambridge University we spread our interests widely, and so, although I guided my research students in the choice of a topic, I was soon to know the humbling experience of receiving tutorials (given sometimes with the condescension of which only the young are capable) in geology, hydrodynamics, astronomy, the theory of magnetism, electromagnetism, etc. The subject was a small one then and we all – my research students and I – received education from, in addition to P. M. S. Blackett, E. C. Bullard and Fisher, Walter H. Bucher, O. T. Jones among geologists, and Vening Meinesz, Harold Jeffreys, Maurice Ewing, Keith Bullen, and Francis Birch among geophysicists, and G. I. Taylor, W. L. Bragg, N. F. Mott, and J. W. Mitchell among physicists took a lively and fatherly interest in our efforts.

In universities today, schemes are afoot to improve university lecturing by 'training' – a wonderful word – and doubtless training for being a supervisor of research students will soon be instituted. By the standards such schemes would establish, I was a poor supervisor of research students (and doubtless a poor lecturer to undergraduates) and doubtless would have received a low mark on any of the questionnaires with which we are now familiar, but I am comforted by the fact that, if so, I am in a most distinguished company including J. J. Thomson of whom it was said that he took his fledgling student into an empty room and said "Why don't you try to measure ..." – and then he mumbled something his student didn't understand. Three years later J. J. came back to a laboratory full of complicated vacuum systems and heard the proud student relate the fascinating results of his work only to have his professor say "Very interesting but not what I told you to do." The famous Dean of St. Paul's, W. R. Inge, once said that "religion is caught not taught" and I am inclined to think that of the other three objectives of Cambridge University contained in its statutes – "education, learning and

---

* The absence of administrators in this list will seem incomprehensible but they rarely impinged on the consciousness of the research worker then.

research" – the same can be said. The fascinating and important contributions in this volume by very well known geophysicists, who have honoured me in this collection, may perhaps attest to my being a little infectious.

S. K. RUNCORN

PALEOMAGNETISM OF GABBROS OF THE EARLY PROTEROZOIC BLACHFORD LAKE INTRUSIVE SUITE AND THE EASTER ISLAND DYKE, GREAT SLAVE LAKE, NWT: POSSIBLE EVIDENCE FOR THE EARLIEST CONTINENTAL DRIFT

E. IRVING

Pacific Geoscience Centre, P.O. Box 6000, Sidney, B.C., Canada V8L 4B2

A. DAVIDSON and J. C. McGLYNN

Geological Survey of Canada, 601 Booth Street, Ottawa, Canada K1A 0E8

Abstract. The Caribou Lake gabbro, part of the Blachford Lake Intrusive Suite accurately dated at -2186 ± 10 mA, has a predominant $NW-/SE+$ magnetization with a mean, irrespective of sign, of $D = 119°$, $I = 50°$, $\alpha_{95} = 5°$ and a palaeopole $14°$ N, $064°$ W, $A_{95} = 5°$; it has not proved possible to determine if the magnetization is primary. The Easter Island dyke, less well-dated in the range -2200 to -2500 Ma, has a predominant $WNW+$ magnetization, whose mean, when corrected for an $8°$ tilt, is $D = 288°$, $I = 46°$, $\alpha_{95} = 5°$ and palaeopole is $32°$ S, $2°$ W, $A_{95} = 5°$; the magnetization is probably primary. A vertical magnetization (D), not significantly different from the present field, occurs sporadically in both units and is considered to be Late Phanerozoic in age. Palaeopoles from the Caribou Lake gabbro and the Easter Island dyke, together with those already known from Early Proterozoic intrusives of the Archaean Slave Structural Province, roughly define a swath (the Slave Track) which maps the motion of the Slave Province relative to the geomagnetic axis during this interval. The corresponding array of palaeopoles (the Superior Track) from the Superior Structural Province does not fall in the same place. Hence it would appear that Slave and Superior were not in their present relative positions in the Early Proterozoic in disagreement with arguments that have been made for a fixed supercontinent during much of the Proterozoic. Mid-Proterozoic paleomagnetic signatures indicate that Slave and Superior had assumed their present relative position by about -1750 mA. These Early Proterozoic relative motions are the earliest for which there is palaeomagnetic evidence.

1. Introduction

New palaeomagnetic results from the Canadian Shields are described. It is appropriate that a volume in honour of S. K. Runcorn should contain a contribution on this topic because it was he who first initiated palaeomagnetic study of Canadian Shield rocks. This first study, begun in June 1951 and made in collaboration with the senior author, was of the Torridonian sandstone of northwestern Scotland. In the Precambrian, this part of Scotland belonged to an enlarged Canadian Shield. It was through this study, and through essentially contemporaneous work in South Africa on the Pilansberg dykes by Gough (1956), that the existence of a geomagnetic field in the Precambrian was established, that serial reversals in pre-Tertiary rocks were discovered, and that palaeodirections in older rocks systematically oblique to the present axis were established (Runcorn, 1955). The latter phenomenon is what we now call apparent polar wandering (apw), the palaeomagnetic signature of continental drift, and it is with this that our paper is mainly concerned.

Earth Physics Branch Contribution No. 1111.

Geophysical Surveys 7 (1984) 1-25. 0046-5763/84/0071-0001$03.05.

E. IRVING ET AL.

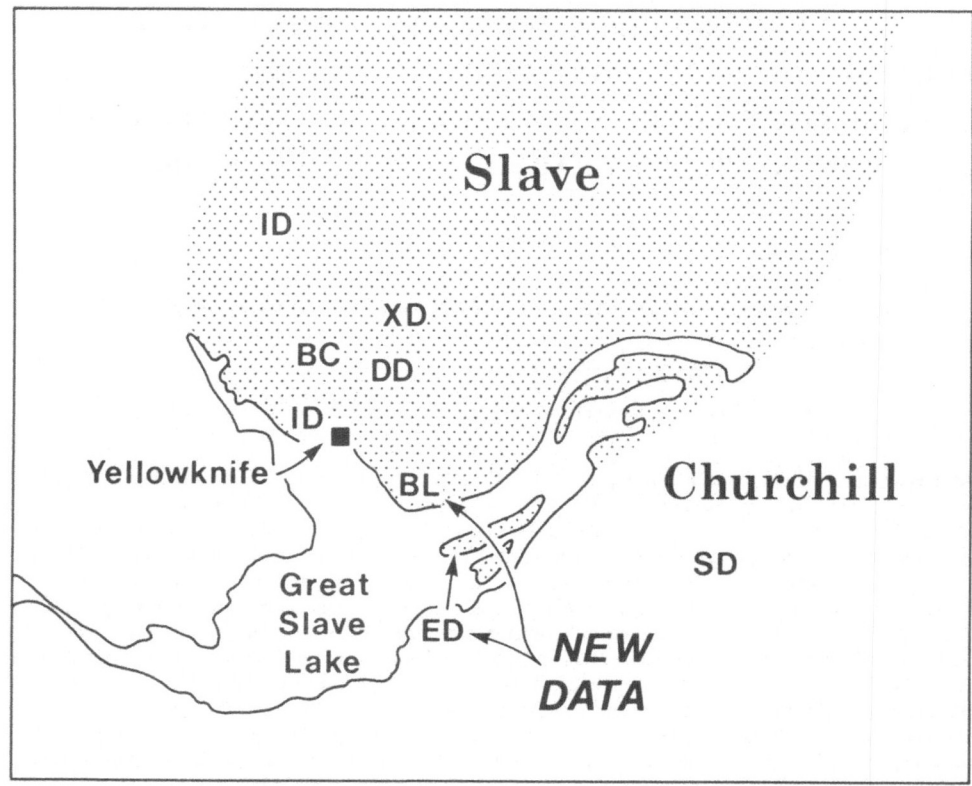

Fig. 1.   Early Proterozoic intrusions of the Slave Structural Province that have been studied palaeomagneti-
cally. BC Big Spruce Complex (Irving and McGlynn, 1976); ID Indin dykes, XD 'X' dykes, DD Dogrib
dykes (McGlynn and Irving, 1975); SD Sparrow dykes (McGlynn *et al.*, 1974); CL Caribou Lake gabbro of
the Blachford Lake Intrusive Suite and ED Easter Island dyke (herein).

The rock-units concerned are Early Proterozoic. They are intrusives into the
Archaean Slave Structural Province in the northwestern part of the Canadian Shield
(Figure 1, see also Figure 15). Units whose palaeomagnetism already has been
described are, the Big Spruce Complex (Irving and McGlynn, 1976), and the Dogrib,
Indin and 'X' dykes (McGlynn and Irving, 1975). Our new data are from gabbros of the
Blachford Intrusive Suite and the Easter Island dyke, and, together with earlier results,
they may be used to suggest the path of apparent polar wander (apw) for the Slave
Province during the Early Proterozoic. Such a path may eventually aid correlation and
the study of thermal history, but of more immediate interest is the possibility of
comparing it with the contemporaneous apw path from the Superior Structural
Province to determine if these two Archaean blocks have undergone relative motions,
as would be expected if the intervening Hudsonian orogen is the product of plate
tectonic processes.

The problem is of general importance because if Early Proterozoic relative motions
could be demonstrated they would be the oldest yet recorded – they would be the

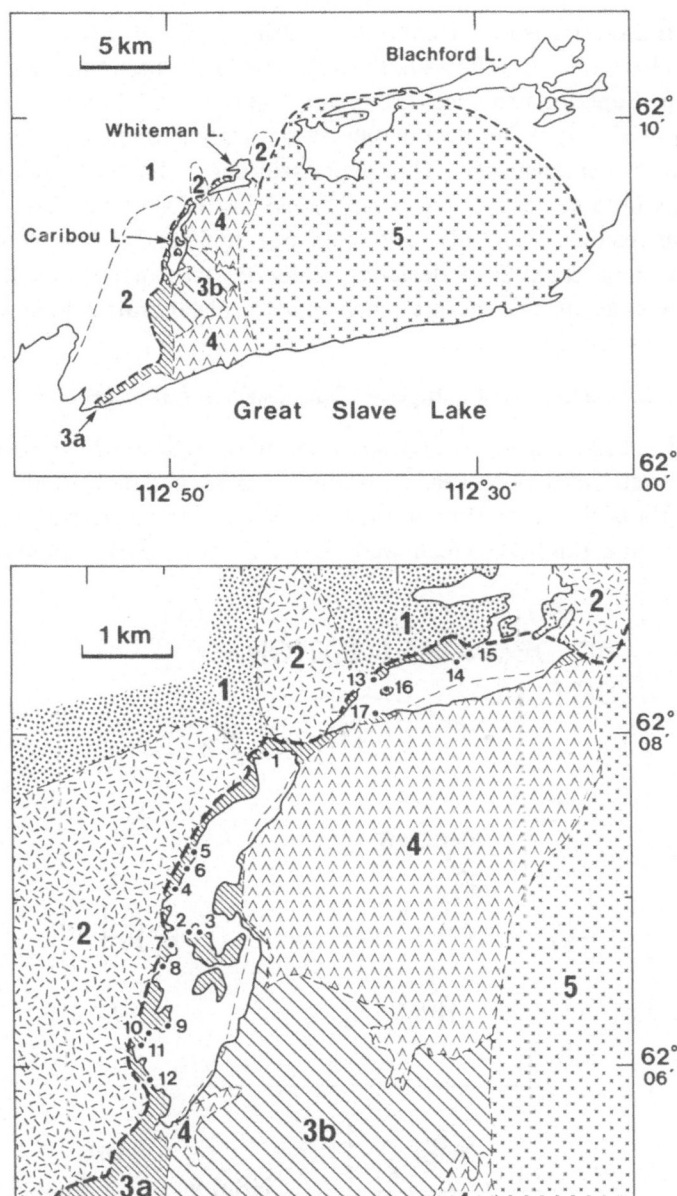

Fig. 2. Above sketch map of the Blachford Lake Intrusive Suite and below of the Caribou Lake gabbro showing sampling sites. The rock-units, numbered in chronological, order are as follows: Proterozoic: Blachford Lake Intrusive Suite: (5) peralkaline granite; (4) metaluminous granite, quartz syenite; (3) Caribou Lake gabbro; (3a) gabbro phase; (3b) leucoferrodiorite phase. Archaean: Slave Province: (2) granite, granodiorite; (1) Yellowknife Supergroup.

earliest quantitative measure of continental drift. Early Proteozoic or Archaean
motions of Archaean blocks relative to the pole (apw) have been observed palaeomag-
netically for the Superior Province (Irving and Naldrett, 1977) and for the Kaapvaal
craton (Layer et al., 1984), but drift of one craton relative to one another so long ago
has not yet been demonstrated. This paper will show that although well-defined
magnetizations (and corresponding palaeopoles) can be obtained from these Early
Proterozoic intrusions, their ages are often not sufficiently well-defined; that is,
palaeomagnetism is capable of detecting Early Proterozoic continental drift but
definitive answers await the accurate geochronological calibration of apw paths.

2. Geology and Sampling of the Caribou Lake Gabbro

The Blachford Lake Intrusive Suite is a large, composite pluton which was emplaced at
high crustal level within Archaean crystalline schists and granitoid rocks at the
southern margin of the Slave Province (Figure 2a). It consists mostly of peralkaline
granite and syenite (unit 5), which were found to have only soft magnetizations

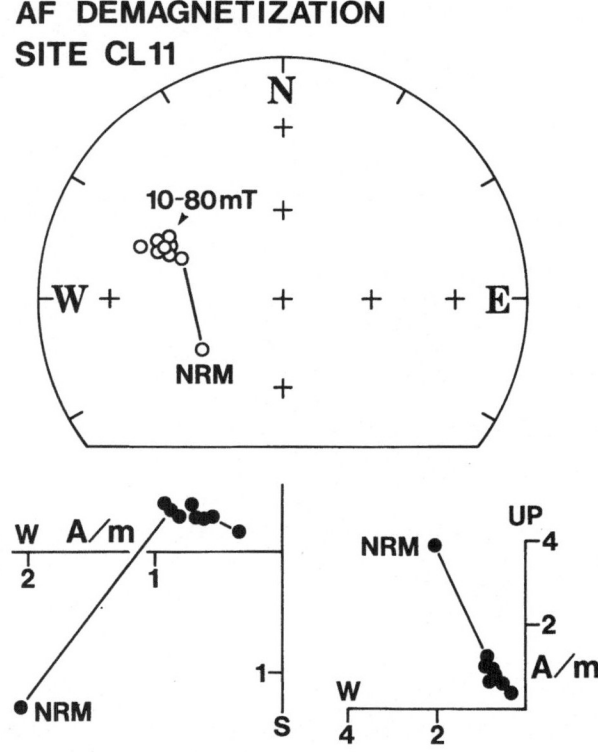

Fig. 3.   Changes in direction (above) and orthogonal plots (below) produced by alternating field (af)
demagnetization of gabbro of Caribou Lake gabbro, site 11. NW— magnetization. Open symbols in the
stereonet indicate negative (upward) inclination, solid symbols positive inclination. 1 millitesla = 10 oersted.

scattered about the direction of the present earth's field (PEF) and contained no discernible record of the Precambrian field. On its western side, earlier plutonic phases of metaluminous granite and quartz syenite (unit 4) intrude a somewhat older gabbroic phase, the Caribou Lake gabbro (unit 3), that is chilled against Archaean rocks. This gabbro is an integral part of the Blachford Lake Intrusive Suite (Davidson, 1978, 1982). It contains stable magnetizations strongly inclined to the PEF. There are no age determination for the gabbro itself, but there are for other members of the suite, all essentially coeval. Six K–Ar ages of alkali amphiboles from the youngest peralkaline unit average −2130 ± 50 Ma, and a whole-rock K–Ar age determined for a partially-melted hornfels inclusion contained within it is −2150 ± 50 Ma. Biotite and hornblende K–Ar ages determined for the geologically older metaluminous granites to the west are −2166 ± 47 Ma and −2127 ± 79 Ma respectively. Rb–Sr isochron ages of −2092 ± 50 Ma and −2130 ± 38 Ma have been obtained for two separate bodies of metaluminous granite, and a U–Pb zircon age from the geologically oldest of these is −2186 ± 10 Ma (Wanless et al., 1979). The Caribou Lake gabbro is not older than about −2200 Ma since K–Ar ages of muscovite and biotite in Archaean granite adjacent to its western contact are −2166 ± 47 Ma and −2109 ± 47 Ma respectively. Hence the U–Pb determination is probably a good approximation to its age. Sampling sites of the Caribou Lake gabbro are shown in Figure 2b.

3. Remanent Magnetization of the Caribou Lake Gabbro

The initial directions of magnetization (nrm) are scattered, sometimes with upward directions, sometimes steeply dipping downwards not far from the PEF. An example of the former is shown in Figure 3. After alternating demagnetization (af) in 10–80 mT the direction becomes well-grouped toward the northwest with upward indication. This is referred to as the $NW-$ magnetization. The orthogonal plots show excellent linear decay to the origin. The magnetization is simple, consisting of a large soft PEF magnetization, readily demagnetized, and a hard $NW-$ magnetization that can be cleanly isolated.

Another example is shown in Figure 4, this time obtained by thermal demagnetization. The soft PEF magnetization is now much smaller in magnitude, and, after its removal at about 200 °C, there is again excellent linear decay on the orthogonal plots, apparently to the origin. At 650 °C, however, there is a dramatic change to a downward direction. This magnetization is referred to as D. D is very small and hence produces little deflection of the orthogonal plots. At 700 °C the directions in this and all specimens studied are random.

The third example (Figure 5) is typical of magnetizations which are initially grouped around the PEF owing to a very large soft magnetization (PEF) which is readily removed by heating to 300 °C. During heating between 300 and 550 °C (the approximate blocking temperature of magnetite) there is an apparent end-point, but upon further heating an underlying D magnetization is revealed. This time D is not negligible in magnitude and to obtain an accurate estimate of the direction of the $NW-$

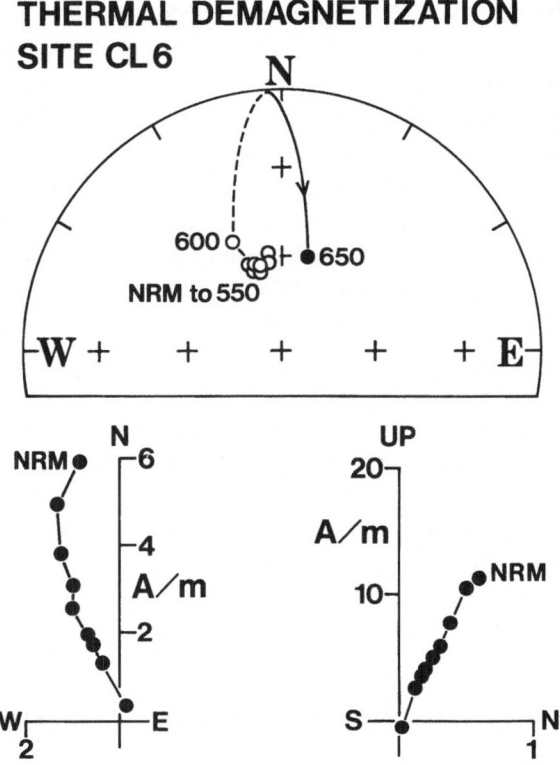

Fig. 4.   Changes in direction (above) and orthogonal plots (below) produced by thermal demagnetization of
Caribou Lake gabbro, site 6. Small *PEF*, *NW*− and a very small *D* magnetizations are present.

magnetization (R in Figure 5) its effect must be removed. The 'end-point' is not real and
does not itself give an accurate estimate of *NW*−.

All specimens have been studied in this detailed way and an attempt made to separate
the magnetizations present. At sited 7 and 8 the stability spectra overlapped strongly
and no magnetizations could be accurately isolated. The directions of all magnetizations observed are plotted in Figure 6, and their averages listed site by site in Table I.
Predominant are *NW*− magnetizations and magnetizations directed to the southeast
and downward (*SE*+). Both have unblocking temperatures generally in the range 500
to 600 °C and may be ascribed to magnetite.

When summarizing the results there are weighting problems. A given magnetization
may occur in several specimens from a collecting site, sometimes only in one, or it may
be absent altogether. To accord specimens unit weight would emphasize too much
those sites at which there were many observations of the same magnetization. To give
sites unit weight would emphasize too much those sites with only one observation. We
suggest that if there are three or more physically satisfactory observations of a given
magnetization at a site it can be considered reasonably well-established. In Table I the
NW−, *SE*+ magnetizations have been averaged first including all results, and then by

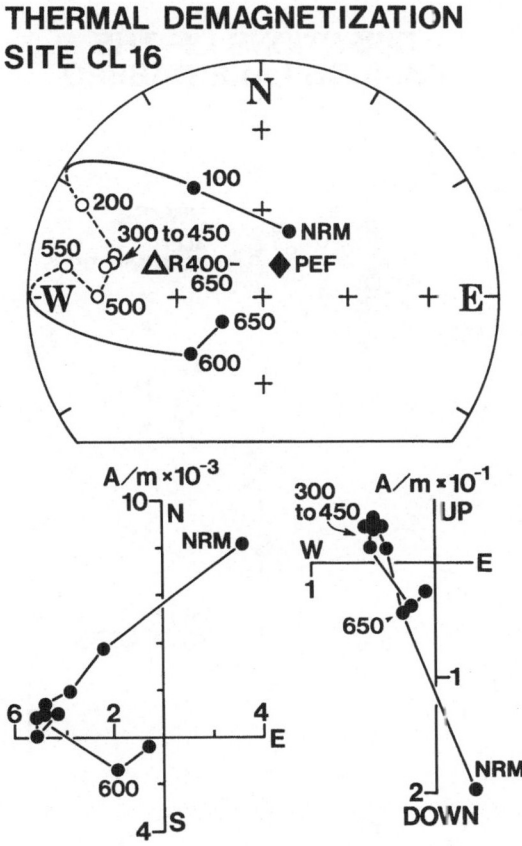

Fig. 5. Changes in direction (above) and orthogonal plots (below) produced by thermal demagnetization of Caribou Lake gabbro, site 16. Large *PEF* and substantial *NW−* and *D* magnetizations are present.

including only those sites for which s, the number of observations, exceeds three. We consider the latter the better procedure. The means of the *NW−* and *SE+* groups are statistically antiparallel (Table II). *D* magnetizations occur in 19 specimens, about 20% of specimens studied, and their mean is not significantly different from the present field (29, 82). *D* magnetization only becomes visible after removal of *NW−* or *SW+*. It has unblocking temperatures in the range 650 to 700 °C and may be ascribed to hematite. Its origin is discussed later.

A subordinate magnetization, directed to the southeast and downward, with significantly steeper indication than *SE+*, occurs at one site (*C* Table I). It is identical with the widespread late Hudsonian overprint found throughout the Canadian Shield, and sometimes called the Coronation overprint (Reid *et al.*, 1981) and which was presumeably acquired about −1750 Ma (Schutts and Dunlop, 1981; Irving and McGlynn, 1981).

The *NW−/SE+* magnetization may have been acquired during cooling of the

DIRECTIONS OF THE CARIBOU LAKE GABBRO

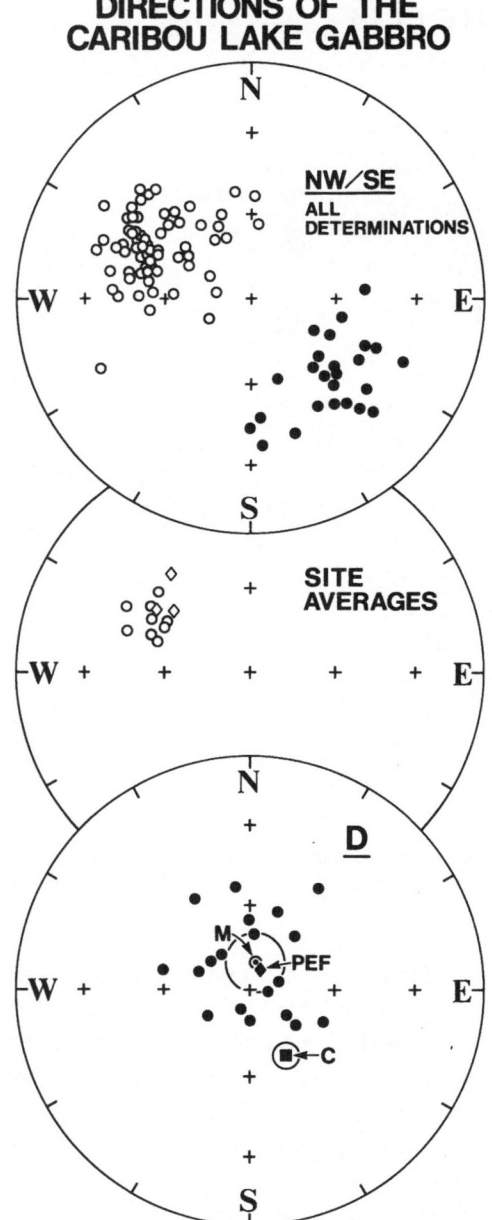

Fig. 6. Above, directions of $NW-$ and $SE+$ magnetizations observed from the Caribou Lake gabbro. Centre, mean site directions $(s > 2)$ irrespective of sign. Below, D directions with mean and error circle showing coincidence with the present field (PEF). C is the Coronation overprint. All with respect to present horizontal.

TABLE I

Results from Caribou Lake gabbro of the Blachford Lake Intrusive Suite

Site	Comp.	$c(s)$	$D°,$	$I°$	k	$\alpha_{95}°$
1	$SE+$	5(8)	142,	44	39	9
2	$NW-$	5(10)	301,	-55	32	9
2	D	1	145,	50	–	–
3	$NW-$	2(2)	295,	-59	–	–
3	$SE+$	4(6)	128,	55	18	16
3	C	4(6)	148,	76	27	13
3	D	1	175,	80	–	–
4	$NW-$	2(2)	336,	-47	11	–
5	$SE+$	5(8)	123,	49	25	11
5	D	3(3)	129,	88	17	31
6	$NW-$	5(8)	304,	-48	14	15
9	D	1	010,	60	–	–
9	$NW-$	5(7)	288,	-55	85	7
9	D	2(2)	340,	73	45	–
10	$NW-$	5(5)	298,	-55	16	20
10	D	2(2)	21,	57	19	–
11	$NW-$	6(9)	299,	-50	43	8
11	D	2(2)	19,	79	10	–
12	$NW-$	3(4)	312,	-46	26	18
12	D	1	113,	62	–	–
13	$NW-$	2(2)	314,	-45	5	–
13	$SE+$	1	136,	48	–	–
13	D	2(2)	357,	63	10	–
14	$NW-$	3(5)	298,	-39	73	9
14	D	1	322,	14	–	–
15	$NW-$	3(4)	289,	-43	11	29
15	D	1	122,	74	–	–
16	$NW-$	2(2)	296,	-39	18	–
16	$SE+$	2(2)	177,	42	209	–
16	D	1	239,	73	–	–
17	$NW-$	6(7)	291,	-52	34	11
17	$SE+$	2(2)	128,	42	22	–
17	D	2(2)	13,	69	10	–

Notes. c is the number of oriented drill cores and s the number of specimens cut from them. D, I are the declination and inclination of the mean direction relative to present horizontal, k Fisher's estimate of precision, $_{95}$ the error ($P = 0.05$). The data have been obtained as follows: site 1 $SE+$, average 400 to 550 °C (2), ave. 30 to 80 mT(5), R 30–60 mT(1); site 2$NW-$ ave. 50 to 80 mT(5), 500–650 °C(5); site 2D, 650 °C; site 3$NW-$, ave. 40–80 mT(2); site 3$SE+$ ave. 50–100 mT(2), R 500–650 °C(4); site 3C R 100–50 mT(2), R 200–500 °C(2); site 5$SE+$, ave. 30 to 80 mT(5), ave. 400 to 550 °C(2); site 5$SE+$, ave. 30 to 80 mT(5), ave. 400 to 550 °C(1), 500–650 °C(2); site 5D, 650 °C; site 6$NW-$, ave. 30 to 60 mT(3), ave. 400 to 500 °C(1), R 20–50 mT(2), R 550–650 °C(2); site 6$NW-$, R 30–60 mT(6), R 500–650 °C (2); site 6D, 650 °C; site 9$NW-$, ave. R 30–80 mT(5), R 550–650 °C(1), ave. 300 to 550 °C(1); site 9D, 650 °C; site 10$NW-$, ave. 30 to 80 mT(5); site 10D, 650 °C; site 11$NW-$, ave. 30–80 mT(8), R 500–600 °C(1); site 11D, 650 °C; site 12$NW-$, 30 to 80 mT(3), 550–650 °C(1); site 12D, 650 °C; site 13$NW-$, ave. 30 to 80 mT(1), R 20–50 mT; site 13$SE+$, ave. 20 to 50 mT; site 13D, 650 °C; site 14$NW-$ ave. 30 to 50 mT(3); ave. 300–400 °C(1), R 550–650 °C(1); site 15$NW-$, ave. 40 to 60 mT(2), R 500–650 °C(1); site 15D, 650 °C; site 16$NW-$, ave. 40 to 80 mT(1), R 550–650 °C(1); site 16$SE+$, ave. 30 to 50 mT(2); site 16D, 650 °C; site 17$NW-$, ave. 500 to 600 °C(1), ave. 30–50 mT(4); R 500–650 °C(2); site 17D, 650 °C. No coherent magnetizations were found at sites 7 and 8.

TABLE II

Summary of results from the Easter Island dyke and the Caribou Lake gabbro

	Sites(s)	$D°$,	$I°$	k	$\alpha_{95}°$	lat°, long°($dm°, dp°$)
Easter Island Dyke						
in situ	15*(146)	281,	42	74	5	–
tilt corrected (ED)	15*(146)	288,	46	74	5	32S, 022W(06, 03)
average of site-poles	15*			68	5	32S, 002W
Hearne dyke remagnetization (HC)	1 (06*)	310,	−6	18	16	15S, 061W
Caribou Lake Gabbro						
$NW-$ sites	13*(67)	302,	−49	58	5	13N, 064W(7, 5)
$NW-$ sites $s > 2$	9*(59)	298,	−49	113	5	15N, 061W(7, 4)
$SE+$ sites	6*(27)	139,	48	31	12	7N, 078W(16, 11)
$SE+$ sites, $s > 2$	3*(22)	132,	50	89	13	10N, 072W(18, 12)
$NW-$ and $SE+$ combined, $s > 2$	12*(81)	301,	−50	92	5	14N, 064W(18, 12)
average of site-poles (CL)	12*			71	5	14N, 064W
C Coronation overprint	1 (06*)	149,	76	27	13	37.4N, 095.4W(6, 4)
D magnetization (CLD)	12 (19*)	15,	79	14	9.4	81N, 073W(18, 17)

Symbols as in Table I. The palaeopole is given on the right. The letters in brackets in the first column are the palaeopole labels used in Figures 15 and 16. Asterisk indicates unit-weight.

Blachford Intrusive Suite (-2180 Ma). If it is a thermoremanent magnetization it is unlikely to be younger than the muscovite and biotite ages of -2166 ± 47 and -2109 ± 47 Ma observed in the western contact aureole because its unblocking temperature (500 to 600 °C) is higher than the argon blocking temperature in micas. If it is a chemical remanent magnetization it could substantially post-date the time of intrusion.

4. Geology and Sampling of the Easter Island Dyke

The Easter Island dyke is composed predominantly of alkali gabbro with a syenitic differentiate to the east of Figure 7. It intrudes Archaean basement and is overlain by the Great Slave Supergroup. Its age is post-Archaean (-2500 Ma) and pre-Great Slave Supergroup. The Compton laccoliths in the upper part of the Great Slave Supergroup yield K–Ar ages of -1865 Ma (Wanless *et al.*, 1979). Hence the age of the Easter Island dyke lies between these limits. The oldest K–Ar dates from the dyke itself are -2170 Ma (Leech *et al.*, 1966) and -2200 Ma (Burwash *et al.*, 1963). The dyke is 200 to 300 m wide and 20 km long. Its margins are sub-vertical. The overlying Great Slave Supergroup dips 8° to the northeast.

5. Remanent Magnetization of the Easter Island Dyke

The initial directions in the gabbro are scattered, but are generally directed downward

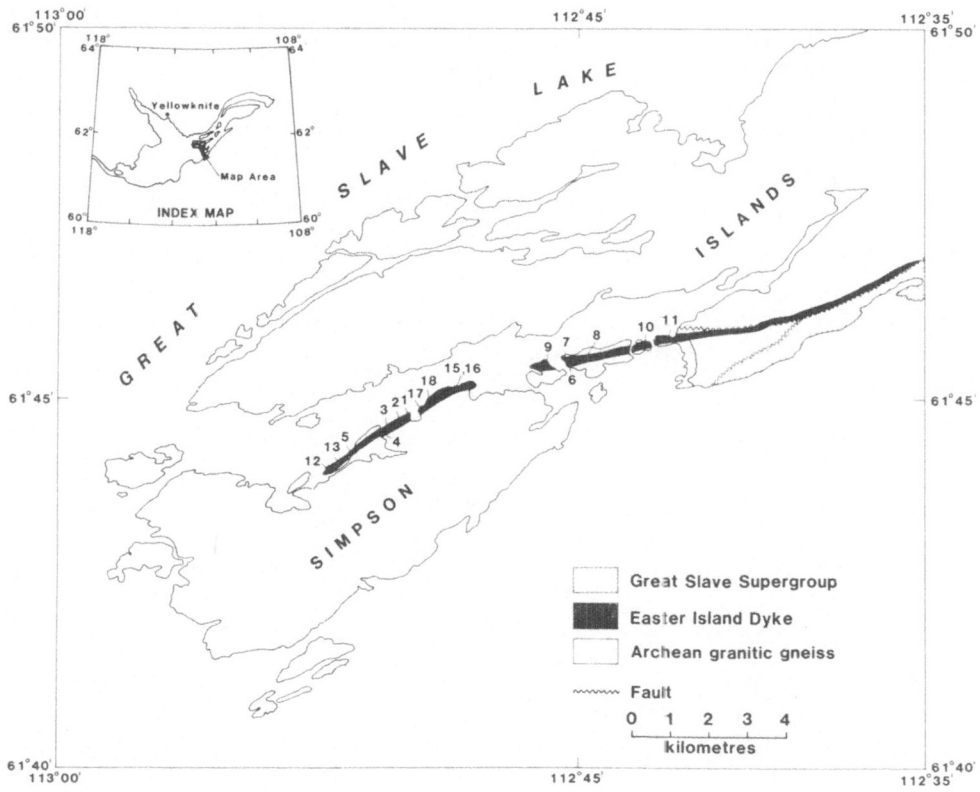

Fig. 7. Easter Island dyke and sampling sites. Site 8 is in baked gabbro, site 16 a granite contact. Site 14 is in syenite and is 2 km off the map to the east.

to the north and west. The directions from the syenite (site 14) are random and no coherent magnetizations were isolated. During AF treatment the magnetization of the gabbro becomes directed toward the west with intermediate inclination (Figure 8). This is referred to as the $WNW+$ magnetization. The decay curve is smooth. After treatment in about 50 mT the intensity is reduced one hundred times, and at higher fields the directions begin to scatter as the magnetization becomes increasingly randomized. Orthogonal plots show perfect linear decay to the origin above about 15 mT (Figure 9). Thermal demagnetization yields square-shouldered decay curves and the directions remain coherent up to 575 °C (Figure 10). At higher temperatures the intensity is reduced to a few per cent of its initial value and becomes scattered, usually but not always randomly. Sometimes there is a systematic tendency to become steeply downward (Figure 10). A similar effect can occasionally be observed during af demagnetization (Figure 8).

Apart from a soft component magnetization that is readily removed at low temperatures and in low alternating fields, the remanent magnetization of the Easter Island dyke consists overwhelmingly of a single hard $WNW+$ magnetization. It has

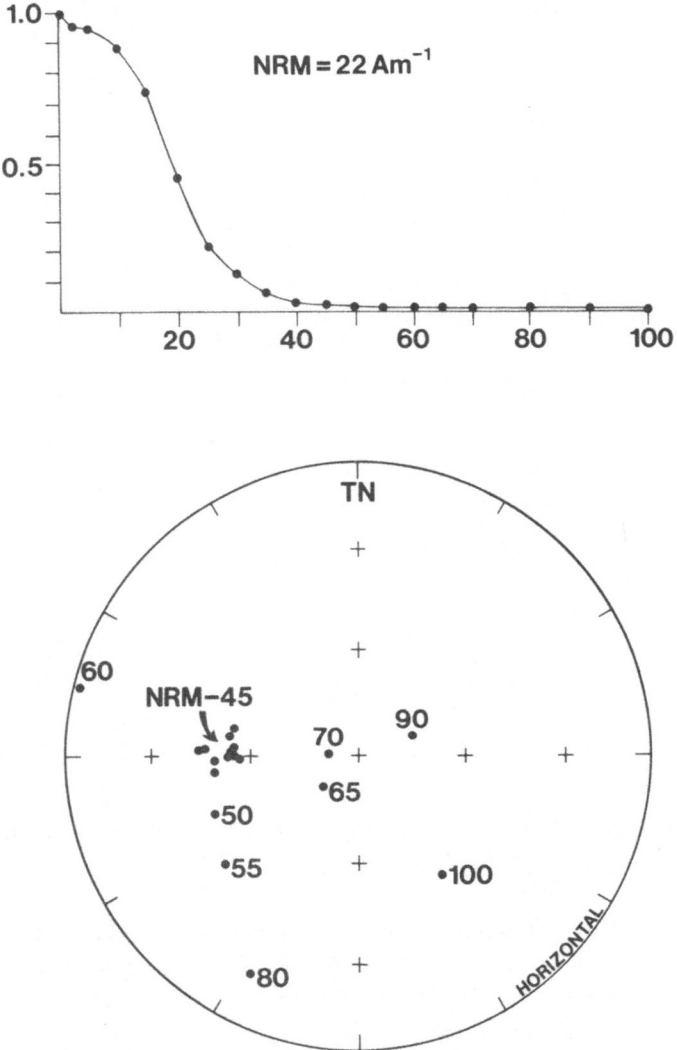

Fig. 8. Changes of intensity (above) and direction (below) of magnetization of gabbro of the Easter Island dyke, site 1, during af demagnetization. Solid (open) symbols denote downward (upward) inclination.

unblocking temperatures in the range 500 to 600 °C, and remanent coercive force typically between 10 and 50 mT occasionally as high as 100 mT. In addition, there is sometimes a very small D magnetization, no more than a per cent of the nrm, that is uncovered by heating to 600 °C or above.

The directions of the $WNW+$ magnetizations observed are plotted in Figure 11. The mean directions at each site are listed in Table III, and the overall mean *in situ* and corrected for tilt in Table III.

TABLE III

Results from Easter Island dyke

Site	$c(s)$	$D°,$	$I°$	k	$\alpha°_{95}$	Cleaning
1	5(11)	283,	46	233	3	30 mT, 575 °C
2	5(10)	288,	47	77	6	30, 40 mT, 525 °C
3	5(10)	293,	46	40	8	30, 40 mT, 575 °C
4	5(11)	279,	42	73	5	30, 40 mT, 575 °C
5	6(11)	278,	43	59	6	30, 40 mT, 575 °C
6	5(9)	282,	36	118	5	30, 40 mT
7	6(11)	287,	38	52	6	30, 40 mT
9	5(10)	269,	44	64	6	30, 40 mT
10	4(7)	294,	32	87	7	40 mT
11	5(9)	294,	29	29	10	40 mT
12	5(10)	275,	51	131	4	40 mT
13	5(10)	275,	43	68	6	30, 40 mT
15	5(10)	268,	44	73	6	40 mT
17	4(7)	282,	38	32	11	40 mT
18	6(10)	262,	40	156	4	40 mT, 550 °C

The treatment used is shown on the right. Symbols as in Table I. Sites plotted in Figure 7 but not listed here are where special studies have been undertaken, as detailed in Section 3.

The uniformity of the $WNW+$ magnetization and its high stability indicate that it was acquired in the direction of the palaeofield when the dyke first cooled. To test this assumption two contacts were studied. Samples of pink gneissic granite were taken at site 16 where the dyke cuts the Archaean basement (Figure 7). The intensities are low (less than 10^{-3} Am^{-1}) and the directions widely scattered. Neither thermal nor af demagnetization produced any coherence. Like many other similar attempts to study baked rocks at intrusive contacts in the Slave Province (McGlynn and Irving, 1975), this contact test failed. Even when reheated by Proterozoic intrusions the Archaean basement, for the most part, seems incapable of retaining a memory of the Precambrian field. At the second contact, samples of gabbro were obtained where a thinner younger Hearne dyke, 10 m wide, cuts the much wider Easter Island dyke (site 8, Figure 7). All samples were taken within 40 cm of the sharp contact. The reheated gabbro of the Easter Island dyke has a highly stable magnetization. After removal of soft magnetizations in low alternating fields there is excellent convergence to the origin on the orthogonal plots (Figure 13) and a true end-point (Figure 12). Thermal demagnetization also yields excellent true end-points with unblocking temperatures mainly between 550 and 600 °C (Figure 14). This very stable magnetization is directed to the northwest with low inclination. Correction for the tilt of the Easter Island dyke causes no significant change. The mean direction *in situ* is 310°, 2°. Corrected for tilt it becomes 310, $-6°$. This direction is very close to that observed in the X dykes (Figures 1, 15). The magnetization was probably acquired at the time this particular Hearn dyke was

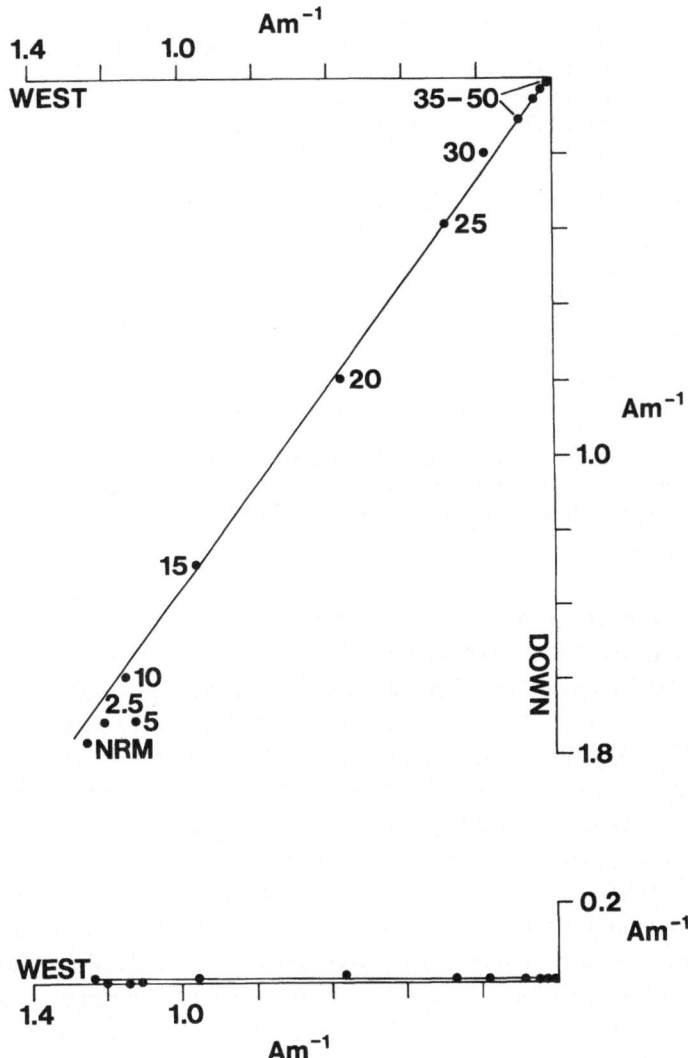

Fig. 9.   Orthogonal diagrams showing linear decay of magnetization of gabbro at site 1, Easter Island dyke.
Above vertical east-west plane and below horizontal plane. Fields are given in mT. The experimental points
for fields greater than 50 mT are close to the origin and indistinguishable at this scale.

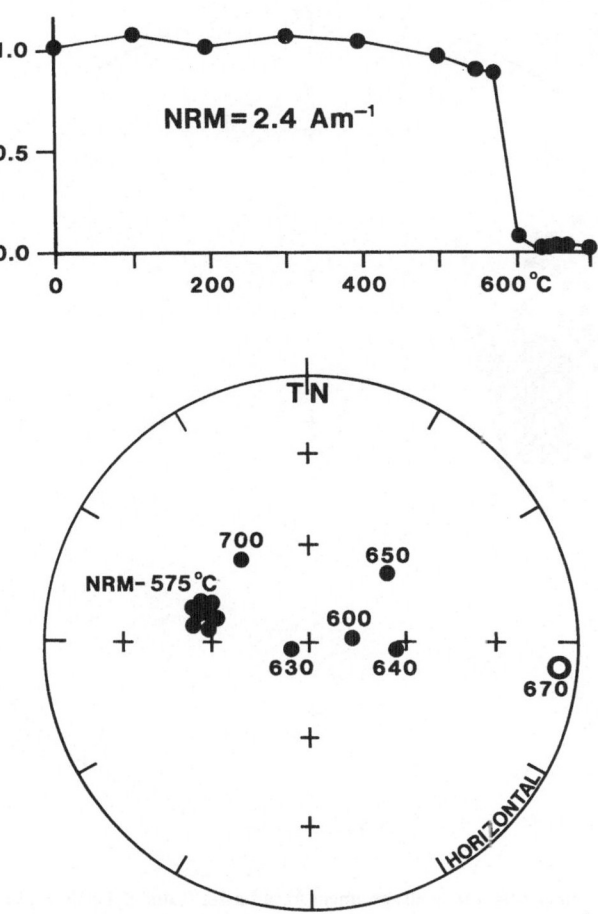

Fig. 10. Changes of intensity (above) and direction (below) of magnetization at site 1, Easter Island dyke, during thermal demagnetization 700 °C.

intruded and cooled, and hence post-dates the magnetization of the unheated Easter Island dyke. The later, therefore, in all probability, predates the intrusion of the Hearne dyke. The Hearn dyke itself is much less stable than the gabbro it bakes. Its magnetization is complex and, despite extensive demagnetization studies, it has not proved possible to resolve it.

Because of very limited geochronological data, the age of the magnetization of the Easter Island Dyke can only be defined within wide limits. The unblocking temperatures are higher than those for argon-loss in micas so the magnetization is probably at least as old as the oldest K–Ar date, − 2200 Ma. It must be post-Archaean so the age limits are − 2200 to − 2500 Ma.

E. IRVING ET AL.

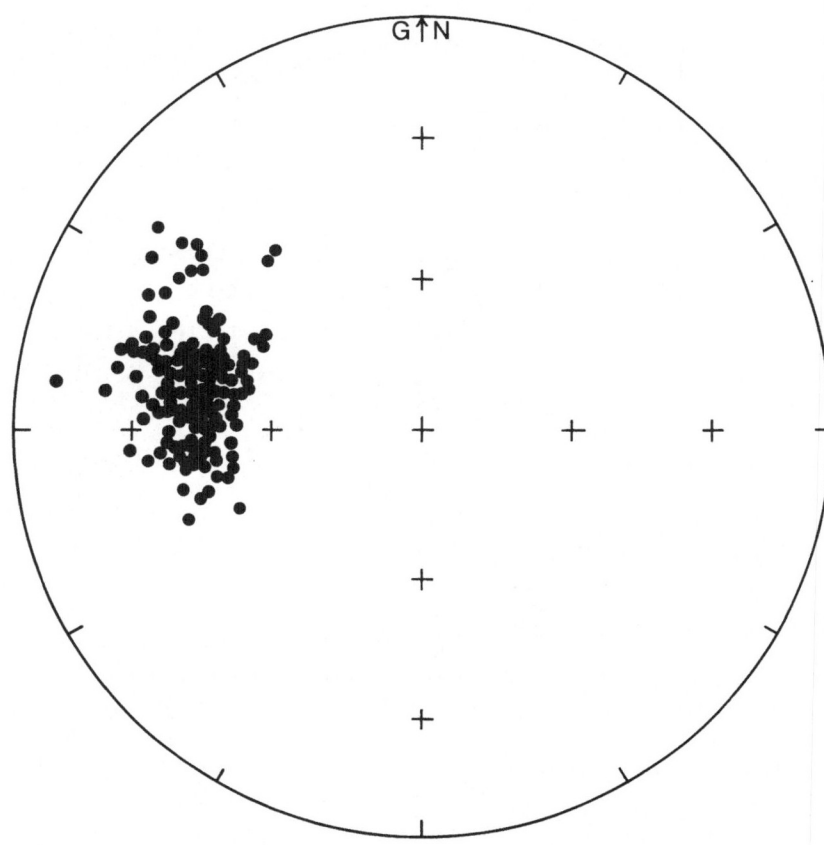

Fig. 11. Directions of magnetization in all specimens from Easter Island dyke after the cleaning specified in Table III. Perimeter is the present horizontal.

6. Early Proterozoic Slave Track

Paleopoles from Early Proterozoic rocks of the Slave Province are plotted in Figure 15. Excepting the D palaeopoles, they form an array stretching from the South Atlantic to the Caribbean. We refer to this as the Slave Track. Most evidence indicates that the trend is from southeast (older) to northwest (younger), but this is not certain. The Dogrib dykes (palaeopole DD) and the Easter Island dyke (ED) are the oldest of the intrusives studied, certainly older than the Indin dykes (palaeopole ID) which cut the Dogrib dykes. The palaeopole for the baked gabbro (HC) is to the northwest of that for the Easter Island dyke, which it post-dates. The northern end of the Slave Track is reasonably well fixed by the Indin dykes (-2050 ± 86 Ma, Gates and Hurley, 1973). Its older limit is tentatively fixed by the Easter Island dyke palaeopole as older than -2200 Ma. Dogrib dykes also falls in the age-range of -2200 to -2500 Ma and their palaeopole is consistent with this proposal. The estimated age of the Big Spruce

Complex is -2066 ± 40 (Rb/Sr (Martineau and Lambert, 1974) and its dominant magnetization yields a palaeopole (BC2) in excellent agreement with the Indin dykes and Caribou Lake gabbro. One of its subordinate magnetization, BC3, yields a palaeopole in the middle of the Slave Track. It would seem therefore that the age of the Slave Track is older than -2050 Ma, probably in part as older as -2200 Ma, and it is certainly post-Archaean. If the trend in age is truly from southeast to northwest, as indicated in Figure 15, and if the NW-/$SE+$ magnetization of the Caribou Lake gabbro is primary, then the particular Hearne dyke (site 8) that intrudes the Easter Island dyke should predate the Blachford Lake Intrusive Suite. But in the north, other members of the Hearn dyke swarm are known to cut the suite. Hence the preferred age-trend of Figure 15 is wrong, or the Hearn dykes are not a single swarm but comprises an older

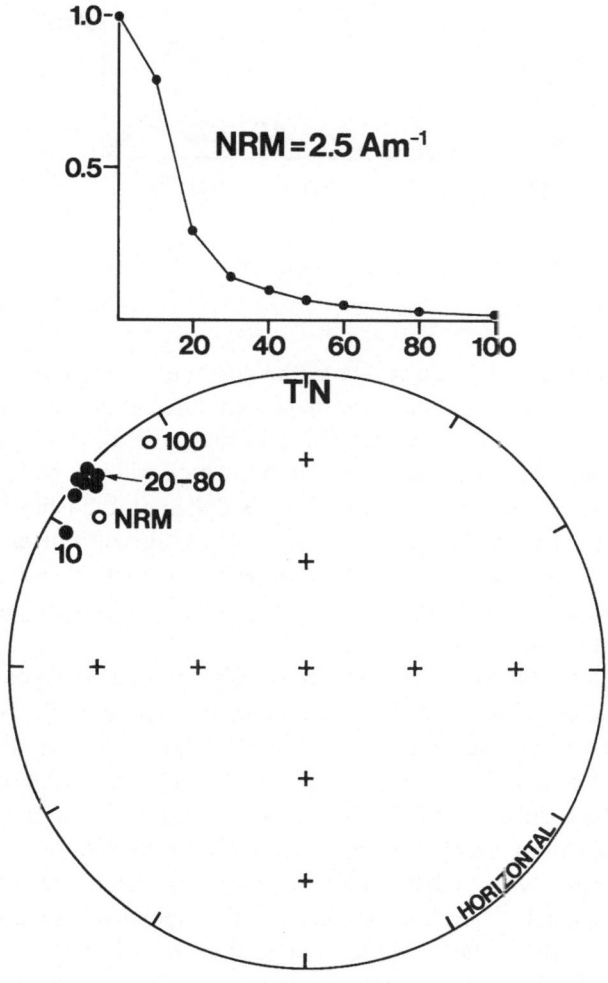

Fig. 12. Af demagnetization of gabbro of Easter Island dyke remagnetized by Hearn dyke, site 8.

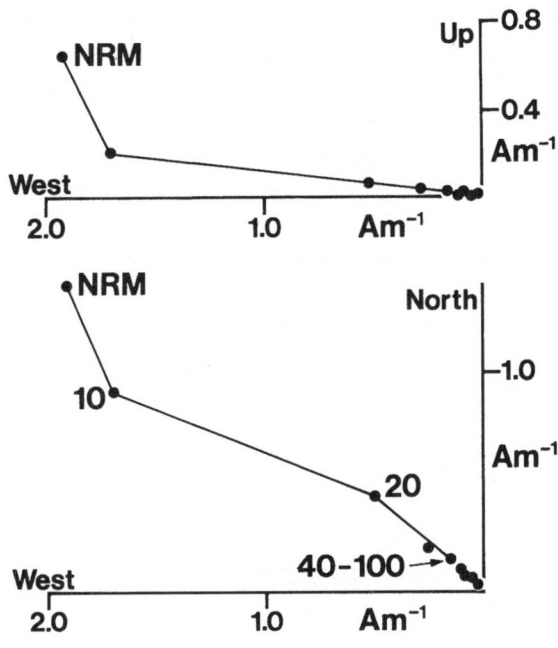

Fig. 13. Orthogonal plots of data of Figure 12.

swarm to the south and a younger one to the north, or the $NW/NW\text{-}/SE+$ magnetization of the Caribou Lake gabbro is secondary acquired at the time of intrusion of the
Indin dykes. Clearly, the accuracy with which the magnetizations are presently dated is
insufficient to unambiguously resolve the trend, which will not be firmly established
until better age-determinations of the Dogrib dykes and Easter Island dyke are
available, and until a palaeomagnetic and age study of the entire Hearne dyke swarm
has been made. In the meantime all that can be said is that the Slave Track is early
Proterozoic (probably in the range -2200 to -2050 Ma) it is roughly $50°$ long, and
that it is situated as shown in Figure 15.

The palaeopole for the Sparrow dykes (SD) is also plotted on Figure 15 at the north
end of the Slave Track. These dykes intrude sedimentary rocks of the Nonacho Group,
and are situated within what is now the Churchill Province just south of the exposed
southern margin of the Slave Province (Figure 1). SD is indistinguishable from the
paleopoles for the Indin dykes the Caribou Lake gabbro, and from the BC2 palaeopole
of the Big Spruce Complex. All three rock-units contain reversals. Contact tests for the
Indin and Sparrow dykes indicate that the magnetizations date from the times of
intrusion. Whole-rock $^{39}Ar/^{40}Ar$ studies on the Sparrow dykes were indecisive
(McGlynn et al., 1974). They yield ages ranging from -2000 to -1700 Ma with a
preference stated by the authors for the younger values. Therefore we originally placed
the Sparrow dyke palaeopole, not on the Slave Track, but on a younger section of the

path, the Coronation Loop. These magnetic similarities however could indicate that the true age of the Sparrow dykes is, in fact, closer to the older than the younger limit of the $^{39}Ar/^{40}Ar$ results. This would imply that the Nonacho Group and the gneissic 'Hudsonian' basement upon which they rest, is older than -2000 Ma, a conclusion consistent with the fact that muscovites from it have yielded K/Ar ages as old as -2300 Ma (Burwash and Baadsgaard, 1962). Indeed these gneisses may be a re-activated southerly extension of the Archaean Slave Province. A further implication is that there has been no very large relative movement between the Nonacho Basin and the Slave Province since the Sparrow and Indin dykes were intruded. Obviously, it is far from certain that the Sparrow dykes palaeopole should be assigned to the Slave Track.

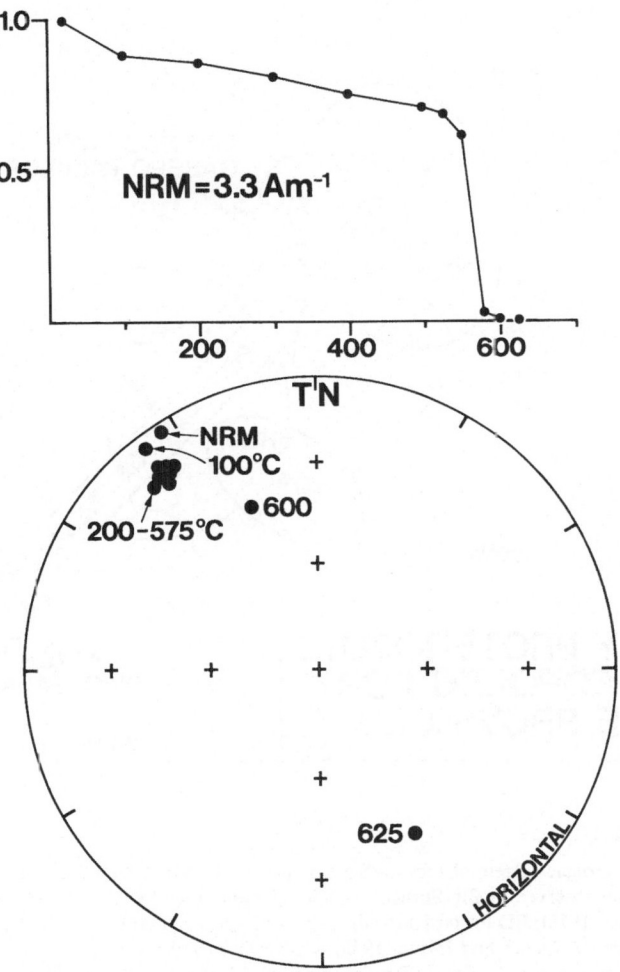

Fig. 14.   Thermal demagnetization of Easter Island dyke gabbro remagnetized by a Hearne dyke, site 8.

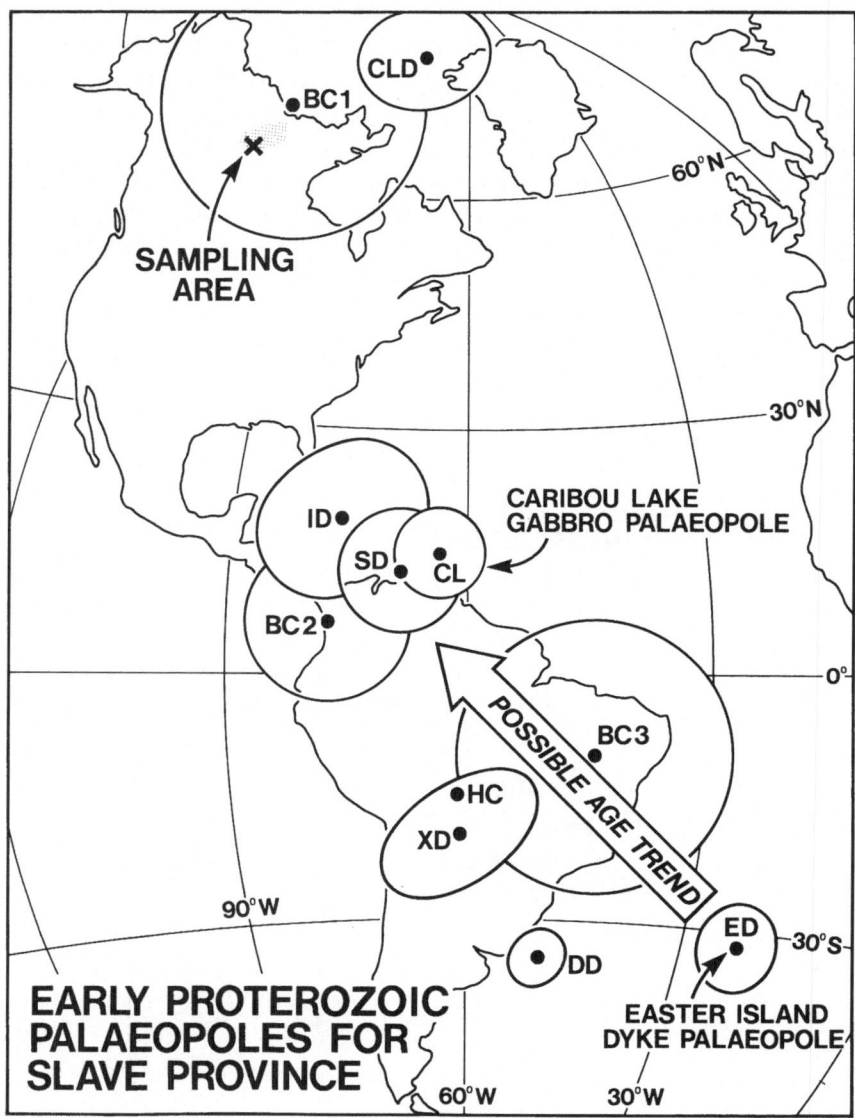

Fig. 15.   Early Proterozoic palaeopoles for the Slave Structural Province. BC1, BC2, BC3 palaeopoles from
three magnetizations observed in Big Spruce Complex (Irving and McGlynn, 1976); DD Dogrib dykes
(McGlynn and Irving, 1975); ED Easter Island dyke, Table 2 herein; ID Indin dykes (McGlynn and Irving,
1975); XD 'X' dykes (McGlynn and Irving, 1975); CL Caribou Lake gabbro mean of $SE+$ and $NW-$
magnetizations, Table II herein; CLD from D magnetization of the Caribou Lake gabbro, Table II herein. SD
is from the Sparrow dykes of the Nonacho Basin (McGlynn et al., 1974). The arrow denotes the suggested age
trend of the Slave Track. See text for discussion of age uncertainties.

It is entirely possible, as we have previously assumed, that it lies on a younger segment of the apw path, and that its position on the older Slave Track is coincidental. An accurate age determination for the Sparrow dykes is needed.

The D magnetizations are not significantly different from the PEF and have the same direction and properties as the D magnetization (BC1 Figure 15) of the Big Spruce Complex. D is always small in magnitude compared to other magnetizations present. Similar magnetizations have been observed from younger Precambrian rocks of Victoria and Baffin Island (Palmer et al., 1983; Christie and Fahrig, 1983). They could be produced by incipient weathering, or they could have been produced by repeated cycling through the Morin transition since they occur in areas subject to repeated permafrost during the Pleistocene glaciations, – an interesting unsolved rock magnetic problem.

7. Slave-Superior Comparison

Earlier Proterozoic palaeopoles for the Superior Structural Province are plotted in Figure 16 and define a Superior Track, essentially that drawn by Irving and Naldrett (1977). Several palaeopoles, derived particularly from the Nipissing diabase from what we consider to be secondary magnetization, fall in latitudes of about 20 °S (documented in Irving and McGlynn, 1981), are not included. Palaeopoles such as TS and GG5, that fall to the side of the track, are assumed to have been displaced by local tectonic rotations such as are common in other fold belts (Irving and McGlynn, 1981). The older part of the Superior Track may be as old as -2500 Ma (Matachewan dykes, Figure 16). Its younger end is fixed by the well-dated Abitibi dykes (-2150 ± 25 Ma, Hanes and York, 1979), an age much the same as that of the Slave Track, and indistinguishable from the Blachford Lake Intrusive Suite. The reconstruction of neither the Superior nor Slave Track is unique. As they are drawn in Figure 16 they seem to us to be the simplest and most satisfactory way of explaining the age relationships of the various magnetizations as they are presently known.

The positions of the tracks are different. This implies that the Slave and Superior Provinces were in different relative positions during the Early Proterozoic. That they had amalgamated by about -1750 Ma is almost certain, because the widespread overprints that have been observed in the Slave (Irving and McGlynn, 1976; Schutts and Dunlop, 1981), Churchill (Reid et al., 1981; McGlynn et al., 1974) and Superior (Larochelle, 1966; Pullaiah and Irving, 1975; Schultz and Dunlop, 1981) Provinces yield concordant palaeopoles (review in Schutts and Dunlop, 1981 and Irving and McGlynn, 1981). But when, precisely, in the interval -2100 (approximate younger limit of tracks) to -1750 Ma (approximate age of Coronation overprint) the Slave and Superior Provinces achieved their present relative positions (when the Hudsonian Orogen became stabilized) is unknown.

The reconstruction of Figure 16 requires the existence of a wider gap than at present between the Slave and Superior Provinces in the Early Proterozoic. This interpretation is similar in principle to that given by Burke et al. (1976), Cavanaugh and Seyfert

(1978), but is simpler. Two simple segments of apw path for the two provinces replace a complex pattern of loops proposed by these. This simplification is made possible by assuming that the southerly group of palaeopoles from the Nipissing diabase (the *N1* group of Morris (1979) see Irving and McGlynn (1981, Figure 23.10)) are secondary (Mid-Proterozoic) and that the *D1* magnetizations are late Phanerozoic.

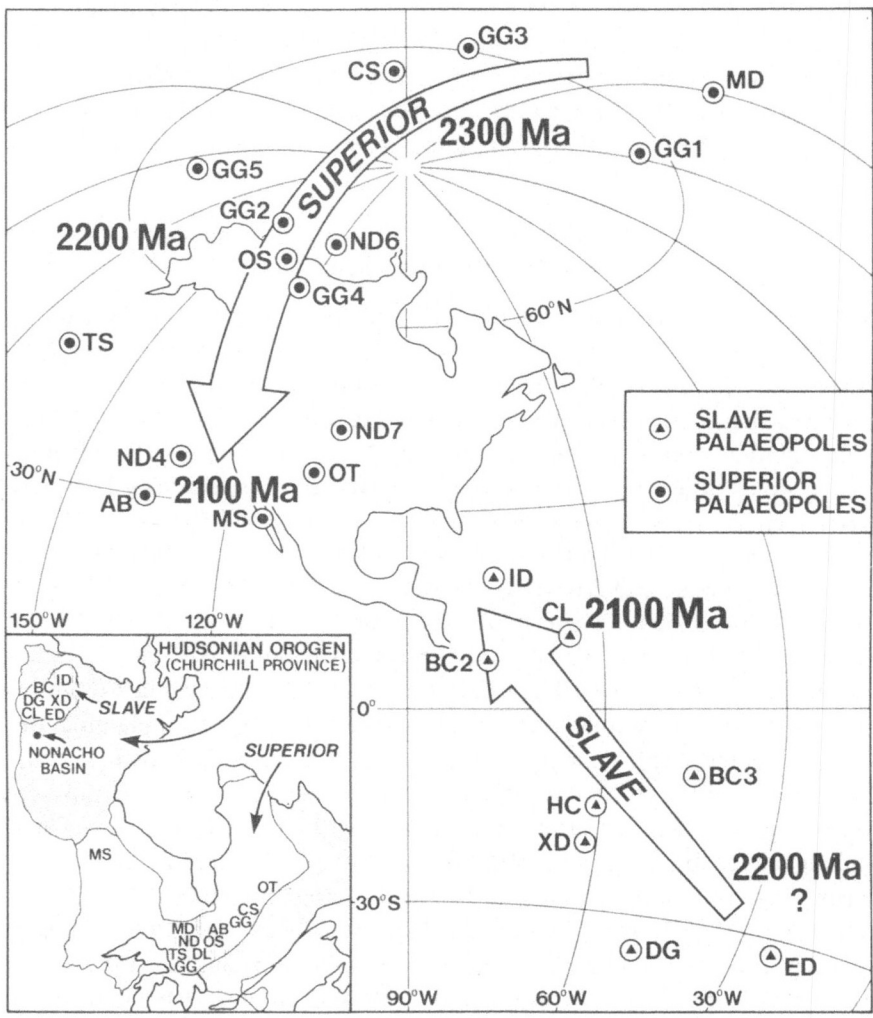

Fig. 16. Tentative Superior and Slave Tracks compared. Slave palaeopoles labelled as in Figure 15. Superior palaeopoles as follows: AB Abitibi dykes (Irving and Naldrett, 1977); CS Chibougamau sills (Ueno and Irving, 1975); GG1 to 5 Gowganda Formation (Symons, 1975; Morris, 1979; Roy and Lapointe, 1976); MD Matachewan dykes (Irving and Naldrett, 1977); MS Molson dykes (Ermanovic and Fahrig, 1975); ND Nipissing diabase (ND4, Symons, 1970; ND6, Symons, 1975, ND7, Roy and Lapointe, 1976)); OS Otto Stock (Pullaiah and Irving, 1975); OT Otish gabbro (Fahrig and Chown, 1973), TS Thessalon Volcanics (Symons and O'Leary, 1978).

This interpretation differs radically from that given earlier (Pullaiah and Irving, 1975; McGlynn *et al.*, 1975; Irving and McGlynn, 1976; Roy and Lapointe, 1976) in which a single south-trending track (then called Track 5) was constructed for the Early Proterozoic for the entire Canadian Shield. We did this because of the excellent agreement between the palaeopole for the Otto Stock (OS) in the Superior Province and the BC1 palaeopole (derived from *D* magnetization which we then assumed primary) of the geochronometrically indistinguishable Big Spruce Complex in the Slave Province. The possibility that *D* may be much younger was discussed at the time but regarded as improbable (Irving and McGlynn, 1976). Our reasons for believing that our initial interpretation of *D* was mistaken have been presented in detail elsewhere (Irving and McGlynn, 1981).

It seems therefore that there is ground for supposing that the Slave and Superior Provincies were separated in the Early Proterozoic and that they had come to their present relative positions by the Middle Proterozoic. That is relative continental drift was occurring then as now. The hypothesis that there was a fixed supercontinent, embracing all continental crust, in existence through the Proterozoic (Piper, 1982) seems incorrect. Piper would connect all palaeopoles of Figure 16 in one continuous highly complex apparent polar wander path. The separate clustering of paleopoles from Early Proterozoic rocks of the Slave and Superior provinces apparently belies this interpretation. However, although the spacial resolution of palaeomagnetic observations is fully adequate to detect Early Palaeozoic drift their timing is not sufficiently accurate to provide a convincing case. The present results therefore, although suggestive of drift do not provide a definitive measure of it. That can only come about through more detailed geochronological studies.

Acknowledgement

Jean Hastie and Elizabeth Tanczyk made most of the laboratory measurements and we are very grateful to them for this work. Jean Roy provided a meticulous review.

References

Burke, K., Dewey, J. F., and Kidd, W. S. F.: 1976, 'Precambrian Palaeomagnetic Results Compatible with the Wilson Cycle', *Tectonophysics* **33**, 287–299.

Burwash, R. A. and Baadsgaard, H.: 1962, 'Yellowknife and Nonacho, Age and Structural Relationships', *Roy. Soc. Can. Spec. Publ.* **4**, 22–30.

Burwash, R. A., Baadsgaard, H., Campbell, F. A., Cumming, G. L., and Folinsbee, R. E.: 1963, 'Potassium-Argon Dates of Dykes-Systems, District of Mackenzie, N.W.T. Trans, *Canad. Inst. Min. Metall.* **66**, 303–307.

Cavanaugh, M. D. and Seyfert, C. K.: 1977, 'Apparent Polar Wander Paths and the Joining of the Superior and Slave Provinces During Early Proterozoic Time', *Geology* **5**, 207–211.

Christie, K. W. and Fahrig, W. F.: 1973, 'Paleomagnetism of the Borden Dykes of Baffin Island and its Bearing on the Grenville Loop', *Canad. J. Earth Sci.* **20**, 275–289.

Davidson, A.: 1978, 'The Blachford Lake Intrusive Suite: An Aphebian Alkaline Plutonic Complex in the Slave Province, Northwest Territories', *Curr. Res. Geol. Survey Canada*, paper 78-1A, 119–127.

Davidson, A.: 1982, 'Petrochemistry of the Blachford Lake Complex near Yellowknife, Northwest Territories', in Y. T. Maurice (ed.), *Uranium in Granites*, Geol. Surv. Canada, Paper 81-23, 71-79.

Ermanovic, I. and Fahrig, W. F.: 1975, 'The Petrochemistry and Paleomagnetism of the Molson Dikes, Manitoba', *Canad. J. Earth Sci.* **12**, 1564-1575.

Fahrig, W. F. and Chown, E. H.: 1973, 'The Paleomagnetism of the Otish Gabbro from North of the Grenville Front, Quebec', *Canad. J. Earth Sci.* **10**, 1556-1564.

Gates, T. M. and Hurley, P. M.: 1973, 'Evaluation of the Rb–Sr Method Applied to the Matachewan, Abitibi, Mackenzie and Sudbury Dyke Swarms in Canada', *Canad. J. Earth Sci.* **10**, 900-919.

Gough, D. I.: 1956, 'A Study of the Palaeomagnetism of the Pilansberg Dykes', *Mon. Not. Roy. Astron. Soc. Geophys. Suppl.* **7**, 196-213.

Hanes, J. A. and York, D.: 1977, 'A Detailed $^{40}Ar/^{39}Ar$ Age Study of an Abitibi Dike from the Canadian Superior Province', *Canad. J. Earth. Sci.* **16**, 1060-1070.

Irving, E. and McGlynn, J. C.: 1976, 'Polyphase Magnetization of the Big Spruce Complex, Northwest Territories', *Canad. J. Earth Sci.* **13**, 476-489.

Irving, E. and McGlynn, J. C.: 1981, 'On the Coherence, Rotation and Palaeolatitude of Laurentia in the Proterozoic', in A. Kroner (ed.), *Precambrian Tectonics*, Elsevier, pp. 561-598.

Irving, E. and Naldrett, A. J.: 1977, 'Paleomagnetism in Abitibi Greenstone Belt and Abitibi and Matachewan Diabase Dykes: Evidence of the Archean Geomagnetic Field', *J. Geol.* **85**, 157-176.

Larochelle, A.: 1966, 'Palaeomagnetism of the Abitibi Dyke Swarm', *Canad. J. Earth Sci.* **3**, 671-683.

Layer, P. W., Kroner, A., and McWilliams, M. O.: 1984, 'Paleomagnetism of the Archean Usushwana Complex, Southern Africa', *J. Geophys. Res.* (to be publ.)

Leech, A. P.: 1963, 'Potassium-Argon Dates of Basic Intrusive Rocks of the District of Mackenzie, N.W.T.', *Canad. J. Earth Sci.* **3**, 389-412.

Martineau, M. P. and Lambert, R.: 1974, 'The Big Spruce Lake Nepheline-Syenite Carbonatite Complex, N.W.T.', (abstr.) *Geol. Assoc. Canada Ann. Meeting Program* p. 59.

McGlynn, J. C. and Irving, E.: 1975, 'Paleomagnetism of Early Aphebian Diabase Dykes from the Slave Structural Province, Canada, *Tectonophysics* **26**, 23-38.

McGlynn, J. C., Hanson, G. N., Irving, E., and Park, J. K.: 1974, 'Paleomagnetism and Age of Nonacho Group Sediments and Associated Sparrow Dykes, District of Mackenzie', *Can. J. Earth Sci.* **11**, 30-42.

McGlynn, J. C. and Irving, E.: 1981, 'Horizontal Motions and Rotations in the Canadian Shield during the Early Proterozoic', in F. H. A. Campbell (ed.), *Proterozoic Basins of the Candian Shield, Geol. Surv. Canada*, paper No. 81-10, 183-190.

McGlynn, J. C., Irving, E., Bell, K., and Pullaiah, G.: 1975, 'Palaeomagnetic Poles and the Proterozoic Supercontinent', *Nature* **255**, 315-319.

Morris, W. A.: 1979, 'A Positive Contact Test between Nipissing Diabase and Gowganda Argillite', *Canad. J. Earth Sci.* **16**, 607-611.

Palmer, H. C., Baragar, W. R. A., Fortier, M., and Foster, J. H.: 1983, 'Paleomagnetism of Late Proterozoic Rocks, Victoria Island, Northwest Territories, Canada, *Can. J. Earth Sci.* **20**, 1456-1469.

Piper, J. D. A.: 1982, 'The Precambrian Paleomagnetic Record – The Case for the Proterozoic Continent', *Earth Plan Sci. Lett.* **59**, 61-89.

Pullaiah, G. and Irving, E.: 1975, 'Paleomagnetism of the Contact Aureole and Ate Dykes of the Otto Stock, Ontario and its Application to Early Proterozoic Apparent Polar Wandering', *Can. J. Earth Sci.* **12**, 1609-1618.

Reid, A. B., McMurry, E. W., and Evans, M. E.: 1981, 'Paleomagnetism of the Great Slave Supergroup, Northwest Territories, Canada: Multicomponent Magnetization of the Kahochella Group', *Can. J. Earth Sci.* **18**, 574-583.

Roy, J. L. and Lapointe, P. L.: 1976, 'The Paleomagnetism of Huronian Red Beds and Nipissing Diabase: Post-Huronian Igneous Events and Apparent Polar Wandering for the Interval − 2300 to − 1500 m.y. for Laurentia', *Can. J. Sci.* **13**, 749-773.

Runcorn, S. K.: 1955, 'Rock Magnetism – Geophysical Aspects', *Phil. Mag. Supp. Adv. Phys.* **4**, 244-291.

Schutts, L. D. and Dunlop, D. J.: 1981, 'Proterozoic Magnetic Overprinting of Archaean Rocks in the Canadian Superior Province', *Nature* **291**, 642-645.

Symons, D. T. A.: 1970, 'Paleomagnetism of Nipissing Diabase, Ontario', *Geol. Surv. Canada* 70-63, 1-29.

Symons, D. T. A.: 1975, 'Huronian Glaciation and Polar Wander from the Gowganda Formation, Ontario', *Geolog* **3**, 303-306.

Symons, D. T. A. and O'Leary, R. J.: 1978, 'Huronian Polar Wander and Paleomagnetism of the Thessalon Volcanics', *Canad. J. Earth Sci.* **15**, 1141-1150.

Ueno, H. and Irving, E.: 1976, 'Paleomagnetism of the Chibougamau Greenstone Belt, Quebec, and the Effects of Grenvillian Post-Orogenic uplift', *Precambrian Res.* **3**, 303–315.
Wanless *et al.*: 1979, 'Geol. Survey Canada', 79–2.

A SYSTEMATIC APPROACH TO RADIOMETRIC AND PALEOMAGNETIC STUDIES IN A MOBILE OROGENIC BELT: I. THE WANING PHASE OF ACTIVITY IN THE SOUTHERN APPALACHIANS OF SOUTH CAROLINA

ROBERT E. DOOLEY*

The Geology Department The University of South Carolina

M. MANZONI

Laboratorio di Geologia Marina, C.N.R. University of Bologna

and

A. E. M. NAIRN

Earth Sciences and Resources Institute, The University of South Carolina

Abstract. The survey of radiometric and paleomagnetic work on the mafic rocks of South Carolina is consistent with, and amplifies the studies on the acidic rocks of the southeast by Ellwood (1982). The westerly post-early Mesozoic tilt of the southeastern Appalachians proposed by Dooley and Smith (1982) over most of the Piedmont balances out the post-late Paleozoic southeastern tilt of Ellwood (1982). Only in the Elberton–Sparta block is the tilting important and here the interpretation proposed is of a greater initial tilt (approximately 25–30°) reduced by the post-early Mesozoic tilt.

There is no evidence of displaced terrains as far as the King's Mountain, Charlotte, and Slate belts are concerned at least since 300 m.y. ago and perhaps as early as 350 m.y. ago. The anomalous paleomagnetic data from the Kiokee belt is best interpreted as due to tectonic displacements associated with the late Paleozoic event described by Secor and Snoke (1978) and Snoke *et al.* (1980).

The paleopoles of the mafic rocks are in agreement with paleopoles on the North American apparent polar wander path (APWP) at about 300 m.y. The resolution of K–Ar apparent ages of 350 m.y. or older will require $^{40}Ar/^{39}Ar$ studies and such age relationships are critical to the reasonable application of tilt corrections in the southern Appalachians.

1. Introduction

Paleomagnetism has gone through a series of new developments in the past two decades that has enabled researchers to study rocks from terrains which have not always been a part of the stable cratonic continental masses to which they are now attached. Paleomagnetic studies of rocks from such areas, until recently, have been avoided because of their questionable value in the establishment of polar wander paths (PWP) for stable cratonic masses used in the reconstruction of continents in the development of theories of continental drift and global plate tectonics (Runcorn, 1962; Bullard *et al.*, 1965; Wilson, 1972; Creer, 1967, 1968; Dewey and Bird, 1970; Smith and Hallam, 1970; Duff, 1980; amongst others). Apparent polar wander paths (APWP) have now been established for the various continents through much of geologic time

---

* Robert E. Dooley is currently at the Department of Geosciences, Northeast Louisiana University, Monroe Louisiana 71209, U.S.A.

(see for example: Van Alstine and de Boer, 1977; Irving, 1977, 1979; Irving and Irving, 1982; Watts, 1982; Briden and Duff, 1981; Bachtadse *et al.*, 1983; Thompson and Clark, 1982). While such APWP's continue to be revised and modified, an increasingly reliable body of pole positions is becoming available from rocks lying outside the cratonic continental areas, in mobile orogenic belts (Van der Voo and Channell, 1980).

Tectonothermal activity in mobile orogenic belts has been shown to be quite complex spatially and temporally (Lowry, 1964; Rodgers, 1967, 1970; Hatcher, 1972; Williams, 1976, 1978). Tectonic motions within orogenic belts such as tilting, rotation, thrusting, strike-slip faulting, and folding have been exhaustively documented (in the Appalachian orogen, see: Rodgers, 1967, 1970; Hatcher, 1972; Butler and Ragland, 1969; Secor and Snoke, 1978), and thermal events of regional dynamothermal metamorphism, retrogressive metamorphism and burial metamorphism as well as cooling events such as exhumation of metamorphic and igneous terrains are throughly documented (in the southern Appalachians see: Hurst, 1970; Hadley, 1964; Butler and Ragland, 1969; Fullagar, 1971; Dallmeyer, 1975, 1978; Zimmerman, 1979). The tectonothermal history of mobile orogenic belts traditionally has been studied mainly by the recognition that metamorphic isograds and mineral pairs have been used to establish the variables of temperature and pressure within an entire orogen throughout the duration of tectonic events. A relative time scale of the various tectonic and thermal events within an orogen has been established using metamorphic isograds in conjunction with structural data (for example see: Cloos, 1964; Rogers, 1967, 1970; Hurst, 1970; and Harte and Johnson, 1969). The limitations of such studies is that metamorphic isograds prove to be useful only over small geographic areas and it is difficult to establish their synchronous nature (Winkler, 1976). Furthermore, such eve its as burial metamorphism and exhumation through erosion and isostatic uplift of orogenic terrain can obscure the occurrence of cogenetic metamorphic mineral pairs.

Radiometric geochronology has been extensively applied to determine the time at which closure of isotopic systems have occurred, thus delineating the times of partial melting, crystallization, intrusion, deformation, and unburial (depending on which isotopic technique is used, for examples in the southern Appalachians see: Hadley, 1964; Kulp and Eckleman, 1961; Butler and Ragland, 1969; Dallmeyer *et al.*, 1981; Wright and Seiders, 1977; Snoke *et al.*, 1980; Zimmerman, 1979; Hurst, 1970; Fullagar and Kish, 1981; Harper and Fullagar, 1981; Fullagar, 1981). If isotopic diffusion is blocked in an isothermal process then the age of the isotopic system ideally represents cooling of such a system through a particular isotherm and such age data have been used to contour areas of similar age ranges to suggest a relationship between the age contours and isotherms (Hadley, 1964; Hurst, 1970; Dallmeyer, 1978; York, 1978). York (1978) coined the term 'Chrontour' lines to represent such contours relating age and temperature to their intersection with the present topographic surfaces. Blocking temperatures of the various isotopic systems vary (depending upon the geologic material dated) from very high-temperatures, greater than 800-1000 °C, (for example, the Nd–Sm isotopic system where the time of isotopic fractionation dates the time the rock was in the magmatic state rather than the time of crystallization) to very low-temperature, 150-450 °C (as seen in K–Ar isotopic systems and in fission track dating techniques where annealing requires temperatures of only 100-150 °C) (Dallmeyer, 1975, 1978; Allegre *et al.*, 1982; Harrison and MacDougal, 1980a, b; Zimmerman, 1979). K–Ar isotopic systems are particularly useful in combination with paleomagnetic data because the range in 'blocking' temperatures of magnetic minerals is similar (York, 1978; Berger and York, 1979). Caution must be taken however in the use of isotopic age data (particularly with K–Ar age data) because of problems with partial resetting of a system (loss of radiogenic daughter atoms), such as in unburial or in thermal pulses of only local extent, and because of problems with incorporation of excess argon (in excess to the radiogenic daughter atoms) in a crystallization or cooling igneous rock from surrounding country rocks, from country rocks at greater depths or excess ^{40}Ar present as initial argon at the site of partial melting (Evans and Tarney, 1964; Dooley, 1977; Dooley and Wampler, 1983; Dalrymple *et al.*, 1975; Giletti, 1971; Wilson, 1972; Hebeda *et al.*, 1973; Claesson, 1976). Incremental release spectra of the ^{40}Ar/^{39}Ar release technique have been extremely valuable in evaluating the conditions of closure of the K–Ar isotopic system and with care can be used to accurately establish the age of blocking temperatures of various mineral phases in a rock (Lanphere and Dalrymple, 1971, 1976; Dallmeyer, 1975, 1978; Harrison and MacDougal, 1980a, b).

While isotopic systems offer temporal relationships of thermal events in igneous and metamorphic rocks within orogenic belts (under certain circumstances), it is paleomagnetism that can uniquely offer a solution for both spatial and temporal relationships of thermal events in such rocks, revealing the thermal and structural history of orogenic belts with one analytical technique (Van der Voo and Channell, 1980). Blocking temperatures of magnetic minerals (where there is a range of domain size in magnetic particles) occur through a thermal process similar to that for closure of isotopic systems in various mineral phases (York, 1978). Thus a magnetic direction may record a remanence of the rock when it crystallized and cooled or it may record one or more secondary thermal (or chemical) overprints which occurred at a later time. In rocks from mobile belts these secondary blocking temperatures may be carried by several 'hard' magnetic components with overlapping coercivity spectra. Ancient areas undergoing isostatic uplift (for example: Precambrian Shields) are continually being uplifted through various isotherms and may therefore record the polar wander path over a substantial portion of geologic time (Morgan, 1976; McWilliams and Dunlop, 1978; Morgan and Briden, 1981; Watts, 1982).

Magnetic systems containing multiple magnetic components, whether primary or secondary, cannot readily be resolved on the conventional Zijderveld diagrams because of the problems of magnetic streaking. The method of least squares has been used to fit remagnetization circles to such magnetic streaking and if the degree of overlap of coercivity spectra is sufficiently small then the intermediate magnetic directions can be resolved and various methods have been described by several researchers (Roy and LaPointe, 1976; Halls, 1976, 1978, 1979; Stupavsky and Symons, 1978; Williamson and Robertson, 1976; Hoffman and Day, 1978; Dunlop, 1979; Park and Roy, 1979; Park, 1981; French and Van der Voo, 1977). There remain however many cases where even such techniques are unsuccessful.

With the development of these techniques to resolve multiple component magnetic systems and of theoretical curves showing the dependence of blocking temperature and magnetic intensity on cooling rates it is possible to consider metamorphic rocks as a potentially valuable subject for paleomagnetic study, and under ideal circumstances, upper and lower age limits may be placed on deformation in orogenic belts (Pullaiah *et al.*, 1975). However, such studies are not without problems because of the complexity of multiply folded metamorphic rocks. In metamorphic terrains, fluid pressures have an important effect on the development of metamorphic fronts. Such catalysts may also impart a secondary chemical remanence to the rock or in a later event, such as hematization of rocks through the oxidizing effects of ground water percolation, both can occur as a post-cooling disturbance of the magnetic systems, but such processes are poorly understood, particularly in resolving the time at which such alteration may occur. An additional complication is the growth of new metamorphic mineralogy in metamorphic rocks under variable metamorphic conditions. Furthermore rocks with certain magnetic mineralogies and grain sizes (multiple domain material) are not necessarily suitable for paleomagnetic investigation (Merrill, 1981). Thus the major task in studying metamorphic rocks is to find a suitable magnetic mineralogy and of sufficiently small grain size (single domain) to have recorded a fossil remanence (Watts, 1982). Even as problems still exist with the accurate recording of the paleomagnetic field, metamorphic rocks with multiple tectonothermal histories may further obscure this record. Such factors will always impart limitations to paleomagnetic study, and such data must be carefully assessed as to its meaning, if meaningful at all.

Given the above mentioned limitations (and there are likely to be more undiscovered limitations) of the various geologic, radiometric, and paleomagnetic techniques, this paper addresses such problems with the first stage of a systematic study of a complex orogenic belt, the southern Appalachians, applying all three techniques, geologic synthesis, conventional K–Ar geochronology, and paleomagnetic study of igneous and metamorphic rocks in the orogen. Such a study must consider various tectonic models that have been proposed for the formation of such mobile regions of the earth's crust, and must also recognize that the various zones into which an orogenic belt can be divided may reflect differences in thermal and/or tectonic history. Thus ideas such as microplate tectonics, large-scale thrusting with accretion of island arcs to continents, large-scale strike-slip faulting of exotic terranes with displacements of greater than 1000 km, tilting of orogenic areas due to domal upwarping and continental splitting after an orogenic phase, and differential erosion of orogenic surfaces (and subsequent tilting therein) must all be considered and tested in a synthesized model of tectonothermal history of an orogenic belt.

As the first step in a systematic study in the southern Appalachians of South Carolina, the mafic igneous rock intrusions that span much of the orogenic history of the southern Appalachians (particularly the younger phases) have been examined and an attempt has been made to resolve paleomagnetic directions for the waning stages of the Alleghenian (Hercynian) orogeny of the Appalachian cycle including initial results from some metamorphic rocks as well as the characterization of a possible early Mesozoic tilt correction for the updoming event that accompanied the opening of Atlantic Ocean and led to continental splitting. These data may be combined with already existing data from granitic igneous plutons (Ellwood, 1982) which together with radiometric age data provide points of reference to the APWP for this area of the southern Piedmont through much of the last 300–350 m.y. The succeeding stage will involve more detailed studies of the complex magnetization of mafic rocks recording earlier stages in the tectonic history of the southern Appalachians.

2. Regional Geology

The Appalachian mountains was originally a sedimentary wedge forming a long linear belt which was transformed into a mountain range during the Appalachian orogenic cycle. The cycle was characterized by three particularly strong orogenic periods associated with deformation, metamorphism and intrusion of magmas, the Taconic (middle to late Ordovician), the Acadian (middle Devonian) and the Alleghanian (late Carboniferous) orogenies. These affected the northern and southern Appalachians to differing degrees (Rodgers, 1967, 1970).

Taconic orogenic movements have been recognized predominantly in the northern Appalachians (for example see: Boucot et al., 1964; Rodgers,1967) and locally in the southeastern portion of the Carolinas and Virginia (Overstreet and Bell, 1965; Hatcher, 1972). Acadian orogenic activity has been found primarily in the northern Appalachians (Canada and New England; Boucot, 1962) and is virtually absent in the southern Appalachians. Alleghanian movements (Hercynian) have been found in the Maritime Provinces and along the southeast side of the northern Appalachians (for example see: Webb, 1963) and to a limited extent in the southern Appalachians (Secor and Snoke, 1978; Snoke et al., 1980).

Exposed rocks of the Appalachian fold mountains occupy the northwestern half of South Carolina. There, a number of belts have been recognized which can be traced into or through the adjoining states of Georgia and North Carolina (Figure 1). The belts have been made familiar by the work of Overstreet and Bell (1965), Hatcher (1972), Butler (1972) and more recently by Secor and Snoke (1978) as well as several other researchers. To the southeast the folded rocks of the Appalachians disappear below the younger, flat lying sediments of the coastal plain. The existence in subsurface of down-faulted early Mesozoic basins and Paleozoic metamorphic and igneous rocks similar to those existing elsewhere along the chain has been recognized through gravity and magnetic surveying (Long et al., 1976; Daniels and Zeitz, 1980) and proved by drilling near Charleston (Gohn et al., 1977, 1978).

The rocks of the southern fold belt are in general poorly exposed and weathering has penetrated deeply in many areas. To this must be added the paucity of information on the age of the rocks. A recently discovered middle to upper Cambrian trilobite (in North Carolina) and trilobite fauna (in South Carolina) within the Carolina Slate belt

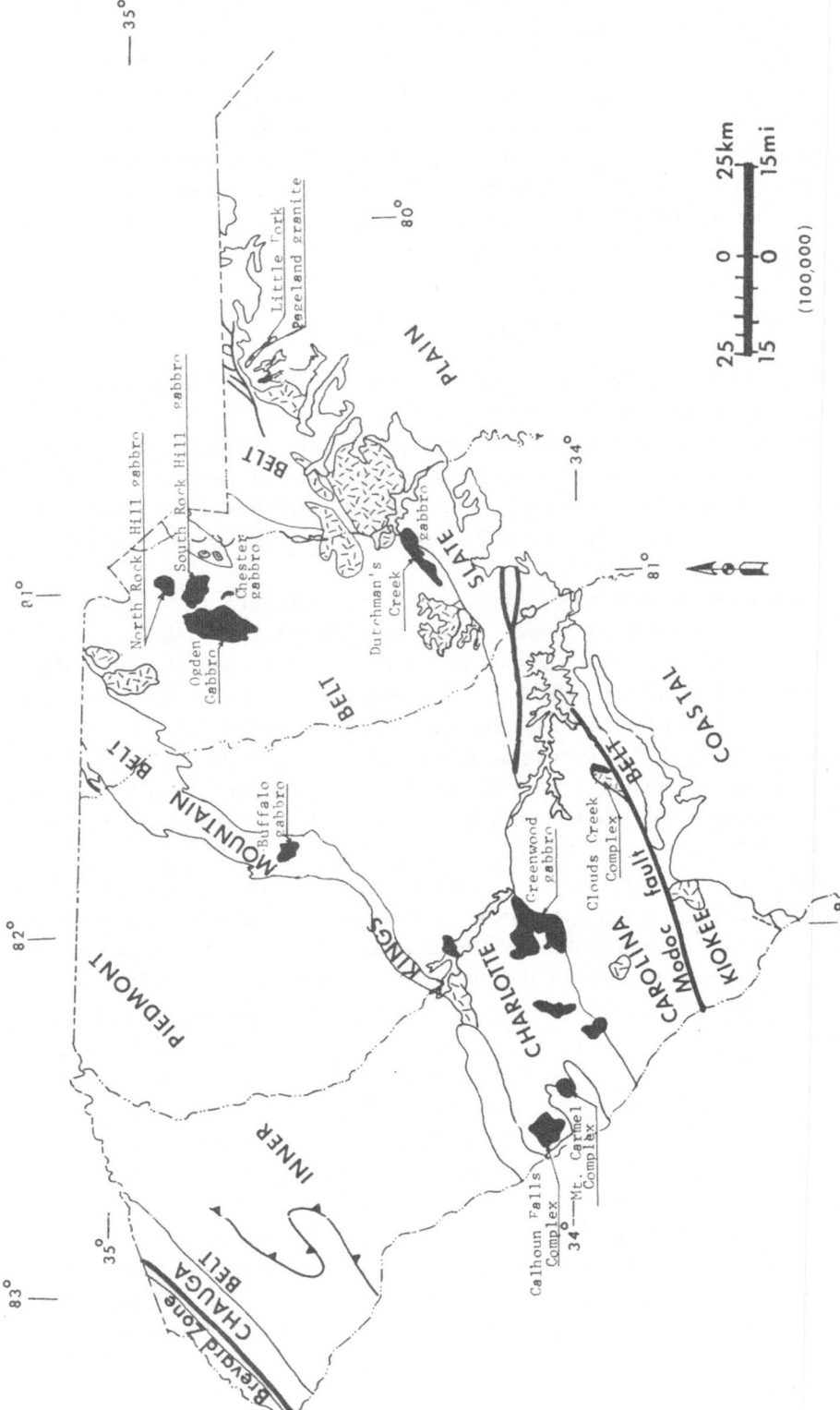

Fig. 1   Geologic Map showing gabbrioc and granitic intrusions in South Carolina (diagrammatic).

dramatically spotlights the almost total absence of paleontological control (St. Jean, 1973; Sara Samson, pers. comm.). This lack of control in consequence has led to an almost total dependance upon radiometric age determinations, made on igneous bodies whose field relationships are well known, as the basic method of assigning ages to the various tectonic events to have affected the area (Fullagar, 1971; Butler, 1972; Fullagar and Butler, 1979).

Significant new advances in the understanding of the geology of the southern Appalachians such as the recognition of an important late Paleozoic tectonothermal event (Secor and Snoke, 1978; Snoke et al., 1980), the interpretation of COCORP seismic profiles in light of thin-skinned tectonics (Cook et al., 1979; Hatcher, 1972; Iverson and Smithson, 1982), as well as the recognition of displaced terranes in the northern Appalachians (Kent and Opdyke, 1978) re-emphasizes the need for accurate temporal and spatial relationships in the southern Appalachians.

In an attempt to provide some of this required data an investigation of the basic igneous rocks, to a large extent ignored up to this point (see: Butler and Ragland, 1969), was begun. Conventional K–Ar age determinations provide provisional information on temporal relationships but, because of argon loss or contamination, verification through $^{40}Ar/^{39}Ar$ methods is needed. Paleomagnetic studies from which spatial information may be derived are not too numerous. Until recently (Bell et al., 1980; Ellwood, 1982; Dooley and Smith, 1982; Dooley, 1983; Brown and Barton, 1983; and Phillips, 1983) little paleomagnetic information had been published on South Carolina rocks in particular or on the southern Appalachian rocks in general apart from progress reports (Edinger et al., 1979; Nairn et al., 1978; Ellwood et al., 1980; Brown and Barton, 1980). The results from the basic rocks discussed in this paper and interpreted within the framework of existing age data, are consistent and complimentary to the recently published works of Ellwood (1982) and Barton and Brown (1983).

The earliest conventional K–Ar data are by Long et al. (1959), and Kulp and Eckleman (1961) with apparent ages in the range 250–350 m.y. old found where a young mineral age belt (250 m.y. old) traversing the Piedmont of Georgia and South Carolina (Long et al., 1959; see also: Dallmeyer, 1978; Hurst, 1970) was identified. The interpretation of their data has centered around uplift and cooling as a result of unburial (Hadley, 1964; Dallmeyer, 1978) rather than around a pervasive thermal event (Long et al., 1959; Hurst, 1970). Butler (1972) also suggested that the localized occurrence of 250–350 m.y. old mica ages from metamorphic rocks in the Inner Piedmont and Charlotte belts was a result of cooling. This contrasts with the suggestion of a Hercynian metamorphism in the Kiokee belt by Snoke et al. (1980).

Radiometric studies summarized by Fullagar and Kish (1980), Harper and Fullagar (1980), and Fullagar (1981) has led to the recognition of four periods of igneous activity: (1) 255 m.y. ago in the Kiokee belt, (2) 320–280 m.y. ago in the Charlotte, Slate, and Kiokee belts, (3) 415–390 m.y. ago in the Charlotte belt only and (4) 545–495 m.y. ago both in the Charlotte and Slate belts.

Only three conventional K–Ar determinations on post-metamorphic mafic intrusions in the southern Piedmont exist (Medlin, 1969; Butler and Ragland, 1969). These are mineral ages in the range from 386–390 m.y. for the Concord syenite (in North Carolina) which intrudes the post-metamorphic gabbro of the Mecklenburg Complex (Butler and Ragland, 1979). The post-metamorphic gabbroic bodies themselves have been studied little.

The age of the early Mesozoic diabase dikes in eastern North America is relatively well known through the work of Sutter and Smith (1979), Smith and Noltimier (1979), Dooley and Wampler (1983), Dooley and Smith (1982). In the southeast two periods of igneous activity were identified (195 and 180 m.y.)* similar to those in the northeast (Sutter and Smith, 1979) and there exists the possibility of a third period at 205 m.y. One consequence of the work of Dooley and Wampler (1983) was the recognition of contamination by excess ^{40}Ar and an attempt has been made to relate this to crustal thickness.

In the present study the results of radiometric and paleomagnetic study of mafic igneous and metamorphic rocks of the southern Appalachians are reported. The mafic rocks of the southern Appalachians which were studied using radiometric and paleomagnetic techniques can be divided into two groups, the early Mesozoic intrusives and the late Paleozoic mafic intrusives (predominantly gabbros). The amphibolites of the Kiokee belt were only studied paleomagnetically, for radiometric results were already available (Secor and Snoke, 1978; Snoke et al., 1980). Studies of the early Mesozoic intrusives have indicated age results in the 180–205 m.y. range, while the Kiokee belt metamorphism has been identified as in the age range 240–300 m.y. (Permian to late Carboniferous). The conventional K–Ar apparent ages from the late Paleozoic gabbros appear to occur in two age ranges; 499–1565 m.y. and from 285–426 m.y. At least the older group of ages are thought to have been influenced by the presence of excess ^{40}Ar, since they appear older than the rocks they intrude. Thus, it is rocks of post-Taconic age and their implications in the interpretation of the tectonic history of the southern Appalachians particularly as it affects the South Carolina Piedmont which are considered here.

3. Radiometric Data from Mafic Rocks in South Carolina

Analytical procedures for conventional K–Ar analyses appear in Dooley and Wampler (1983) and Dalrymple and Lanphere (1969). Descriptions of the mass spectrometric analysis and the argon extraction line can be found in Dooley and Wampler (1983) and thus will only briefly be described here. Fresh rock samples were crushed and split using a jaw crusher and aluminum sample splitter. Potassium was measured in one aliquot by atomic absorption spectrophotometry. Argon isotopes

---

* Values for the decay constants and isotopic abundance of ^{40}K recommended by the IUGS subcommission on Geochronology in 1976 (Steiger and Jaeger, 1977) differ from those that were in general use until recently, the difference is such that K–Ar apparent ages near 200 m.y. are 2.3% greater when the new values are used. In referring to earlier work in which the newly recommended values were not used, we have recalculated the radiometric ages using the new values.

TABLE I

Conventional K–Ar data for igneous rocks of the southern Piedmont

Sample name	Potassium (%)	'Radiogenic' Argon[a] (%)	(nmol g^{-1})	Calculated age (m.y.)[b]	Locality	Description of material
(A) Early Mesozoic Intrusives						
Diabase Dikes						
CD46–3–6	0.174	77.32	0.0616	204.6±4[c]	South Carolina	Whole rock diabase
CD38–4–6	0.345	85.16	0.1864	287.5±6		Whole rock diabase
CD56–5–1	0.196	78.09	0.0726	201.8±4		Whole rock diabase
CD56–6–2	0.280	65.88	0.0994	193.9±5		Whole rock diabase
CD54–7–1	0.162	76.08	0.0565	190.7±4		Whole rock diabase
CD52–8–2	0.169	82.33	0.0597	193.2±4		Whole rock diabase
CD45–10–2	0.152	67.56	0.0540	194.2±5		Whole rock diabase
CD46–11–3	0.269	73.28	0.9590	194.7±4		Whole rock diabase
CD46–12–3	0.208	75.65	0.0739	194.2±4		Whole rock diabase
CD54–13–6	0.300	58.90	0.1140	206.7±5		Whole rock diabase
CD54–14.7	0.231	67.8	0.0774	183.6±4		Whole rock diabase
Corehole 1	–	–	–	109.0±4		Whole rock lava
Corehole 1	–	–	–	94.8±4		Whole rock lava (Gohn, 1977)
Rhyolite Dikes					North Carolina	
RMWD1	2.814	97.03	0.9757	187.8±4		whole rock rhyolite
Dup	2.806	97.08	0.9538			
RBFT10B	4.340	93.77	1.4914	178.0±10		sanadine separate
Dup	4.710	93.63	1.4371			(Delorey *et al.*, 1982)
(B) Post-Metamorphic Late Paleozoic Gabbros					South Carolina	
Buffalo Pluton						
CDU18–16–3	0.233	93.34	0.6259	1107±25		whole rock gabbro
Dup	0.237	92.70	0.6190			
CDU18–16–4	0.184	91.95	0.7654	1464±50		whole rock gabbro
Dup	0.173	92.30	0.7757			
Dutchman's Creek Pluton						
DC1C	0.583	92.19	0.4278	396±30		whole rock gabbro
Dup	0.597	91.65	0.4895			
DC1HB	0.097	77.85	0.0967	499±25		hornblende separate
DC2	0.685	83.65	0.5097	385±12		whole rock gabbro
South rock Hill Pluton DD11–1–4	0.335	81.48	0.2262	353±11	South Carolina	Whole rock gabbro
Ogden Pluton DD8–1–4	0.109	90.98	0.4394	1494±45		Whole rock gabbro
Chester Pluton DD23–2–6	0.248	96.95	0.6006	1034±30		Whole rock gabbro
Calhoun Falls Pluton DC39–1–4	1.679	97.39	1.1606	360±11		Whole rock gabbro

Table I (continued)

Sample name	Potassium (%)	'Radiogenic' argon[a] (%)	Calculated (nmol g^{-1})age (m.y.)[b]	Locality	Description of material
Mt. Carmel Pluton					
CC40-1-2	0.309	96.63	0.4394	676 ± 1[c]		Whole rock gabbro
Pageland Gabbro					
LF–13–BC	0.877	93.01	0.5000	293 ± 6		Whole rock sample
Dup	0.915	93.83	0.4886			
LF–13–BC–BT					
	3.903	94.56	2.1503	285 ± 20		Biotite separate
Dup	4.501	96.01	2.3265			
Speumarrow Pluton				North Carolina	
DN–1	0.498	89.91	0.3482	364 ± 11		Whole rock gabbro
Chatham Co Pluton					
DN2–9	0.115	76.79	0.1742	712 ± 25		Whole rock gabbro
Bear Poplar Pluton					
DN9–3	0.736	96.61	0.6129	426 ± 13		Whole rock gabbro
(C) Late Paleozoic Diorites					
CD46–1–5	1.485	93.95	0.8038	288 ± 10		Whole rock diorite
CD46–2–2	1.240	95.41	0.6148	265 ± 10		Whole rock diorite

[a] The term 'Radiogenic argon' is used to represent the conventional formula for calculating an apparent age (Dalrymple and Lanphere, 1969). It is put in quotation marks because it is known that several of these samples contain excess argon and are not radiogenic in the sense that the conventional formula requires.

[b] The calculated age is regarded as an apparent age and is calculated using the decay constants and isotopic abundances of Steiger and Jaeger (1977): $n^{40}K/nK = 0.000\,116\,7$; $\lambda_\epsilon = 0.581 \times 10^{-10}\,y^{-1}$; $\lambda_\beta = 4.9162 \times 10^{-10}\,y^{-1}$.

[c] The analytical precision is an estimate at the 66 % confidence level of reproducability for duplicate analyses carried out in another study (Dooley and Wampler, 1983). Further description of this estimate can be found there.

were released from another aliquot by melting the sample in a vacuum furnace purging contaminants before being transferred into the mass spectrometer in the static mode. The argon isotope ratios were used to calculate the amount of 'radiogenic' and 'atmospheric' argon by a conventional formula (Dalrymple and Lanphere, 1969, p. 56–57). The results of these analyses are listed in Table I. Most of these data result from conventional K–Ar analyses of single whole rock samples. There are however a few mineral separate analyses (RBFT 10; LF–13–BC–BT; DC1–HB) and some duplicate analyses (the Buffalo Pluton CDU–18–16; CDU–18–17; LF–13–BC–BT; RBFT 10) in a few instances. The large error limits on the mineral separates is thought to be the result of sample inhomogeniety and such error limits are greater than our average analytical precision. The estimate of analytical precision is based on repeated analyses of diabase samples and standard samples (Dooley and Wampler, 1983) and is taken to be 2% at the 66% confidence level.

4. Paleomagnetic Data from South Carolina Mafic Rocks

The sampling and measurement techniques used have been described by Dooley and Smith (1982) and Dooley (1983). Results from the gabbroic bodies have been presented in Dooley (1983), while the early Mesozoic diabase dike results from South Carolina diabase (Dooley and Smith, 1982) add to the growing volume of data on these young intrusions. There are also detailed studies on two sites from an amphibolite unit in the Kiokee belt. The results from these rocks and from several sites of acidic dikes from the central Piedmont of North Carolina (of early Mesozoic age) are summarized in Table II.

TABLE II

Paleomagnetic results for mafic rocks in the south Carolina Piedmont

N	K	alpha 95[a]	Long° E	Lat° N	Reference	Location
(A) Early mesozoic diabase dikes						
25	–	7.7	80.1	64.9	Watts, 1975	North Carolina Ga. S.C.
5	22.0	6.3	83.0	62.0	Bell *et al.*, 1980	South Carolina
10	14.5	8.9	83.9	66.1	Dooley and Smith, 1982	South Carolina
Mean pole position for diabase dikes						
50	48.9	3.1[a]	86.1	66.4		
Early mesozoic rhyolite dikes						
6	36.0	11.3	122.9	63.7	Delorey *et al.*, 1982	North Carolina
(B) Late paleozoic mafic igneous rocks with single characteristic magnetization						
Buffalo Pluton						
6	39.2	10.8	121.5	39.4	Dooley, 1982	South Carolina
Dutchman's Creek Pluton						
19	32.1	6.0	122.3	37.8	Dooley, 1982	South Carolina
Clouds Creek Gabbro						
4	35.5	15.6	137.9	45.5	Dooley, 1982	South Carolina
Little Fork Gabbro						
2	201.9	17.7	103.2	32.1	Dooley, 1982	South Carolina
Jenkinsville diorites						
2	–	–	103.2	33.3	Dooley, 1982	South Carolina
Mean pole position for late paleozoic mafic igneous rocks						
6	26.7	10.2[a]	120.8	38.9	Dooley, 1982	

N = Number of sites.

K = Fisherian (1953) precision parameter.

alpha 95 = semi-angle of the cone of confidence.

[a] A_{95} = circle of confidence using site vectors as pole positions in calculating mean pole positions.

5. Discussion

5.1. POST OROGENIC EARLY MESOZOIC IGNEOUS ACTIVITY

The youngest exposed igneous rocks in the southern Piedmont of South Carolina are the diabase dikes of late Triassic to early Jurassic age (Table I) for there is no evidence of Cenozoic activity such as is found in Virginia (Fullagar and Bottino, 1969; Core *et al.*, 1974; Wampler and Dooley, 1975; Ressetar and Martin, 1976). Acidic dikes in North Carolina formerly regarded as possibly Cretaceous age (Spence and McDaniels, 1979) are actually early Jurassic in age, following a paleomagnetic and radiometric (K–Ar and Rb–Sr) study (Delorey *et al.*, 1982).

The large number of early Mesozoic diabase dikes is only becoming apparent as a result of detailed mapping together with magnetic and aeromagnetic surveys (Bell *et al.*, 1980; Daniels and Zeitz, 1980; Ragland, 1982; Ragland *et al.*, 1983), for as the diabase rapidly breaks down under the prevailing climatic regime, outcrops are frequently reduced to boulder trains. These early Mesozoic dikes are some of the most widely studied igneous rocks of the southern Appalachians geochemically, paleomagnetically, and radiometrically (for example see: Weigand and Ragland, 1970; Deininger *et al.*, 1975; Watts, 1975; Dooley, 1977; Dooley and Wampler, 1978; Meadows, 1978; de Boer and Snider, 1979; Bell *et al.*, 1980; Dooley and Wampler, 1983) and to this the diabase dikes of South Carolina are no exception (Dooley and Smith, 1982). Two pulses of mafic activity are reported in the northern Appalachians (Smith and Noltimier, 1979) at 180 and 195 m.y. (all radiometric dates are calculated or recalculated using the decay constants and isotopic abundances of Steiger and Jaeger, 1977; see footnote[*]). In South Carolina the 195 m.y. ages are dominant but the 180 m.y. ages are also found and there is a suggestion of perhaps an older group near 205 m.y. (Dooley and Smith, 1982). Verification of these latter groups must await $^{40}Ar/^{39}Ar$ incremental release studies. There are two K–Ar age determinations of tholeiitic flows from under the coastal plain of South Carolina (109 and 95 m.y.) where the ages are suspect, for argon loss may have occurred from these highly altered and weathered rocks (Gohn *et al.*, 1977, 1978; Gottfried *et al.*, 1978). The most reliable age determination for southern diabase is perhaps that of Lanphere (1983) who produced a $^{40}Ar/^{39}Ar$ plateau age of a basalt flow at Clubhouse Crossroad test hole No. 2 of 184 ± 3.3 m.y.

The apparent ages for the South Carolina diabase dikes do not range as high as those examined in Georgia (Dooley and Wampler, 1983). They are however consistent with the trend to progressively higher apparent ages from the southeast to the northwest across the Piedmont (see Figure 2). Apparent ages plotted as a histogram (Figure 3) show a peak in the 180-200 m.y. range, similar to that recorded for both Georgia (Dooley and Wampler, 1983) and Liberia (Dalrympler *et al.*, 1975).

The remanence directions yield a geomagnetic pole position not significantly different from the 190 m.y. pole position (Irving and Irving, 1982). As Dooley and Smith (1982) point out, this direction assumed no tilting of the Appalachian surface.

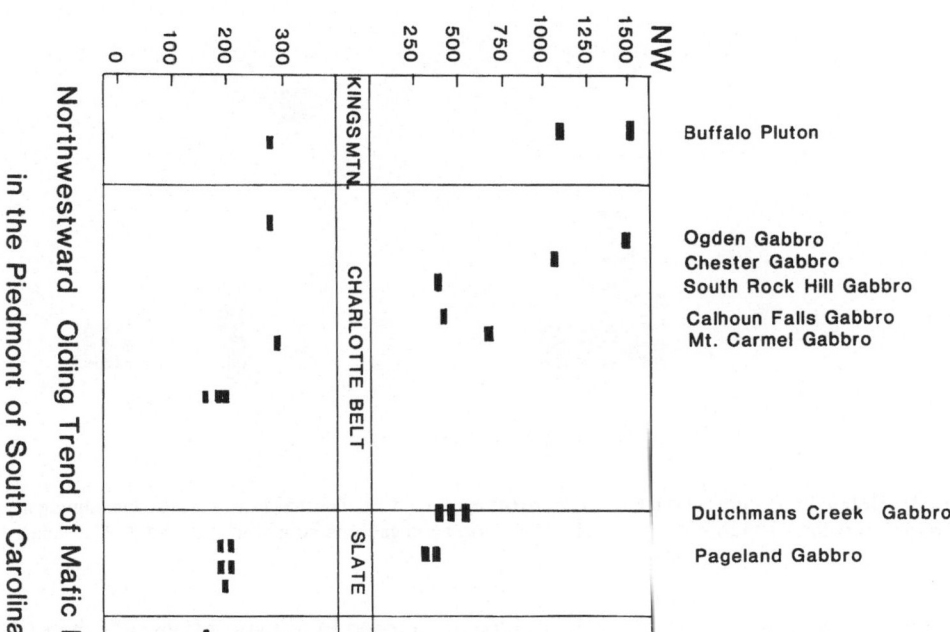

Fig. 2. Diagrammatic cross section across the South Carolina Piedmont with Conventional K-Ar apparent ages plotted as a function of position in the Piedmont with a comparison to the metamorphic time line (a) early Mesozoic diabase and (b) post-metamorphic gabbros.

Two tilt models can be tested, intrusion when the surface was updomed prior to Atlantic rifting (May, 1971) or subsequent tilting following isostatic rebound and erosion of the Appalachians, by applying a modest 10° tilt correction. The result of this test, while not offering definitive proof, of the first model appears to have ruled out the second. Ellwood *et al.* (1978) and Ellwood (1982) have proposed a 15° southeasterly post-Carboniferous tilting of the Elberton-Sparta crustal block, in a direction opposite to the proposed post-early Mesozoic tilt of Dooley and Smith's (1982) but in the adjacent Piedmont the 'Atlantic block', they considered no tilt correction necessary for their group 'A' granitic rocks. Most of the gabbros of Dooley (1983) were collected

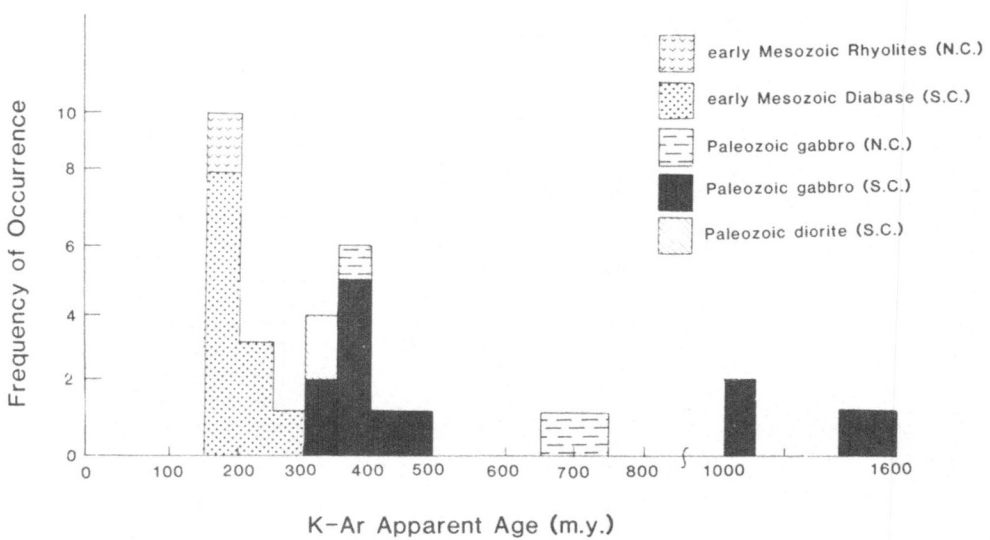

Fig. 3   Histogram of conventional K–Ar apparent ages for data reported in this study, including early Mesozoic diabase and rhyolites, late Paleozoic diorites and gabbros from North and South Carolina.

from sites occurring within the Atlanta block proposed by Ellwood (1982). If their tilt correction proposed is applied to the whole of the area it would virtually balance out the tilt correction of Dooley and Smith's (1982), explaining why no correction was considered necessary by Ellwood (1982). Such tilting is presumably in opposing direction and would leave Ellwood's interpretation for his group 'A' rocks unchanged. In light of the post-Mesozoic tilt the simplest interpretation is of two tilts which essentially balance out with the net result similar to that as if no tilt correction were applied. In the Elberton-Sparta block, on the other hand, the difference between the two tilts (of 15°) requires a steeper initial tilting for the Elberton-Sparta block, an observation that is similar to that of Ellwood's (1982).

5.2. Late Paleozoic mafic intrusions

The conventional K–Ar apparent ages listed in Table I must be considered provisional since in most cases there are no duplicate analyses and $^{40}Ar/^{39}Ar$ incremental studies have not been completed. As the histogram of these apparent ages illustrates (Figure 3) there are two groups; a 386–426 m.y. old group and an anomalously older 499–1595 m.y. old group. This latter group of apparent ages is anomalous as they are generally older than the metasediments into which they are intruded (St. Jean, 1973; Stromquist and Sundelius, 1969; Sara Samson, pers. comm.). The source of the anomalously old ages may represent the presence of excess ^{40}Ar as has clearly been documented in other K–deficient mafic rocks of basaltic composition (Dooley and Wampler, 1983); certainly the age trend is similar (Figure 2) though not as clearly defined. The significance of the occurrence of excess ^{40}Ar in mafic rocks in increasing amounts across the Piedmont to the northwest has been related to differences in crustal thickness (Dooley and Wampler, 1983) as suggested by the gravity gradient in the Charlotte-Inner Piedmont belt area (Long, 1979). Other geochemical and geophysical observations point to a distinct difference in the character of the crust at depth across this gradient (Zeitz et al., 1980; Ellwood et al., 1980; Wenner, 1981; Ellwood, 1982). Any further comparisons about the incorporation of excess ^{40}Ar in gabbros relative to that of diabase cannot be reasonably made because a great difference exists between the intrusive and cooling histories of the older (up to 100–150 m.y. older) plutonic gabbros and the younger hypabyssal early Mesozoic diabase dikes.

The younger group of apparent ages of post-metamorphic gabbros (386–426 m.y. old group) is more difficult to interpret. Since these apparent ages are younger than the rocks they intrude there is no a priori evidence to suggest the presence of excess ^{40}Ar on the basis of conventional K–Ar data alone. Also, while some gabbros have shown to contain excess ^{40}Ar (above) the possibility of argon loss must equally be considered for the younger group of ages. Thus, no definite age can be placed on the time of the cooling of post-matamorphic gabbros on the basis of preliminary K–Ar data. Studies using the ^{40}Ar/^{39}Ar technique will be required to determine whether argon loss or the incorporation of excess ^{40}Ar has affected the apparent ages of the post-metamorphic gabbros. Unlike the situation for the early Mesozoic diabase dikes where there is fairly definite knowledge of their age (Dooley and Smith,1982; Sutter and Smith, 1979) and levels of excess ^{40}Ar contamination (Dooley and Smith, 1982; Sutter and Smith, 1979) and further, the levels of excess ^{40}Ar contamination (Dooley and Wampler, 1983) can be estimated, no such comparison of age or estimates of excess ^{40}Ar contamination can be made for the South Carolina gabbros. It is interesting to note that the only other K–Ar conventional data (386 to 390 m.y.) for biotite separates from a diorite intruding the Mt. Carmel pluton (Medlin, 1968, 1969) reinforces the suggested apparent age range of the younger group of gabbros in this study. Finally, several gabbroic bodies from North and South Carolina have been studied using Nd-Sm mineral isochrons (McSween et al., 1982) with an average age of 405 ± 30 m.y. This has been interpreted as the age of emplacement of the Rock Hill and Ogden plutons. Given the present uncertainties of the radiometric age data for the post-metamorphic gabbros we must also turn to geologic and paleomagnetic evidence for additional information of the geologic age of the post-metamorphic gabbros.

Nine of the mafic intrusions for which there is geochronological data have been investigated with the paleomagnetism, five gabbroic and two small meta-dioritic bodies are described by Dooley (1983). These data are listed in Table II. The remaining gabbros in the South Carolina Piedmont possess multiple hard components not yet fully analyzed.

This mean pole position for these seven bodies plots on the North American APWP (Irving, 1977, 1979; Irving and Irving, 1982) in a position close to the 300–310 m.y. pole; with a circle of confidence enclosing poles in the 250–350 m.y. range (Figure 4). It is consistent with other late Carboniferous paleopole positions for North America and in particular for igneous rocks in the southern Appalachians (Brown and Barton, 1980;

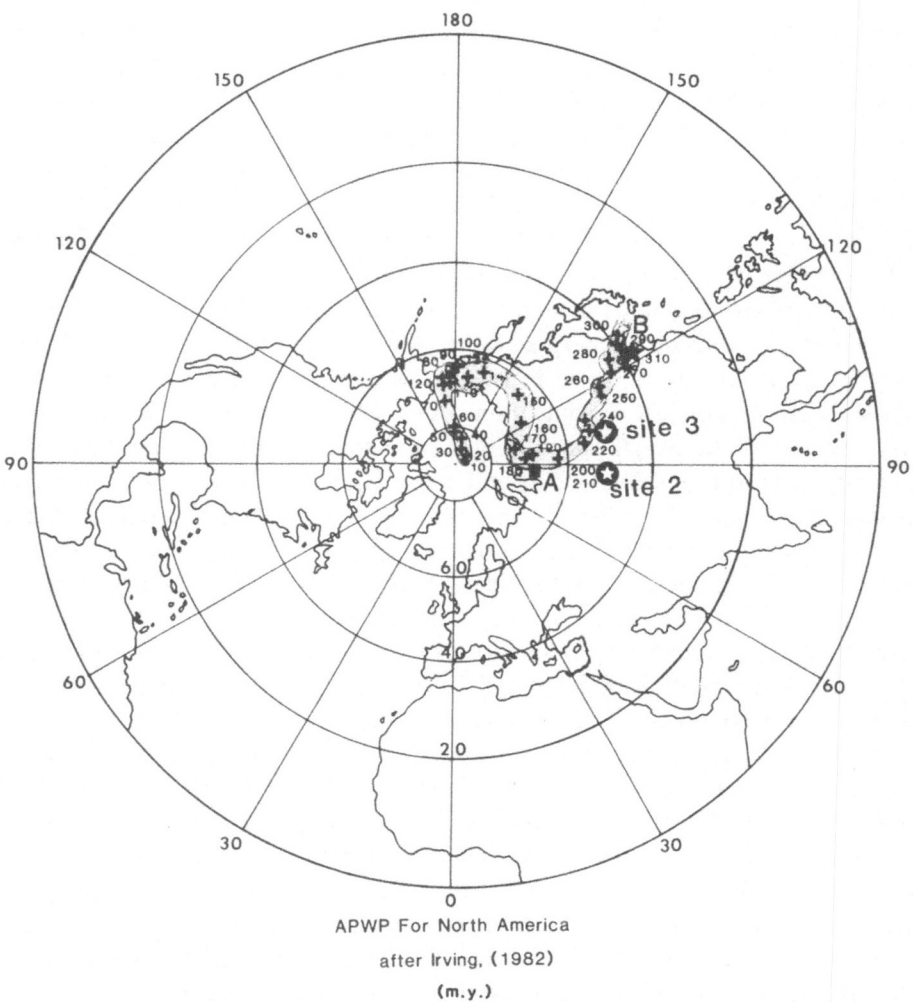

Fig. 4. Mean pole position for the early Mesozoic diabase and the late Paleozoic gabbros relative to the APWP of North America (Irving, 1977, Irving and Irving, 1982).

Barton and Brown, 1983; Watts and Van der Voo, 1979; Ellwood, 1982; Dooley, 1983). If as indicated earlier the post-late Paleozoic tilt correction of Ellwood appears to balance out with the post-early Mesozoic tilt, the conclusion is that the best representation of the age of these rocks from paleomagnetic data is around 300 m.y. This is in agreement with Ellwood's (1982) estimates based upon his group 'A' acidic rocks; it contrasts with the apparent K–Ar ages for the basic rocks, particularly those that do not intrude the Carboniferous acidic plutons. Those gabbros that do intrude such acidic plutons indeed give K–Ar apparent ages around the 300 m.y. range thus it is these pole positions that are compared to the results from gabbros that do not intrude acidic plutons. If for the 'Atlanta Block' the westerly tilt correction is applied, then the age estimate from paleomagnetism shifts to about 350 m.y. The $^{40}Ar/^{39}Ar$ incremental release spectra of mineral separates thus should provide not only age discrimination but aid in tilt correction hypotheses.

5.3. THE KIOKEE BELT AMPHIBOLITES

In South Carolina the Kiokee belt is bounded to the northwest by the Carolina Slate belt but to the southeast it generally disappears under the younger sediments of the Coastal Plain except in the region of the Savannah River where it is bounded by the Belair belt (O'Conner and Prowell, 1978).

The rocks of the Kiokee belt chiefly consist of polyphase deformed amphibolite facies metamorphic rocks and associated granitoid bodies. These have recently been described by Secor and Snoke (1978), and Snoke *et al.* (1980). Protoliths of the metamorphic rocks included felsic to mafic volcaniclastic rocks (Secor and Wagener, 1968; Overstreet and Bell, 1965), which accumulated during late PreCambrian and early Paleozoic time in an island arc environment (Butler and Ragland, 1969; Hatcher, 1972). Comparable rocks of the adjacent Carolina slate belt have not been metamorphosed beyond greenschist grade. The higher grade metamorphism of the Kiokee belt is considered to have occurred in the late Paleozoic during the interval 290–250 m.y. ago (Kish *et al.*, 1979; Secor and Snoke, 1978; Snoke *et al.*, 1980). The effects of this Hercynian event were impressed over those of an earlier low grade metamorphic event, probably Taconic. The Taconic event is the principal phase of deformation of the Slate belt, subsequently an epoch of intrusive activity is defined by radiometric ages of about 300 m.y. (Kish *et al.*, 1979; Fullagar and Butler, 1979). Only subtle Hercynian effects (Snoke *et al.*, 1980) are seen in these Slate Belt rocks.

The boundary between the Kiokee and Carolina Slate belts records a complex history with an early ductile phase of deformation and late stage activity characterized by brittle fracture. According to Secor and Snoke the approximate limit between the two belts is similar to an Abscherungzone (D_2) characterized by a steep metamorphic gradient along which the Modoc fault was later localized.

The work of Kish *et al.* (1979) and Secor and Snoke (1978) relates the phase of high grade metamorphism to the interval 290–250 m.y. ago, their 'Hercynian' (Alleghanian) event. The K–Ar radiometric ages from mineral separates show slight differences consistent with the diachronous closure of the hornblende and biotite phases at different temperatures (530 and 280 °C respectivily). The argon became locked in hornblende at a temperature close to the Curie temperature of magnetite, consequently a paleomagnetic study should reveal a remanence which dates from this blocking temperature. A variety of high grade metamorphic rocks were sampled in the Lake Murray spillway (Figure 5). Of these samples stable results were obtained only from the fine grained amphibolites.

The paleomagnetic study of the amphibolites indicates that consistent direction of magnetization, although scattered, could be obtained after magnetic cleaning in alternating fields from the fine grained varieties (Figure 6). The coarser grained

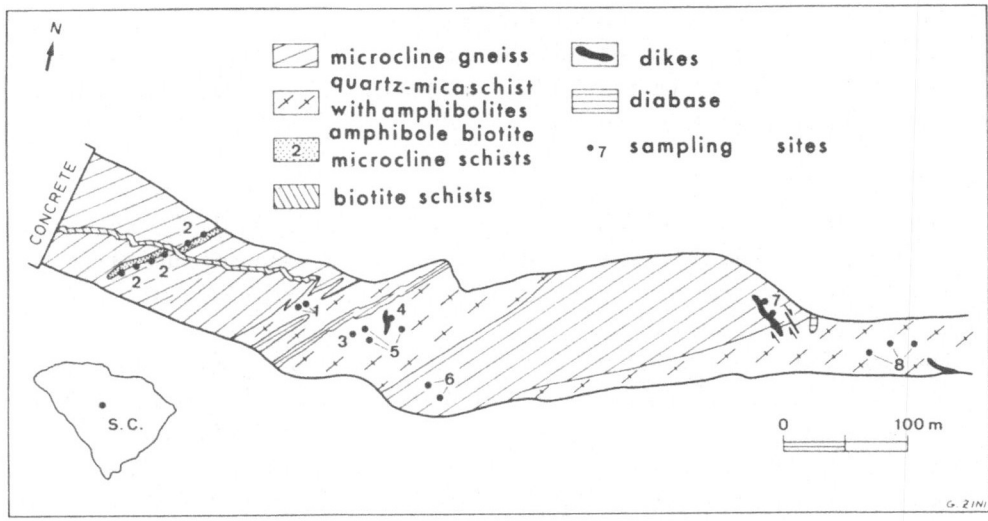

Fig. 5.   Sample distribution within the Lake Murray Spillway.

varieties such as the gneissic samples, had an unacceptably high degree of dispersion even after cleaning. The results from the two sites sampled in Lake Murray spillway near Columbia, South Carolina yielded the following:

Site 2   Biotite amphibole schist (24 samples) declination 185.4°, inclination 0.4°, alpha 95, 8.4°; demagnetizing field, 35 MT, Paleomagnetic pole position of 89.2° E; 55.7° M, dp 4.2°; dm 7.3°.

Site 3   Fine grained amphibolite (9 samples) declination 180.8°; inclination 3.0°; alpha 95 14.5°; demagnetizing field, 35 mT; paleomagnetic pole position 97.3° E; 57.4° N; dp, 8.14°, dm, 14.5°.

The tentative virtual geomagnetic pole position (VGP) for these rocks does not coincide with results from the stable craton for upper Carboniferous or lower Permian times (Figure 7). Since the VGP position most closely approximate the lower to middle Mesozoic North American geomagnetic poles and such an age is inconsistent with geological and radiometric data, a tectonic interpretation is proposed. The better geometric solution to return the observed directions of remanence to a direction consistent with hornblende radiometric dates involves a rotation of 20° about a vertical axis accompanied by a 15° tilt to the north-northwest. This tilt, at least, is similar to that proposed by Dooley and Smith (1982) as a possible tilt correction which could be applied to diabase dikes. The rotation about the vertical axis is visualized as a change in direction during tectonic transport (i.e.: folding and for minor rotation of a thrust slice). As the alternate geometric solution requires a large (80°) angular rotation about a horizontal axis oblique to the trend of the Appalachians it is considered geologically less probable. The direction suggested for granitic rocks in Elberton-Sparta block (Ellwood, 1982), some of which lie in the Kiokee belt region is somewhat different.

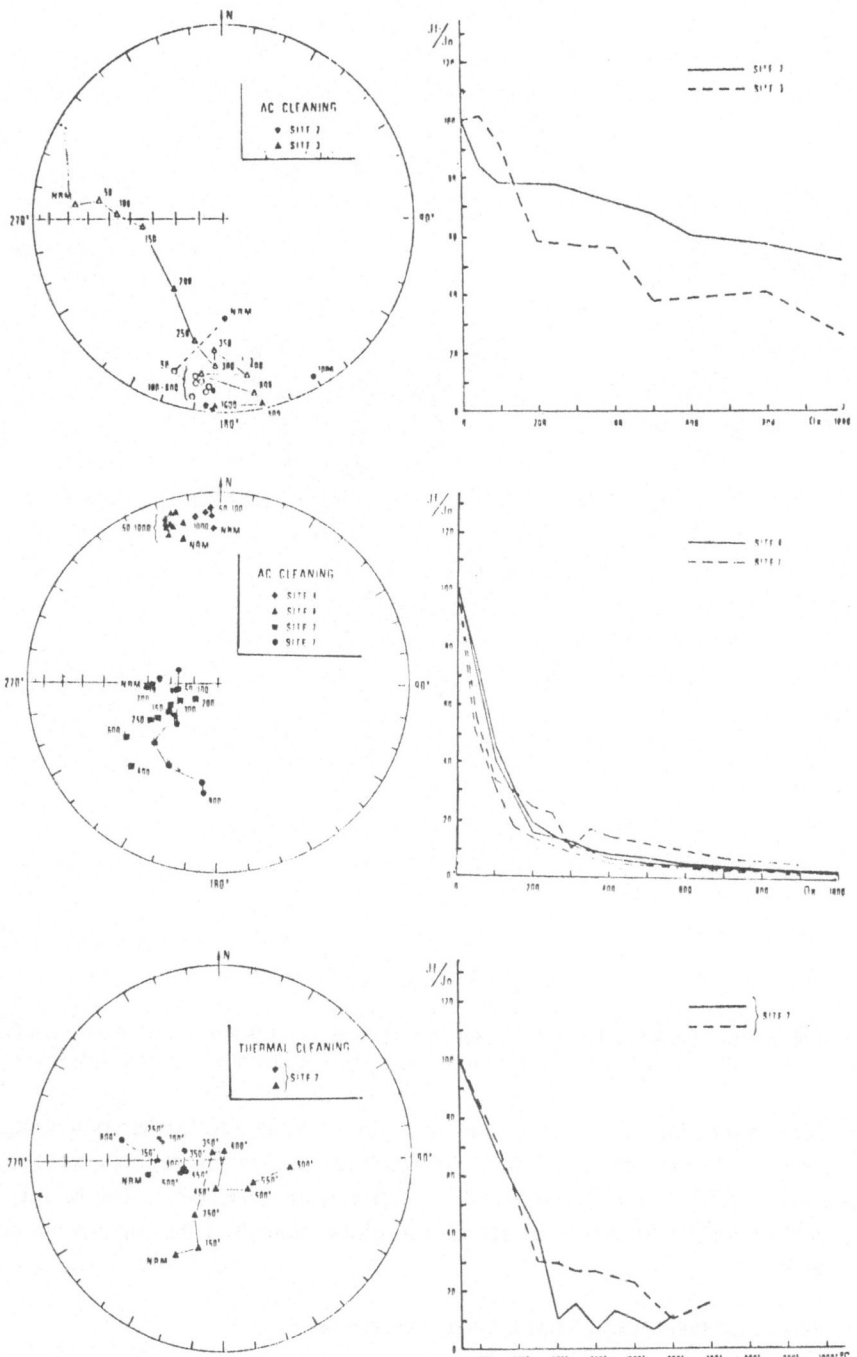

Fig. 6.   AF demagnetization spectra and the effects of magnetic cleaning for selected amphibolite samples.

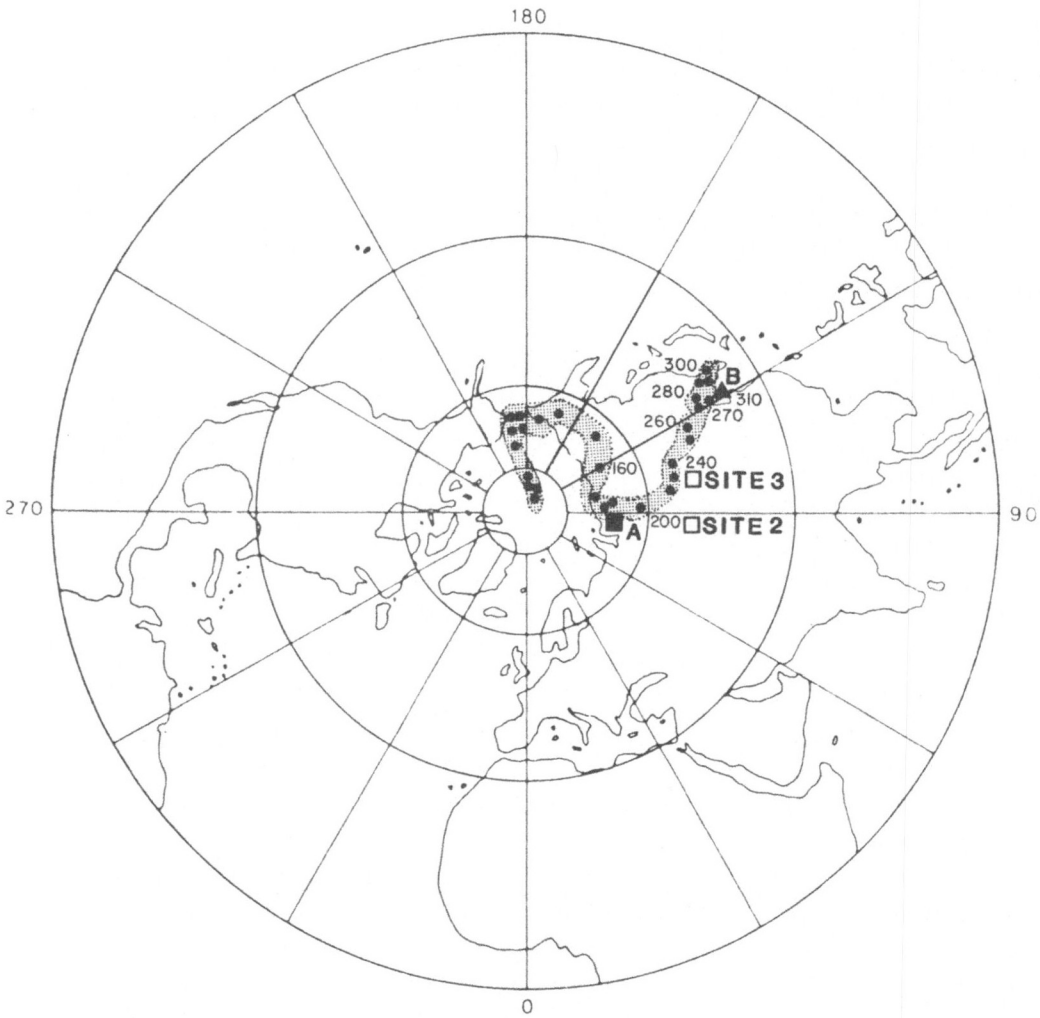

Fig. 7.   Pole position for Amphibolite rocks plotted relative to the APWP for North America in Permian
times. Site 2 and 3 referred to text. B site means pole of late Paleozoic mafic rocks.

These data cannot be reconciled at this preliminary stage and further paleomagnetic
study is required to be certain of the measured directions in this report and to resolve
the tectonic significance of the different tilt corrections observed in the Kiokee belt,
which lies within the Elberton-Sparta crustal block, and the Elberton-Sparta crustal
block itself.

5.4 LARGE SCALE DISPLACEMENT AND DISPLACED TERRANES

The present radiometric and paleomagnetic data from granitic mafic rocks of the South
Carolina Piedmont yield 300 m.y. old stable cratonic pole positions. Significantly, only
the Kiokee belt metamorphic rocks do not fit this pattern. As these rocks appear to be

more highly metamorphosed than the lithologies present in the adjacent Slate belt, the simplest interpretation is that the remanence reflects displacements associated with local scale folding rather than a large degree required for the displaced terrane hypothesis. The 300 m.y. old position reflects the post-magmatic cooling of the gabbros which may or may not be the time of intrusion for these bodies (McSween *et al.*, 1982). Based upon the geomagnetic pole positions of the mafic intrusions, the principal result is to suggest that there is no evidence of displaced terranes in the South Carolina Slate Belt and the Charlotte belts of the Piedmont younger than 300 m.y. old. Ellwood (1982) reached the same conclusion based on his studies of the granitic intrusions.

Large scale horizontal thrust displacements proposed by Cook *et al.* (1979), the magnitude of which has been questioned recently (Iverson and Smithson, 1982) cannot be discriminated with radiometric-paleomagnetic data currently available. Under ideal conditions, that is paleomagnetic data with a maximum degree of precision (alpha 95 of 2°) is difficult to achieve and would detect horizontal displacements only if they were greater than about 250 km. With such precise data and cratonic pole positions calculated then the degree of horizontal displacement would have to be less than the limits of detection, that is 250 km.

The granitic rocks occur in both Grenvillian times and in Paleozoic times. Significant intervals of granitic intrusion, have been identified around 250–300 m.y. (Permian to late Carboniferous), 300 m.y. (Late Carboniferous), and 400–450 m.y. (Ordovician-Devonian).

Generally, the distribution of age maxima for mafic igneous intrusions (using only the data which provisionally appears acceptable) and granitic intrusions tend to alternate (Figure 8). The granitic intrusions correlate well with compressional tectonic events during orogenic episodes of the Appalachians. Whereas, only the early Mesozoic diabase can be related to a tensional tectonic regime prior to and during the opening of the central Atlantic ocean. The tectonic environment during the intrusion of the post-metamorphic gabbros is not well understood, primarily because their age is not well known. If the late Carboniferous pole position correctly dates the time of intrusion (rather than cooling) for the gabbros then at least some of them can be viewed as differentiation products of the Carboniferous granitic intrusions during compressional tectonic regime. However, if some of the gabbros are older (as old as 405 m.y.; McSween *et al.*, 1982) then they may have intruded at a time of relative tectonic quiescence or in a tensional rifting tectonic regime earlier that Mesozoic times within the Appalachian cycle and have only been subjected to minor effects of the Acadian orogeny in the southern Appalachians (Rodgers, 1967, 1970). Further resolution of radiometric and paleomagnetic data is required to resolve this question.

5.5. CORRELATION OF MAFIC AND GRANITIC INTRUSIONS IN THE TECTONIC HISTORY OF THE SOUTHERN APPALACHIANS

Figure 8 is a histogram of the frequency of occurrence of conventional K–Ar apparent ages for mafic intrusions compared with that of Rb–Sr dates of granitic intrusions in South Carolina (Fullagar, 1981). The Rb–Sr ages are based on mineral and whole rock

R. E. DOOLEY ET AL.

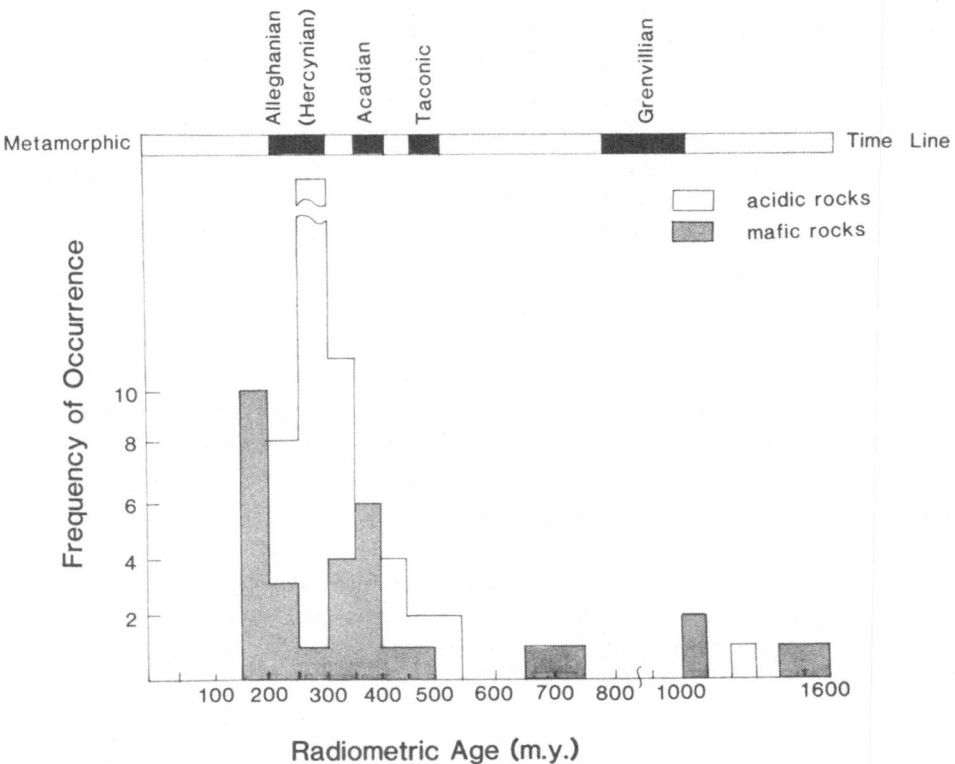

Fig. 8. Histogram of mafic and granitic intrusions in the South Carolina Piedmont taken from this study and Fullagar (1981).

isochron techniques applied to granitic rocks (and thus are not plagued with the problems of Ar loss and initial Ar contamination) and generally represent their correct ages of intrusion and/or cooling. Such results have been used as the basis for assigning ages to the orogenic phases of deformation affecting the southern Appalachians. However, the K–Ar apparent ages for mafic rocks generally have proven to have problems with excess ^{40}Ar contamination and/or argon loss in both the early Mesozoic diabase and the late Paleozoic gabbros. The maximum for the histogram for diabase rocks (Figure 8, stipled pattern), in the 150–200 m.y. interval, is comprised of two pulses of tholeiitic igneous intrusions, one at 195 m.y. ago and the other at 180 m.y. ago. A second maximum for the mafic rocks occurs between 300–400 m.y. ago for the late Paleozoic gabbros suggesting that their intrusion and/or cooling ages may be broadly within this range. In fact, recent Nd–Sr isochron ages of 405 ± 30 m.y. for several gabbros from North and South Carolina (McSween *et al.*, 1982) suggest that these magmas fractionated rare earth elements upon partial melting and injection into the crust around 400 m.y. ago in a tensional tectonic regime. Thus ages younger than 400 m.y. could reflect cooling of intrusions that were originally intruded at crustal depths of

15 km or more in Silurian-Devonian times. Apparent K–Ar ages greater than 426 m.y. have been found in several gabbros (Table I), generally older than the metasediments they intrude (see above); these clearly are anomalous and cannot be used in a comparison of mafic and granitic radiometric data.

6. Conclusions

It is clear from the survey of data presented here that the combination of radiometric and paleomagnetic data can make impressive contributions to the interpretation of tectonic problems associated with a mobile belt such as the southern Appalachians. Particularly, if a systematic approach toward an orogenic belt using several lines of geological and geophysical evidence is done, then the maximum amount of interpretative history can be gleaned out of data from several different analytical techniques. It is this sort of corroboration that gives added emphasis to the validity of any one analytical technique. Also, limitations of a particular technique can be better judged when data from independant means is available, to which one can make comparisons and judgemental statements as to the validity of age data and paleomagnetic pole positions. Their significance in the relative time frame that has been developed by careful geologic observation can offer perhaps better resolution of temporal and spatial problems in an orogenic mobile belt.

Acknowledgements

We would like to thank Dr J. M. Wampler for his assistance in reviewing this manuscript as well as the use of his radiometric laboratory at the Georgia Institute of Technology. Further helpful reviews were done by Dr Robert Hatcher, Dr Donald Secor, and Dr Art Snoke at the University of South Carolina. Field assistance was provided by Mr Bill Smith, Mr Keith Sprague, and Mr Johnathan Shireman. Partial support for this work was provided by the Department of Geology at the University of South Carolina. Paleomagnetic studies were conducted in the Geology Department at the University of South Carolina.

References

Allegre, C. J., Dupre, B., Richard, P., Rousseau, D., and Brooks, C.: 1982, 'Subcontinental versus Suboceanic Mantle, II. Nd–Sr–Pb Isotopic Comparison of the Continental Tholeiites, with Mid. Ocean Ridge Tholeiites, and the Structure of the Continental Lithosphere', *Earth and Planet. Sci. Lett.* **57**, 25–34.

Bachtadse, V., Heller, F., and Kroner, A.: 1983, 'Paleomagnetic Investigations in the Hercynian Mountain Belt of Central Europe', *Tectonophysics* **91**, 285–299.

Bell, H. III, Book, K. G., Daniels, D. L., Huff, W. E., Jr., and Popenoe, P.: 1980, 'Diabase Dikes in the Haile-Brewer Area, South Carolina, and Their Magnetic Properties', in *Shorter Contributions to Geophysics, 1979*, U.S. Geol. Surv. Prof. Pap. 1123, C1–C18.

Berger, G. W. and York, D.: 1979, '^{40}Ar/^{39}Ar Dating of Multicomponent Magnetizations in the Archean Shelly Lake Granite, Northwest Ontario', *Can. J. Earth Sci.* **16**, 1933–1841.

Boucot, A. J.: 1962, 'Appalachian Siluro-Devonian: Some Aspect of the Variscan Fold Belt', Manchester, University press, 153–188.

Boucot, A. J., Field, M. T., Fletcher, R., Forbes, W. H., Naylor, R. S., and Pavlides, L.: 1964, 'Reconnaissance Bedrock Geology of the Presque Isle Quadrangle, Maine', *Maine Geol. Surv. Quad. Mapping Ser.* **2**, 123.

Briden, J. C. and Duff, B. A.: 1981, 'Pre-Carboniferous Paleomagnetism of Europe North of the Alpine Orogenic Belt', in M. W. McElhinny and D. A. Valencio (eds.), *Paleoreconstruction of the Continents*, American Geophysical Union, Washington D.C.

Brown, L. and Barton, C.: 1980, 'Paleomagnetism of some Paleozoic Intrisive Rocks in the Southern Appalachian Piedmont', *Geol. Soc. of Am. Abs. with Prog.* **12**, 55.

Barton, C. and Brown, L.: 1983, 'Paleomagnetism of Carboniferous Intrusions in North Carolina', *J. Geophys. Res.* **88** (B3), 2327–2335.

Bullard, E. C., Everett, J. E., and Smith, A. G.: 1965, 'The Fit of the Continents around the Atlantic', *Philos. Trans. R. Soc. London*, Ser. A. **238**, 41–51.

Butler, J. R.: 1972, 'Ages of Paleozoic Regional Metamorphism in the Carolinas, Georgia and Tenessee (Southern Appalachians)', *Science* **272**, 319–333.

Butler, J. R. and Ragland, P. C.: 1969, 'A Petrochemical Survey of Plutonic Intrusions in the Piedmont Southeastern Appalachians, U.S.A.', *Contrib. Mineral. Petro.* **24**, 164–190.

Claesson, S.: 1976, 'The Age of the Ottfjallet Dolerites of Sarv Nappe, Swedish Caledonides', *Geol. Foren. Stockh. Forth.* **98**, 370–374.

Cloos, E.: 1964, 'Wedging, Bedding Plane Slips, and Gravity Tectonics', in W. D. Lowry (ed.), *Tectonics of the Southern Appalachians*, VPI Dept. of Geological Sciences, Memoir 1, 63–70.

Core, D. L., Sutter, J. F., and Fleck, R. J.: 1974, 'K/Ar Ages of Igneous Rocks from Eastern West Virginia and Western Virginia', (Abs.): *EOS, Trans. Am. Geophys. Union* **55**, 471.

Cook, F. A., Albaugh, D. S., Brown, L. D., Kaufman, S., Oliver, J. E., and Hatcher, R. D.: 1979, 'Thin-Skinned Tectonics in the Crystalline Southern Appalachians: COCORP Seismic-Reflection Profiling of the Blue Ridge and Piedmont', *Geology* **7**, 563–567.

Creer, K. M.: 1967, 'A Synthesis of World-Wide Paleomagnetic Data', in S. K. Runcorn (ed.), *Mantles of the Earth and Terrestrial Planets*, Inter-Science Publ., London, 351–382.

Creer, K. M.: 1968, 'Paleozoic Paleomagnetism', *Nature* **219**, 246–250.

Dalrymple, G. B. and Lanphere, M. A.: 1969, *Potassium Argon Dating: Principles Techniques, and Applications to Geochronology*, San Francisco, W. H. Freeman and Company, 258 pp.

Daniels, D. L. and Zeitz, I.: 1980, 'Preliminary Aeromagnetic Map of the Charlotte 1 × 2 Degree Quadrangle, North Carolina and South Carolina', U.S. Geol. Surv. Open-file report 80–229, 1 : 250000.

Dalrymple, G. B., Gromme, C. S., and White, R. W.: 1975, 'Potassium-Argon Age and Paleomagnetism of Diabase Dikes and Sills in Liberia: Initiation of Central Atlantic Rifting', *Geol. Soc. Amer. Bull.* **86**, 399–411.

Dallmeyer, R. D.: 1975, '^{40}Ar/^{39}Ar Age Spectra of Biotite from Grenville Basement Gneisses in Northwest Georgia', *Geol. Soc. Amer. Bull.* **86** (12), 1740–1744.

Dallmeyer, R. D.: 1978, '^{40}Ar/^{39}Ar Incremental-Release Ages of Hornblende and Biotite Across the Georgia Inner Piedmont: Their bearing on Late Paleozoic-Early Mesozoic Tectonothermal History', *Amer. Jour. of Sci.* **278**, 124–149.

Dallmeyer, R. D., Hess, J. R., and Whitney, J. A.: 1981, 'Post-Magmatic Cooling of Elberton Granite: Bearing on Late Paleozoic Tectonothermal History of the Georgia Inner Piedmont', *J. Geol.* **89**, 585–600.

de Boer, J. and Snider, G.: 1979, 'Magnetic and Chemical Variations of Mesozoic Diabase Dikes from Eastern North America; Evidence for a Hot-spot in the Carolinas', *Geol. Soc. Am. Bull.* **90**, 185–198.

Deininger, R.W., Dallmeyer, R. B., and Neathery, T. L.: 1975, 'Chemical Variations and K-Ar Ages of Diabase Dikes in East-Central Alabama (abs.)', *Geol. Soc. America, Abs. with Programs* **7** (4), 482.

Delorey, C. M., Dooley, R. E., Ressetar, R., Fullagar, P. D., and Stoddard, E. R.: 1982, 'Paleomagnetism and Isotope Geochronology of Early Mesozoic Rhyolite Porphyry Dikes, Eastern Southern Appalachian Piedmont', *EOS, Trans. Am. Geophys. Union* **63** (18), 309.

Dewey, J. F. and Bird, J. M.: 1970, 'Mountain Belts and the new Global Tectonics', *J. Geophys. Res.* **75** (14), 2625–2647.

Dooley, R. E.: 1977, 'K-Ar Relationships in Dolerite Dikes of Georgia', Atlanta, Georgia Institute of Technology, M.S. thesis, 185.

Dooley, R. E.: 1983, 'Paleomagnetism of some Mafic Intrusions in the South Carolina Piedmont: Part I. Magnetic Systems with Single Characteristic Directions', *Physics of Earth and Plan. Inter.* **31**, 241–268.

Dooley, R. E. and Smith, W. A.: 1982, 'Age and Magnetism of Diabase Dikes and Tilting of the Piedmont', *Tectonophysics* **90**, 283–307.

Dooley, R. E. and Wampler, J. M.: 1983, 'K–Ar Relationships in Diabase Dikes of Georgia: The Influence of Excess Argon on the Geochronology of Early Mesozoic Igneous and Tectonic Events', in Greg Gohn (ed.), *Studies Related to the Charleston, South Carolina, Earthquake of 1886 – Tectonics and Seismicity*, U.S. Geol. Surv. Prof. Pap. 1313, M1–M24.

Duff, B. A.: 1980, 'The Paleomagnetism of Jersey Volcanic and Dykes and the Lower Carboniferous Paleozoic Apparent Polar Wander Path for Europe', *Geophys. J. Roy. Astron. Soc.* **60**, 355–375.

Dunlop, D. J.: 1979, 'On the Use of Zijderveld Vector Diagrams in Multi-Component Paleomagnetic Studies', *Phys. Earth. Plantet. Inter.* **20**, 12–24.

Edinger, P. R., Nairn, A. E. M., and Zupan, A. P. J.: 1979, 'Paleomagnetic Investigations in the Carolinas (Abs/)', Programme with Abs. Meeting, Can. Geophys. Union, Fredericton, N.B., 20.

Ellwood, B. B.: 1982, 'Paleomagnetic Evidence for the Continuity and Independent Movement of a Distinct Major Crustal Block in the Southern Appalachians', *J. Geophys. Res.* **87**, 5339–5350.

Ellwood, B. B., Whitney, J. A., Wenner, D. B., Mose, D., and Amerigian, C.: 1980, 'Age, Paleomagnetism and Tectonic Significance of the Elberton Granite, Northeast Georgia Piedmont', *J. Geophys. Res.* **85**, 2521–2533.

Evans, C. R. and Tarney, J.: 1964, 'Isotopic Age of Assynt Dykes', *Nature* **204** (4959), 638–641.

Fisher, R. A.: 1953, 'Dispersion on a Sphere', *Proc. Roy. Soc. London*, Ser. A. **217**, 295–305.

French, R.B. and Van der Voo, R.: 1977, 'Remagnetization Problems with the Paleomagnetism of the Middle Silurian Rose Hill Formation of Central Appalachians', *J. Geophys. Res.* **83** ((36), 5803–5806.

Fullagar, P. D.: 1971, 'Age and Origin of Plutonic Intrusions in the Piedmont of Southeastern Appalachians', *Geol. Soc. Am. Bull.* **82**, 2845–2862.

Fullagar, P. D.: 1981, 'Summary of Rb–Sr Whole-Rock Ages for South Carolina', *South Carolina Geology* **25**, 29–32.

Fullagar, P. D. and Bottino, M. L.: 1969, 'Tertiary Felsite Intrusions in the Valley and Ridge Province, Virginia', *Geol. Soc. Am. Bull.* **80** (9), 1853–1858.

Fullagar, P. D. and Butler, J. R.: 1979, '325 to 265 m.y. old Granitic Plutons in the Piedmont of the Southeastern Appalachians', *Am. Jour. Sci.* **279**, 161–185.

Fullagar, P. D. and Kish, A.: 1981, 'Mineral Age Traverses across the Piedmont of South Carolina and North Carolina', in Horton, J. W., Butler, J. R., and Milton, D. M. (eds.), *Geological Investigations of the Kings Mountain Belt and Adjacent Areas in the Carolinas*, Carolina Geol. Soc. Field Trip Guidebook: 155–165.

Giletti, B. J.: 1971, 'Discordant Isotopic Age in Biotites', *Earth. Planet. Sci. Lett.* **10**, 157–164.

Gohn, G. S., Higgins, B. B., Smith, C. C., and Owens, J. P.: 1977, 'Lithostratigraphy of the Deep Corehole (Clubhouse Crossroads Corehole 1) Near Charleston, South Carolina', in *Studies Related to the Charleston, South Carolina Earthquake of 1886 – A Preliminary Report*. Geol. Surv. Prof. Pap. 1028-E: 59–70.

Gohn, G. S., Gottfried, D., Lanphere, M. A., and Higgins, B. B.: 1978, 'Regional Implications of Triassic or Jurassic Age for Basalt and Sedimentary Red Beds in the South Carolina Coastal Plain', *Science* **202**, 887–890.

Gottfried, D., Annell, C. S., and Schwarz, L. J.: 1978, 'Geochemistry of Subsurface Basalt from the Deep Corehole (Clubhouse Crossroads Corehole 1) Near Charleston, South Carolina – Magma Type and Tectonic Implications', in *Studies related to the Charleston, South Carolina, Earthquake of 1886 – Preliminary Report*, Geol. Surv. Prof. Pap. 1028-G: 91–113.

Hadley, J. B.: 1964, 'Correlation of Isotopic Ages, Crustal Heating and Sedimentation in the Appalachian Region', in Lowry, W. D. (ed.), *Tectonics of the Southern Appalachians*, Va. Poly Inst. Dept. Geol. Sci., Mem. 1, 33–45.

Halls, H. C.: 1976, 'A Least-Squares Method to Find a Remanence Direction from Converging Remagnetization Circles', *Geophys. J. Roy. Astron. Soc.* **45**, 297–304.

Halls, H. C.: 1978, 'The Use of Converging Remagnetization Circles in Paleomagnetism', *Phys. of Earth Planet. Inter.* **16**, 1–11.

Halls, H. C.: 1979, 'Separation of Multicomponent NRM: Combined Use of Difference and Resultant Magnetization Vectors', *Earth Planet. Sci. Lett.* **43**, 303–308.

Harper, S. B. and Fullagar, P. D.: 1981, 'Rb–Sr Ages of Granitic Gneisses of the Inner Piedmont Belt of Northwestern North Carolina and Southwestern South Carolina', *Geol. Soc. Am. Bull.* **92**, 864–872.

Harte, B. and Johnson, M. R. W.: 1969, 'Metamorphic History of Dalradian Rocks in Glens Clova. Esk, and Lethnot, Angus, Scotland, *Scot. J. Geo.* **5**, 54–80.

Harrison, T. M. and McDougal, I.:1980a, 'Investigations of an Intrusive Contact, Northwest Nelson, New Zealand – I. Thermal, chronological and Isotopic Constraints', *Geochim. Cosmochim. Acta* **44**, 1985–2003.

Harrison, T. M. and McDougal, I.: 1980b, 'Investigations of an Intrusive Contact, Northwest Nelson, New Zealand – II. Diffusion of Radiogenic and Excess ^{40}Ar in Hornblende Revealed by $^{40}Ar/^{39}Ar$ Age Spectrum Analysis', *Geochim. Cosmochim. Acta* **44**, 2005.

Hatcher, R. D., Jr.: 1972, 'Developmental Model for the Southern Appalachians', *Geol. Soc. Am. Bull.* **83**, 2735–2760.

Hebeda, E. H., Boelrijk, N. A. I. M., Priem, H. N. A., Verdurmen, E. A. Th., and Verschure, R. H.: 1973, 'Excess Radiogenic Argon in the Precambrian Avanevero Dolerites in Western Surinam (South America)', *Earth and Planet. Sci. Lett.* **20**, 189–200.

Hoffman, K. A. and Day, R.: 1978, 'Separation of Multi-Component NRM: A General Method', *Earth Planet. Sci. Lett.* **40**, 433–438.

Hurst, V. J.: 1970, 'The Piedmont in Georgia', in Fisher *et al.* (eds.), *Studies of Appalachian Geology: Central and Southern*, New York, Intersci. Publishers, Chapter 26, 383–396.

Irving, E.: 1977, 'Drift of the Major Continental Blocks since the Devonian', *Nature* **270**, 304–309.

Irving, E.: 1979, 'Paleopoles and Paleolatitudes of North America and Speculations about Displaced Terrains', *Can. Jour. Earth Sci.* **16**: 669–694.

Irving, E. and Irving, G. A.: 1982, 'Apparent Polar Wander Paths Carboniferous through Cenozoic and the Assemly of Gondwana', *Geophys. Surveys.* **5**, 141–188.

Iverson, W. P. and Smithson, S. B.: 1982, 'Master Decollement Root Zone Beneath the Southern Appalachians and Crustal Balance', *Geology* **10**, 241–245.

Kent, D.V. and Opdyke, N. D.: 1978, 'Paleomagnetism of the Devonian Catskill Red Beds: Evidence for Motion of the Coastal New England – Canadian Maritime Region Relative to Cratonic North America', *J. Geophys. Res.* **83**, 4441–4450.

Kish, S. A., Butler, J. R., and Fullagar, P. D.: 1979, 'The Timing of Metamorphism and Deformation in the Central and Eastern Piedmont of North Carolina', (abs.): *Geol. Soc. of Amer., Abs. with Prog.* **11**, 184–185.

Kulp, J. L. and Eckleman, F. D.: 1961, 'Potassium-Argon Isotopic Ages on Micas from the Southern Appalachians', *Ann. N.Y. Acad. Sci.* **91**, 408–419.

Lanphere, M.A.: 1983, '$^{40}Ar/^{39}Ar$ Ages of Basalt from Clubhouse Crossroads No. 2 near Charleston, South Carolina', in Gohn G. (ed.), *Studies Related to the Charleston, South Carolina Eaarthquake of 1886 – Tectonics and Seismicity*, U.S. Geol. Surv. Prof. Pap. 1313: B1–B7.

Lanphere, M. A. and Dalrymple, G. B.: 1971, 'A Test of $^{40}Ar/^{39}Ar$ Age Spectrum Technique on some Terrestrial Materials', *Earth Planet. Sci. Lett.* **32**, 141–148.

Lanphere, M. A. and Dalrymple, B.: 1976, 'Identification of Excess ^{40}Ar by the $^{40}Ar/^{39}Ar$ Age Spectrum Technique', *Earth and Planet. Sci. Lett.* **32**, 141–148.

Long, T. L.: 1979, 'The Carolina Slate Belt – Evidence of a Continental Rift Zone', *Geology* **7**, 180–184.

Long, L. E., Kulp, J. L. and Eckleman, F. D.: 1959, 'Chronology of Major Metamorphic Events in the Southeastern United States'. *Am. Jour. Sci.* **257**, 585–603.

Long, T. L., Talwani, Pradeep, and Bridges, S. R.: 1976, Simple Bouguer Anomaly Map of South Carolina: South Carolina Division of Geology', MS–21, scale 1 : 500 000.

Lowry,W. D. (ed.): 1964, 'Tectonics of the Southern Appalachians', Virg. Poly. Inst., Dept. Geol. Sci., Memoir 1, Blacksburg.

May, P. R.: 1971, 'Pattern of Triassic-Jurassic Dikes around the North Atlantic in the Context of Pre-Rift Positions of the Continents', *Geol. Soc. Am. Bull.* **82**, 1285–1292.

McSween, H. Y. Jr., Sando, T. W., and Misra, K.: 1982, 'Gabbroic Plutonism in the Charlotte Belt of the Southern Appalachians', *Geol. Soc. Am. Abs. With Prog.* **14** (7): 567.

McWilliams, M. O. and Dunlop, D. J.: 1978, 'Grenville Paleomagnetism and Tectonics', *Can. J. Earth Sci.* **15**, 687–695.

Meadows, G. R.: 1978, 'Petrology of Mesozoic Age Diabase Dikes in the Georgia Piedmont', M.S. Thesis, Emory University, Atlanta, 70.

Medlin, J. H.: 1968, 'Comparitive Petrology of Two Igneous Complexes in the South Carolina Piedmont', Ph.D. Dissertation, The Pennsylvania State University, 205.

Medlin, J. H.: 1969, 'Petrology of Two Mafic Igneous Complexes in the South Carolina Piedmont. (Abs.)', *Geol. Soc. Am. Abs. with Prog.* **4**, 52.

Merrill, R. T.: 1981, 'Toward a Better Theory of Thermal Remanent Magnetization', *J. Geophys. Res.* **86** (B2): 937–949.

Morgan, G. E.: 1976, 'Paleomagnetism of a Slowly Cooled Plutonic Terrain in West Greenland', *Nature* **259**, 382–385.

Morgan, G. E. and Briden, J. C.: 1981, 'Aspects of Precambrian Paleomagnetism with new Data from the Limpopo Mobile Belt and Kaa-avaal Craton in Southern Africa', *Phys. Earth Planet. Inter.* **24**, 142–168.

Nairn, A. E. M., Howell, D. E., Ressetar, R., and Martin, D. L.: 1978, 'Initial Paleomagnetic Results from the Piedmont of South Carolina (Abs.)', *Geol. Soc. Am. Abs. with Prog.* **10**, 193.

O'Connor, Bruce, J. and Prowell, David, C.: 1978, 'Belair Fault Zone: Evidence of a Tertiary Fault Displacement in Eastern Georgia, *Geology* **6**, 681–684.

Overstreet, W. C. and Bell, H., III: 1965, 'The Crystalline Rocks of South Carolina', *U.S. Geol. Surv. Bull.* **1183**, 129.

Park, J. K.: 1981, 'Analysis of the Multicomponent Magnetization of the Little Dal Group, Mackenzie Mountain, Northwest Territories, Canada', *J. Geophys. Res.* **86** (B6): 5134–5146.

Park, J. K. and Roy, J. L.: 1979, 'Further Paleomagnetic Results from the Seal Group Igneous Rocks, Quebec', *Can. J. Earth Sci.* **16**, 895–912.

Phillips, J. D.: 1983, 'Paleomagnetic Investigations of the Clubhouse Crossroads Basalt', in Gohn, G. (ed.), *Studies Related to the Charleston, South Carolina, Earthquake of 1886 – Tectonics and Seismicity*. U.S. Geol. Surv. Prof. Pap. 1313: C1–C17.

Pullaiah, E., Irving, K., Buchan, L., and Dunlop, D. J.: 1975, 'Magnetization Changes Caused by Burial and Uplift', *Earth Planet. Sci. Lett.* **28**: 133–143.

Ragland, P. C.: 1982, 'Mesozoic Diabase Dikes in Eastern North America. A Review (Abs.)', *Geol. Soc. Am. Abs. with Prog.* **14** (1, 2), 75.

Ragland, P. C., Hatcher, R. D., Jr., and Whittington, D.: 1983, 'Juxtaposed Mesozoic Diabase Dikes Sets from the Carolinas: A Preliminary Assessment', *Geology* **11**, 394–399.

Ressetar, R. and Martin, D. L.: 1976, 'Paleomagnetism of Eocene Intrusions in the Valley and Ridge Province, Virginia and West Virginia', *Can. Jour. Earth Sci.* **17**: 1583–1588.

Rodgers, J.: 1967, 'Chronology of Tectonic Movements in the Appalachian Region of Eastern North America', *Am. J. Sci.* **265**, 408–427.

Rodgers, J.: 1970, 'The Tectonics of the Appalachians', New York, Wiley Intersci., 271.

Roy, J. L. and LaPointe, P. L.: 1976, 'The Paleomagnetism of Post-Huronian Igneous Events and Apparent Polar Path for the Interval −2300 to −1500 Ma for Laurentia', *Can. J. Earth Sci.* **13**, 749–773.

Runcorn, S. K. (ed.): 1962, 'Memories of Alfred Wegener', in *Continental Drift*, Academic Press, New York.

Samson, Sara, Personal Communication: 1983, The Geology Department, The University of South Carolina Columbia, South Carolina.

Secor, D. T. and Wagener, H. D.: 1968, 'Stratigraphy, Structure and Petrology of the Piedmont in Central South Carolina: South Carolina Geological Survey', *Geological Notes*, **12**, 67–84.

Secor, D. T. and Snoke, A. W.: 1978, 'Stratigraphy, Structure, and Plutonism in the Central South Carolina Piedmont', in Snoke, A. W. (ed.), *Geological Investigations of the Eastern Piedmont Southern Appalachians*, Carolina Geol. Soc. Field Trip Guidebook, 61–99.

Smith, A. and Hallam, A.: 1970, 'The Fit of the Southern Continents', *Nature* **225** (5228): 139–144.

Smith, T. E. and Noltimier, H. C.: 1979, 'Paleomagnetism of the Newark Trend Igneous Rocks of the North Central Appalachians and the Opening of the Central Atlantic Ocean', *Am. J. of Sci.* **279**, 778–807.

Snoke, A. W., Kish, S. A., and Secor, D. T.: 1980, 'Deformed Hercynian Granitic Rocks from the Piedmont of South Carolina', *Am. J. Sci.* **280**, 1018–1034.

Spence, W. N. and McDaniel, R. D.: 1979, 'Upper Cretaceous Trachytes of the Northeastern North Carolina Piedmont', *Geol. Soc. Am. Abs. With Prog.* **11** (4): 213.

Steiger, R. H. and Jaeger, E.: 1977, 'Subcommission of Geochronology: Convention of the Use of Decay Constants in Geo- and Cosmo-Chronology', *Earth Planet. Sci. Lett.* **36**, 359–362.

St. Jean, J.: 1973, 'A new Cambrian Tribolite from the Piedmont of North Carolina', *Am. J. Sci.* **273-A**, 196–216.

Stromquist, A. A. and Sundelius, H. W.: 1969, 'Stratigraphy of the Albermarle Group of the Carolina Slate Belt in Central North Carolina', *U.S. Geological Survey Bulletin*, **1274-B**, B1–B22.

Stupavsky, M. and Symons, D. T. A.: 1978, 'Separation of Magnetic Components from AF Step Demagnetization Least Squares Computer Methods', *Jour. Geophys. Res.* **83**, 4925–4931.

Sutter, J. F. and Smith, T. E.: 1979, $^{40}Ar/^{39}Ar$ Ages of Diabase Intrusions from Newark Trend Basins in Connecticut and Maryland: Initiation of Central Atlantic Rifting', *Am. J. Sci.* **279**: 808–831.

Thompson, R. and Clark, R. M.: 1982, 'A Robust Least-Squares Gondwana Apparent Polar Wander Path and the Question of Paleomagnetic Assessment of Gondwana Reconstructions', *Earth and Planet. Sci. Lett.* **57**, 152–158.

Van Alstine, D. R. and de Boer, J.: 1977, 'A new Technique for Constructing Apparent Polar Wander Paths and the Revised Phanerozoic Path for North America', *Geology* **6**, 137–139.

Van der Voo, R. and Channell, J. E. T.: 1980, 'Paleomagnetism in Orogenic Belts', *Rev. of Geophys. and Space Phys.* **18** (2): 455–481.

Wampler, J. M. and Dooley, R. E.: 1975, 'Potassium-Argon Determination of Triassic and Eocene Igneous Activity in Rockingham County, Virginia (Abs.): *Geol. Soc. Am. Abs. with Prog.* **7** (4): 547.

Watts, D. R.: 1975, 'A Paleomagnetic Study of four Mesozoic Diabase Dikes Swarms of the Southern Appalachian Mountains: M.S. Thesis. The Ohio State University. Columbus, Ohio, 115.

Watts, D. R.: 1982, 'A Multicomponent Dual-Polarity Paleomagnetic Regional Overprint from the Moine of Northwest Scotland', *Earth and Planet. Sci. Lett.* **61**, 190–198.

Watts, D. R. and Van der Voo, R.: 1979, 'Palaeomagnetic Results from the Ordovician Moccasin, Bays, and Chapman Ridge Formations of the Valley and Ridge Province, Eastern Tennessee', *Jour, Geoph. Res.* **84**, 645–655.

Webb, G. W.: 1963, 'Occurrence and Exploration Significance of Strike-Slip Faults in Southern New Brunswick, Canada, *Am. Assoc. Pet. Geol. Bull.* **47**, 1904–1927.

Weigand, P. W. and Ragland, P. C.: 1970, 'Geochemistry of Mesozoic Dolerite Dikes from Eastern North America', *Contributions to Mineralogy and Petrology*, **29**, 195–214.

Wenner, D. B.: 1981, 'Oxygen Isotopic Compositions of the Late-Orogenic Granites in the Southern Piedmont of the Appalachian Mountains, U.S.A. and their Relationship to Sub-Crustal Structures and Lithologies', *Earth Planet, Sci. Lett.* **54**, 186–199.

Williams, H.: 1976, 'Tectonostratigraphic Subdivision of the Appalachian Orogen', *Geol. Soc. Amer. Abst. with Prog.* **8** (2): 300.

Williams, H.: 1978, 'Tectonic-Lithofacies Map of the Appalachian Orogen', Memorial University of Newfoundland, Map No. 1, St. John's, Newfoundland.

Williamson, P. and Robertson, W. A.: 1976, 'Iterative Method of Isolating Primary and Secondary Components of Remanent Magnetization Illustrated by Using the Upper Devonian Catombal Group of Australia'. *J. Geophys. Res.* **81** (14), 2531–2537.

Wilson, M. R.: 1972, 'Excess Radiogenic Argon in Metamorphic Amphiboles and Biotites from the Sulifjelma Region, Central Norwegian Caledonides', *Earth Planet. Sci. Lett.* **14**, 403–412.

Winkler, H. J. F.: 1976, 'Petrogenesis of Metamorphic Rocks', fourth ed., Springer-Verlag, New York, 334.

Wright, J. A. and Seiders, V. M.: 1977, 'U–Pb Dating of Zircons from the Carolina Volcanic Slate Belt, Central North Carolina', *Geol. Soc. Am. Abstr. with Prog.* **9**, 197–198.

York, D.: 1978, 'A Formula Describing both Magnetic and Isotopic Blocking Temperatures', *Earth Planet. Sci. Lett.* **39**, 89–93.

Zietz, I. R. T., Hawthorn, H., Williams, and Daniels, D. L.: 1980, 'Magnetic Anomaly Map of the Appalachian Orogen', *Map 2*, University of Newfoundland, St. John's, scale: 1 : 1 000 000.

Zimmerman, R. A.: 1979, 'Apatite Fission Track Age Evidence of Post-Triassic Uplift in the Central and Southern Appalachians (Abs.)', *Geol. Soc. Am. Abstr. with Progr.* **11** (4): 219.

S. K. RUNCORN'S COMMENTARY

Palaeomagnetism has been like the widow's cruse; hardly anyone associated with its development could have predicted the continued throwing up of new results.

The discovery of reversed as well as normal magnetization in the Tertiary, especially Jan Hospers' work on the Icelandic lavas and his demonstration that the mean geomagnetic field was a dipole aligned along the axis of rotation, seemed to many to cast doubt on whether rocks did really faithfully reproduce the direction of the field at the time of their formation. A great deal of ingenious work, theoretical and experimental, went into the problem of self reversals. The discovery by K. M. Creer, E. Irving, and I in the Palaeozoic and Pre-Cambrian, and in the Mesozoic by Blackett, Clegg and colleagues, that in Pre-Tertiary times rocks in Great Britain were magnetized at a strong angle to the present again raised doubts in many minds as to whether rocks did faithfully reproduce the direction of the field at the time of their deposition, or in the case of igneous rocks, their cooling. Again, many ingenious solutions to allow the field to stay put through geological times were offered by appealing to geological and petrological biases on the iron oxide carriers of magnetization. Of course, successful 'field' tests, experiments on tilting and scrambling which nature performed, played a great part in convincing the early workers of the reliability of the record preserved, in igneous and sedimentary rocks, of the changing geomagnetic field. These contributions, by two of the earliest colleagues of mine trained as geologists, who early had 'faith' in the subject, very well illustrate two of the most important problems in terrestrial palaeomagnetism today – the unravelling of the Pre-Cambrian record and the complex motions of the mobile belts of the Earth's crust in the Phanerozoic.

In an entertaining critical last chapter on polar wandering in their classic 'The Earth's rotation', Munk and MacDonald, writing soon after the discovery of the discrepancy between the polar wandering paths for N. America and Europe, and sceptically recounting palaeomagnetists' explanations in terms of continental drift, rather scornfully asked what palaeomagnetists would say when pole positions for one epoch in one continent were found to disagree. They felt they had scored a valid point when they said pieces of the continent would be juggled around. Nairn and colleagues show that this latest use of palaeomagnetism is giving for the great mountain belts results comparable in interest to the displaced terrains of western N. America.

The Pre-Cambrian palaeomagnetic record is one of daunting difficulty with the many possibilities of overprinting of later over earlier magnetizations, but one in which Ted Irving has been fascinated since I went with him to sample the Torridonian in the summer of 1952. One may hazard a guess that the correct theory of continental drift will not be found until the horizontal displacements of the crust during the vast time interval represented by the rocks of Pre-Cambrian is better understood. Although the record of these displacements in the ocean floor have been so beautifully preserved – and interpreted – they represent only the last 2–3% of the Earth's life. The very large proportion of geophysicists and geologists who have plumped for gravity sliding of the plates propelled by the sinking of the plate edges at the trenches, or their sliding from the ocean ridges have, I think, been misled by the very data which finally removed their objections to continental drift. One can predict that palaeomagnetism has perhaps the most important role in understanding the internal dynamics of the Earth yet to play.

PAST AND PRESENT MAGNETISM OF THE MOON

D. W. COLLINSON

*Institute of Lunar and Planetary Science, School of Physics, University of Newcastle upon Tyne,
NE1 7RU, U.K.*

Abstract. The characteristics of the remanent magnetism of lunar samples suggests that it was acquired in a magnetic field on the Moon. The most likely origin of the field is a dynamo process in a molten, electrically-conducting core, but generation of a transient magnetic field during large meteorite impacts cannot be entirely ruled out. The magnetizing process may be thermoremanence, acquired when the rocks cooled through the Curie point of the constituent iron grains which carry the remanent magnetization, or it may involve shock at the time of a meteorite impact, with or without a partial thermoremanence arising from heating.

Evidence from absolute and relative determinations of the ancient field strength from the sample magnetizations strongly favours a global lunar field. This is implied by a trend which shows the field rising to a maximum value of $\sim 100\,\mu\text{T}$ between about 3.9–3.7 by ago and then decaying to 5–10 μT until ~ 3.1 by. Such a systematic variation of field with time is not expected to be derived from magnetizations acquired in transient, impact-generated fields varying randomly in intensity.

Contributory evidence for a dynamo field is provided by measurements of present lunar surface fields, the present very small dipole moment of the Moon and accumulating evidence of variation of the axis of the lunar field with time. Although there is no direct evidence for the existence of a lunar core the relevant observations are consistent with the presence of a core of up to 400 km in radius. There are some difficulties associated with the lunar dynamo mechanism and its energy source but the evidence for a lunar dynamo is accumulating, with important implications for the structure and thermal history of the Moon.

1. Introduction

One of the most remarkable developments in geophysics in the last 25 yr has been the growth of palaeomagnetism, the study of the natural remanent magnetism (NRM) of rocks, and its application to the study of continental drift (later to be embraced by plate tectonics) and the history of the Earth's magnetic field. It may be said without exaggeration that palaeomagnetism provided the compelling, quantitative evidence for continental movements which led to major rethinking not only about the history of the Earth's surface features but also about the Earth's interior. Keith Runcorn has been in the forefront – at times one might have said the battle front – of many aspects of this revolution in the earth sciences, and it was fitting that geophysicists at Newcastle upon Tyne took an early interest in extending palaeomagnetism to another body in the Solar System, namely the Moon.

The Moon prior to the Apollo missions was a body for which palaeomagnetic studies did not appear to have much relevance. Generally considered to have been formed cold and never to have been significantly heated during its history, there seemed little likelihood of the Moon having ever possessed a molten core in which a magnetic field of dynamo origin could have been generated. The mean relative density of 3.34 and an estimated moment of inertia factor of near 0.40 favoured a nearly uniform interior, and the early spacecraft missions measured only a very weak lunar dipole moment, with an upper limit of only about 10^{-5} of that of the Earth. Thus there was some justified scepticism when proposals were made for studying the magnetic properties of the

Geophysical Surveys 7 (1984) 57-73. 0046-5763/84/0071-0051$02.55.

samples to be returned by the Apollo missions. However, this scepticism proved to be misplaced, and lunar palaeomagnetism has proved to be a research field of considerable interest and importance in its contribution to our ideas concerning the history and structure of the Moon.

2. The Remanent Magnetism of Lunar Rocks

From the outset there were difficulties associated with lunar palaeomagnetic measurements. It seemed likely, and in fact proved to be the case, that the samples would not be of bedrock and therefore that the direction of NRM in the samples could not be referred to lunar reference axes. Thus the data which has made terrestrial palaeomagnetism such a powerful tool is not available for the Moon. Also, where the terrestrial palaeomagnetist usually has as many 25 g samples as he wants to work with, his lunar counterpart is usually restricted to a maximum of one or two 5 g chips from a hand sample, and not unusually to one or two 1 g fragments. Since typical intensities of magnetization of lunar rocks are comparable with those of terrestrial sediments, i.e. $10^{-5} - 10^{-6}$ Am^2 kg^{-1} (Figure 1), the demagnetization of a 1 g chip often requires the measurement of very weak magnetic moments.

Three main types of rock were returned by the Apollo missions – basaltic,

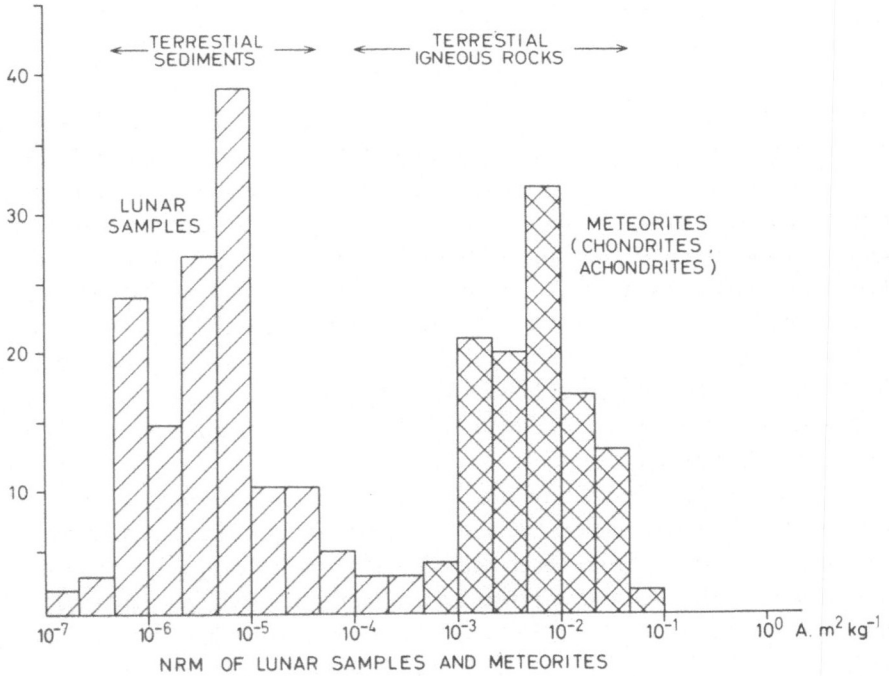

Fig. 1.   Range of intensity of natural remanent magnetization of lunar samples compared with meteorites and terrestrial igneous and sedimentary rocks.

anorthositic and the breccias – and all three types were found to possess a measurable remanent magnetization. The dominant carrier of the NRM is native iron or dilute nickel-iron in a wide range of grain-sizes up to $\sim 100\,\mu$, present to the extent of ~ 0.01– 0.1% by weight in basalts and anorthosites and up to $\sim 1\%$ in breccias.

Preliminary tests showed that the finely-divided iron is prone to oxidation on heating, even in a high vacuum or inert atmosphere, and this led to alternating field (a.f.) rather than thermal demagnetization being the favoured technique for testing the stability of the NRM. The aim was to examine the demagnetization behaviour of the NRM of different samples to determine whether their NRM showed evidence of being acquired in a lunar magnetic field or had been acquired between collection on the Moon and reception in the laboratory.

On a.f. demagnetization, different types of behaviour are observed. Figure 2 shows examples of very stable NRM, both in direction and intensity, suggesting the presence of a single magnetization component. A common type of behaviour is shown in Figure 3, in which a soft secondary NRM is removed by demagnetizing fields in the range 2– 10 mT, revealing a stable, but sometimes weak, primary NRM. Other samples require higher fields before a stable component is revealed (Figure 4) and in a few cases a.f. demagnetization results in an almost random movement of the NRM vector, with no discernible stable end-point (Figure 5).

Where thermal demagnetization has been successfully carried out, similar types of

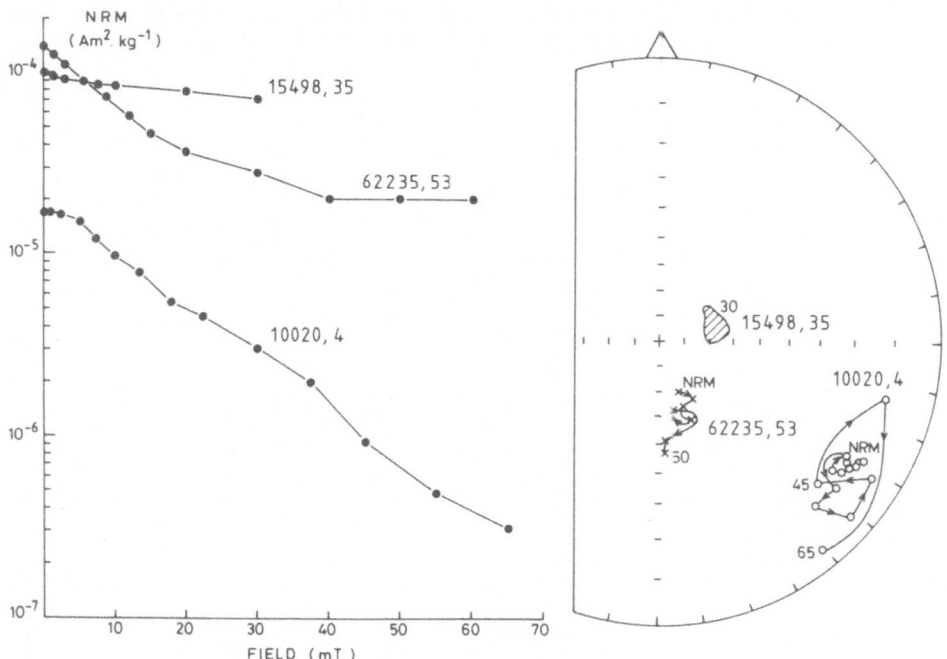

Fig. 2. Examples of stable magnetization in lunar breccia 15498 (Gose *et al.*, 1973) and igneous rocks 62235 (Collinson *et al.*, 1973) and 10020 (Collinson *et al.*, 1972).

D. W. COLLINSON

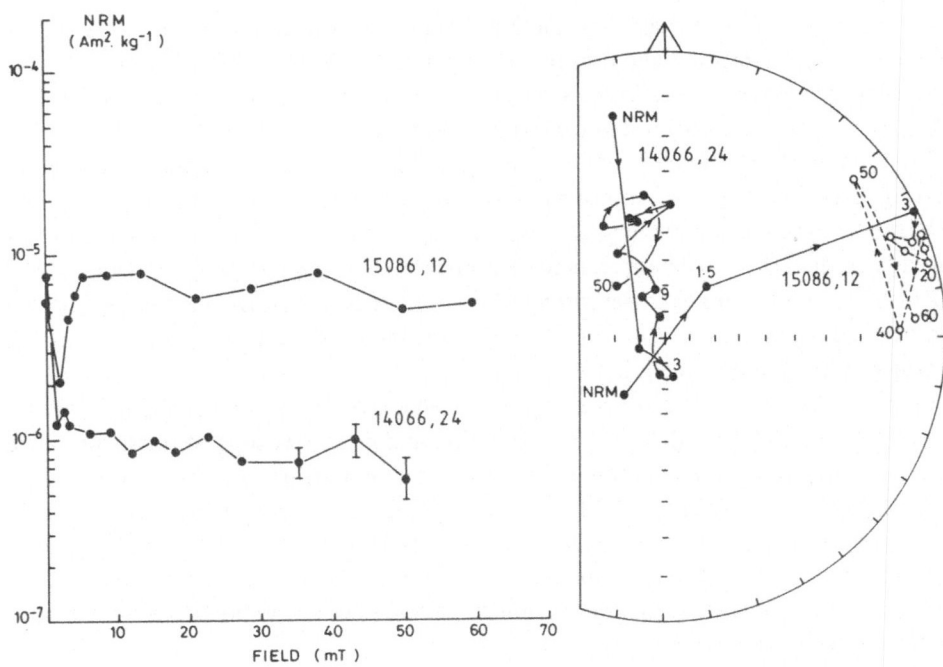

Fig. 3.   Removal of a soft secondary magnetization to reveal a harder, primary NRM in lunar breccias
14066 (Collinson *et al.*, 1972) and 15086 (Collinson *et al.*, 1973).

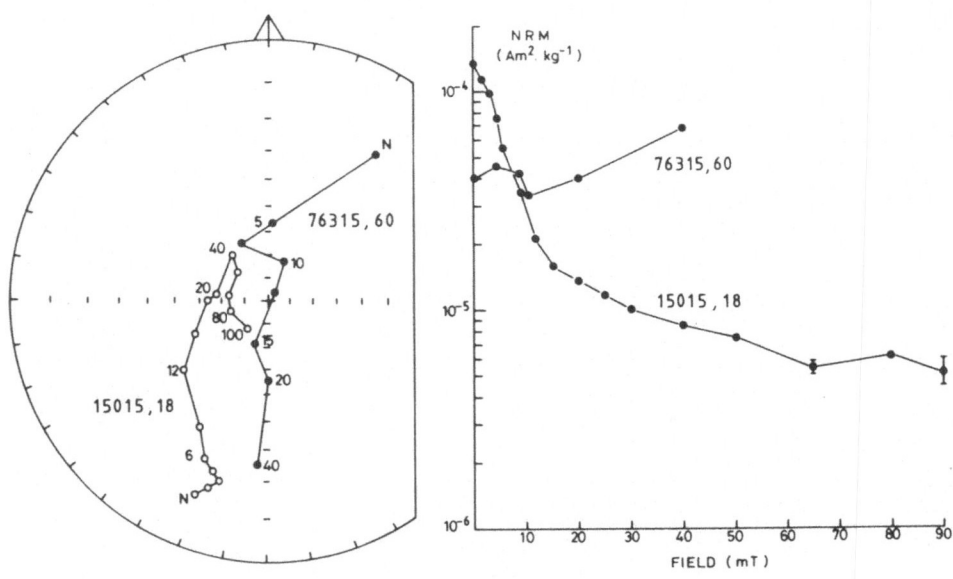

Fig. 4.   Removal of more stable secondary magnetization in two breccias.

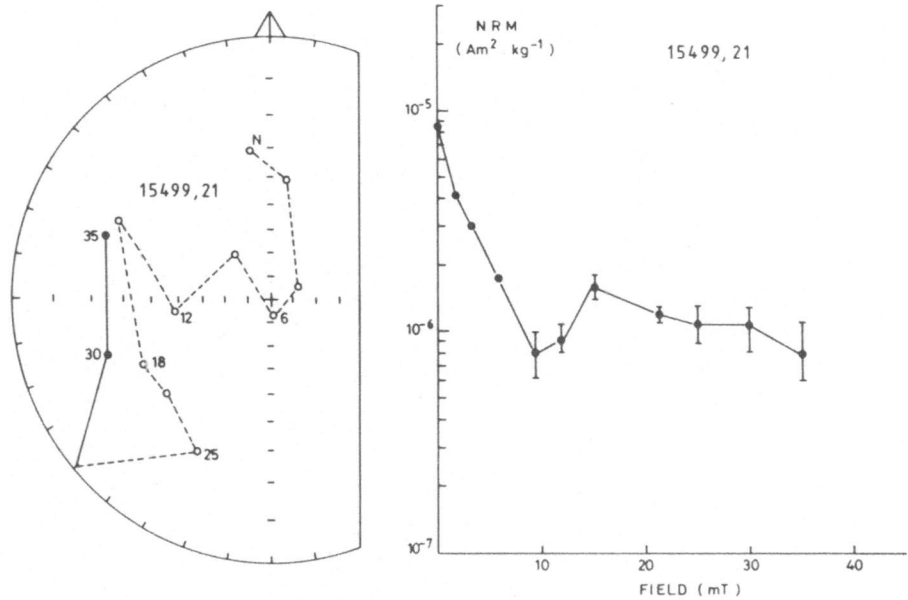

Fig. 5.   Anomalous demagnetization behaviour in igneous rock 15499.18.

behaviour are seen (Figure 6). An important feature in several samples is the persistence of a measurable magnetization up to the Curie temperature of iron ($\sim$ 780 °C).

Thus in many of the lunar samples there is evidence of a primary magnetization, indicated by stability against a.f. and thermal demagnetization and constancy of NRM direction, with or without prior removal of soft magnetization components. The softer, secondary magnetizations appear to arise from two main sources, contaminatory and lunar. The former takes the form of a low-field isothermal remanent magnetization (IRM) acquired in a magnetic field (1–3 mT) in the Apollo spacecraft during return to Earth (Pearce *et al.*, 1973) and a viscous magnetization (VRM) acquired in the geomagnetic field. Secondary magnetizations removed in alternating fields $> \sim$ 20 mT are likely to have been acquired on the Moon, and are discussed further below.

The lunar basalts, like their terrestrial analogues, have cooled from the molten state and the presence in them of an apparently primary NRM was the first indication of the possible existence of an ancient lunar magnetic field in which the samples acquired a thermoremanent magnetization (TRM). This magnetization process is well-documented in terrestrial rocks, and occurs when a rock cools through the Curie temperature (or a lower magnetic 'blocking' temperature) of its magnetic mineral(s) in an ambient magnetic field. The persistence of NRM up to the Curie point of iron (780 °C) during thermal demagnetization of lunar samples is an important pointer to the presence of a TRM.

The accumulation of more data on igneous rocks and the breccias suggesting the presence of an ancient lunar magnetic field led naturally to discussions of the nature and

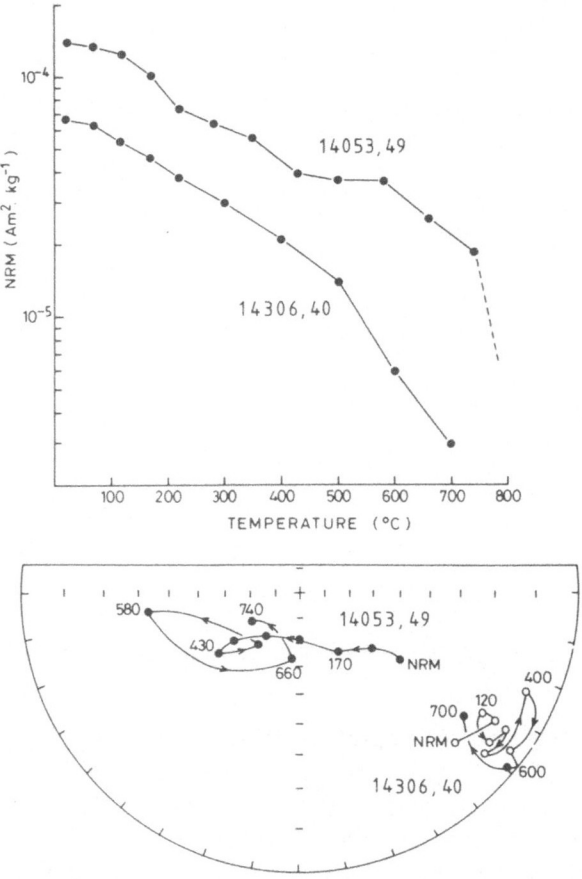

Fig. 6.   Thermal demagnetization of igneous rock 14053 and breccia 14306.

origin of the field. At the same time the observations of NRM in the breccias was suggesting to some investigators another type of magnetization process. These rocks result from the welding together of lunar dust and rock fragments by shock pressure and/or heating during a meteorite impact. Shock experiments on lunar samples (Cisowski *et al.*, 1973, 1974) show that rocks can acquire a magnetization if shocked in an ambient magnetic field, with characteristics similar to the NRM observed in some lunar samples. The ample evidence of extensive meteoritic bombardment of the lunar surface suggested the possible importance of shock remanent magnetization (SRM) acquired in an ambient field. At a later stage in the investigations, the possibility of SRM in a transient magnetic field generated during a meteorite impact was proposed, and this is discussed later in this paper. First we address the question of possible origins of a magnetic field at the surface of the ancient Moon.

3. Origins of a Lunar Magnetic Field

The radiometric ages of the returned lunar samples broadly covers the period 4.0–3.0 × 10⁹ yr ago, and the samples for which palaeomagnetic data were available covered most of this time interval. Preliminary estimates of the strength of the ancient field suggested values at least an order of magnitude greater than those measured at the Apollo 12, 14, 15, and 16 landing sites (Fuller, 1974) where a maximum field of $\sim 360\,\mathrm{nT}$ was observed (Apollo 16).

An enhanced solar wind field, at present of the order of $5\,\mathrm{nT}$, was suggested as a source of the ancient lunar field, but no reasonable mechanism for the amplification required was forthcoming. An alternative idea was that since it was likely that the Moon was much closer to the Earth during the early history of the Earth-Moon system, the Earth's magnetic field at the Moon was substantially greater than at present. However, assuming the geomagnetic field has been of essentially constant strength during its existence, it is easy to show that the Moon would have to be near or within Roche's limit (~ 2.5 Earth radii) to provide a lunar field of $\sim 3000\,\mathrm{nT}$ or more. The estimated time span of the field also requires it to be very close to the Earth for $\sim 10^9$ yr, for which there is no evidence.

Runcorn and Urey (1973) proposed that the whole body of the Moon became magnetized in an early Solar System magnetic field, providing a surface field appropriate to a uniformly magnetized sphere. Subsequent internal heating to above the iron Curie point resulted in removal of the magnetization, after the surface rocks had been extruded and become magnetized in the surface field. When improved lunar palaeointensity data were obtained (Section 4), it was established that the iron content of the Moon, estimated from surface and satellite magnetometer data (Parkin *et al.*, 1973), would have had to be magnetized to near saturation to provide the estimated surface field. There is no evidence of such a strong magnetizing field ($> \sim 50\,\mathrm{mT}$) having existed in the early Solar System.

A more interesting possibility, with important implications for lunar history and structure, is the existence of a molten, electrically-conducting lunar core in which a dynamo process operated to generate the lunar magnetic field, analogous with the origin of the Earth's magnetic field. However, there was considerable scepticism from some lunar scientists concerning this idea, based mainly on lack of positive evidence for a lunar core and certain theoretical difficulties in the generation of a significant magnetic field in the core of a small, slowly rotating body (Daily and Dyal, 1979).

These difficulties associated with the internal generation of a lunar magnetic field led to further interest in the impact theory of lunar magnetism. It was proposed that during a large meteoritic impact a transient magnetic field is generated in the hot cloud of dust and ionized particles surrounding the impact site. The rocks in the vicinity of the impact would then acquire a shock magnetization in this field. The generation of such a transient magnetic field does have some plausibility, and has been considered theoretically by Srnka (1977). There is also some laboratory evidence, obtained from experiments in which laser beams interact with targets (Stamper *et al.*, 1971; Tidman, 1974).

Thus it becomes a matter of considerable interest to obtain evidence which might help to distinguish between the dynamo and impact hypotheses of the origin of the lunar magnetizing field, and a possible key to the problem appeared to lie in the strength of the field and its variation with time.

4. Characteristics of the Lunar Magnetizing Field

It has already been observed that a severe limitation in lunar palaeomagnetism is the lack of oriented samples and consequently any information about the direction of the magnetizing field. Thus it was natural that, from the start, efforts were made at least to determine the strength of the field, which in principle could be obtained from the part of the magnetization vector that was measurable, namely the intensity of magnetization possessed by the samples.

Although the magnetization process occurring in lunar rocks had not been definitely established, the only reasonable approach to determining ancient lunar field intensity was to use the technique commonly used in terrestrial igneous rocks, the Thellier-Thellier method (Thellier and Thellier, 1959). This is based on the assumption that thermoremanence is the magnetizing process, and various modifications of the technique have been used on terrestrial igneous rocks (Collinson, 1982). Assuming that the thermoremanent magnetization acquired by a rock in cooling to 'room' temperature from the Curie point in an ambient magnetic field is proportional to the strength of that field, then

$$\left(\frac{J_{\text{NRM}}}{J_{\text{TRM}}}\right) = \frac{B_a}{B_l}$$

where J_{NRM} and B_a are the intensities of the primary NRM and the magnetic field strength in which it was acquired and J_{TRM} is the intensity of laboratory-induced TRM acquired by the rock in a field B_l. In practice it is usual to make use of the additive property of TRM, and compare the loss of NRM (heating in zero magnetic field) and the gain of partial thermoremanent magnetization (PTRM) acquired in cooling in the field B_l over a series of temperature intervals up to the Curie point. The cumulative loss of NRM is then plotted against the cumulative gain of PTRM for increasing temperatures (Figure 7). If the assumptions on which the method is based are valid and no thermal alteration of the remanence-carrying mineral occurs, a straight line will be obtained, the slope of which is proportional to the ancient magnetic field B_a. Deviations from a straight line can result from the presence of secondary magnetization(s) and physical or chemical alteration of the remanence-carrier due to heating, but it is often possible to select a reasonable temperature range over which linearity is obtained and from which B_a can be determined.

The application of the Thellier-Thellier palaeointensity technique to lunar rocks has not been entirely successful. The fine-grained iron carrying the NRM is extremely susceptible to oxidation and in some samples a magnetic interaction between troilite (FeS) and iron particles closely associated with it occurs at about 250–300 °C, resulting

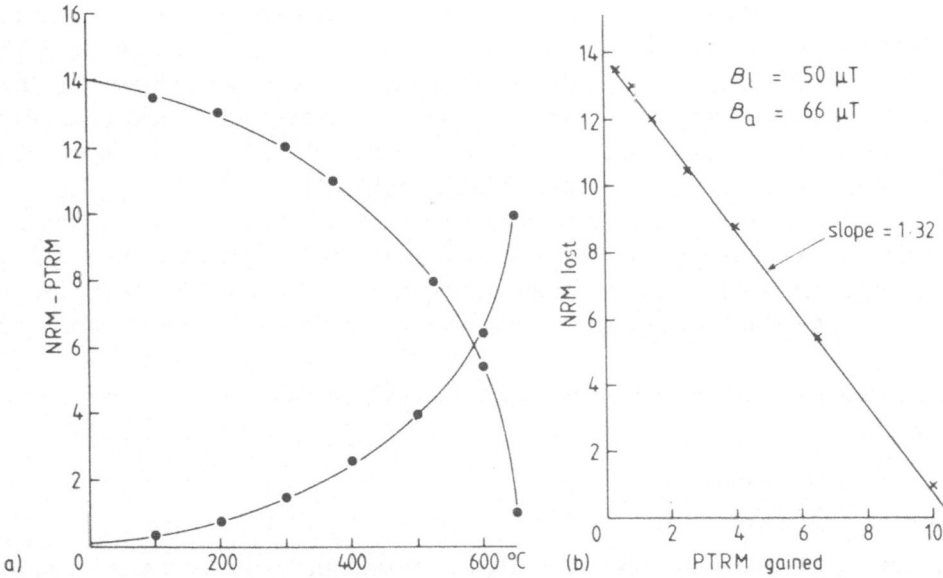

Fig. 7. Ideal behaviour of rock sample in the Thellier palaeointensity method. (a) Plot of NRM lost and PTRM gained after heating to and cooling from the indicated temperatures; (b) Cumulative plot of NRM lost against PTRM gained in $B_l = 50\,\mu T$. Magnetizations in arbitrary units.

in anomalous PTRM acquisition (Pearce *et al.*, 1976). Two samples from which reasonable results were obtained were an Apollo 16 recrystallised basalt, 62235.53 (Collinson et al., 1972) and regolith breccia 15 498.36 (Gose *et al.*, 1972). The former gave a rather well-defined but surprisingly high palaeointensity of $120\,\mu T$ over the temperature range 100–500 °C, and the latter about $2.2\,\mu T$, as derived by the authors, although Stephenson and Collinson (1975) considered that $\sim 7.0\,\mu T$ was a better interpretation of the result. More recently, Sugiura and Strangway (1983) obtained a palaeointensity of $130\,\mu T$ from another chip (.18) of 62235 using the Thellier method, in excellent agreement with the earlier determination. The few other attempts with the Thellier method failed mainly because of anomalous NRM behaviour on thermal treatment. However, an apparently valid palaeointensity value of $82\,\mu T$ has been obtained from an Apollo 11 basalt sample (10017.135) by a variation of the Thellier method, in which the decay during a.f. demagnetization of a laboratory TRM in the sample is compared with that of the NRM over several alternating field intervals (Hoffman, 1979).

The difficulties associated with the Thellier technique encouraged the development of another palaeointensity method, preferably one which involved no heating of the samples. One such method, developed by Stephenson (Stephenson and Collinson, 1974), makes use of the close similarity between anhysteretic remanent magnetization (ARM) and thermoremanent magnetization. ARM is acquired by a rock when it is subjected to a small direct magnetic field (typically 10–100 μT) on which is superimposed a decreasing alternating field. When the alternating field becomes zero the ARM remains in the rock directed along the biasing field direction. In ARM electromagnetic energy replaces the thermal energy in TRM which releases and then allows alignment of the elementary magnetic moments in the ambient field. Like TRM, ARM is generally proportional to the strength of the ambient field in low fields and is additive for successively higher alternating field intervals.

In the ARM palaeointensity method the loss of NRM in an alternating field interval during demagnetization is compared with the gain of ARM in the same interval, these quantities being obtained from the a.f. demagnetization curve and ARM acquisition curve, i.e. ARM intensity plotted against peak alternating field, with constant direct field.

To obtain an ancient field intensity using the ARM method, an expression is used of the form

$$\frac{1}{f}\left(\frac{J_{NRM}}{J_{ARM}}\right) = \frac{B_a}{B_l}$$

where J_{NRM}, B_a and B_l are as before and J_{ARM} is the saturated intensity of ARM acquired in B_l, i.e. the intensity acquired in an infinite alternating field, which can be obtained by extrapolation of the ARM acquisition curve. The factor f is the ratio of TRM intensity to ARM intensity, each acquired in the same ambient (direct) field. This factor has an experimentally determined value, based on tests with synthetic iron samples and lunar rocks, of $\sim 1.34 \pm 10\%$ (Stephenson and Collinson, 1974). Whether this value is valid for all lunar samples is currently in doubt, and is the cause of some discussion concerning the validity of the ARM technique. It was at least encouraging that when the ARM method was applied to another fragment of lunar rock, 62235.53, a palaeointensity of 140 µT was obtained, rather close to the values of 120 and 130 µT obtained with the Thellier technique. Five other lunar samples gave apparently valid palaeointensities with this method (Collinson and Stephenson, 1977; Stephenson et al., 1977).

The possible significance of the palaeointensity data is indicated in Figure 8, in which the ancient field intensities derived from rock samples are plotted against their radiometrically-determined ages. In spite of the limitations and lack of accuracy in the determinations, there is some evidence of a steady decrease of lunar field intensity with time. If true, such a trend argues strongly for an intrinsic lunar magnetic field and against any process that would provide random magnetizing fields, such as would arise from meteorite impacts. The high field of 130 µT derived from samples 62235.53 and 62235.18 by two independent methods is an important result in that it eliminates the Runcorn-Urey hypothesis of a permanently magnetized Moon (Section 3), leaving the dynamo-generation of the lunar field as a strong contender for its origin.

It is clearly highly desirable to extend palaeointensity determinations to many more samples, whose ages span the time interval covered by the lunar rocks, in order to obtain more evidence of any variation of ancient field intensity with time. However, the difficulties associated with the Thellier technique and some scepticism concerning the

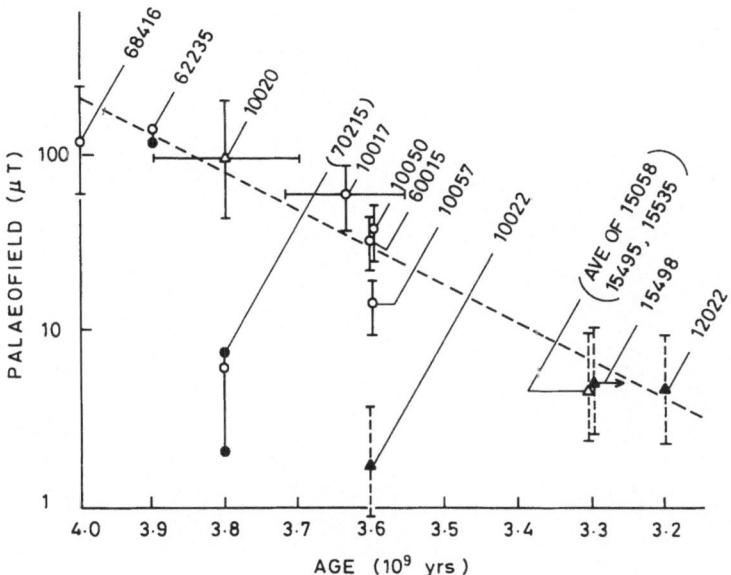

Fig. 8.   Intensities of the ancient lunar field derived by the ARM method (open symbols) and Thellier method (full symbols), plotted against the age of the samples.

validity of the ARM procedure caused problems in extending the work in this way, and Cisowski *et al.* (1977) examined a less rigorous but more widely applicable procedure for estimating relative lunar palaeointensities.

If it is assumed that the iron grains in lunar rocks have a similar size distribution, similar magnetic properties and have undergone a magnetizing process in which the resulting NRM is proportional to the ambient field strength, then the observed primary NRM will depend only on the field strength and the proportion of iron in the rocks. Direct measurement of metallic iron content is not possible, but the saturated isothermal remanent magnetization (IRM$_s$) in different samples should provide a satisfactory relative measure of iron content.

Thus by taking the ratio of primary NRM intensity to saturation remanence, the primary remanence is 'normalized' to constant iron content, and the resulting ratios obtained from different samples should provide a reasonable estimate of relative ancient field strengths in which the samples acquired their NRM. To avoid spurious results due to the presence of secondary magnetizations, Cisowski *et al.* used the NRM values after 10 mT a.f. demagnetization.

Although dependent on several assumptions, not all of which (e.g. a uniform magnetizing process) appear to be entirely valid, a recent extensive series of measurements by Cisowski involving some 70 lunar samples of all types showed an interesting trend (Cisowski *et al.*, 1983). This trend is shown in Figure 9 in which the main feature of interest is the apparent rapid rise of the field approximately 3.9–4.0 by ago and its decay to an intermediate value some 250 my later this field then persisting until ∼ 3.0

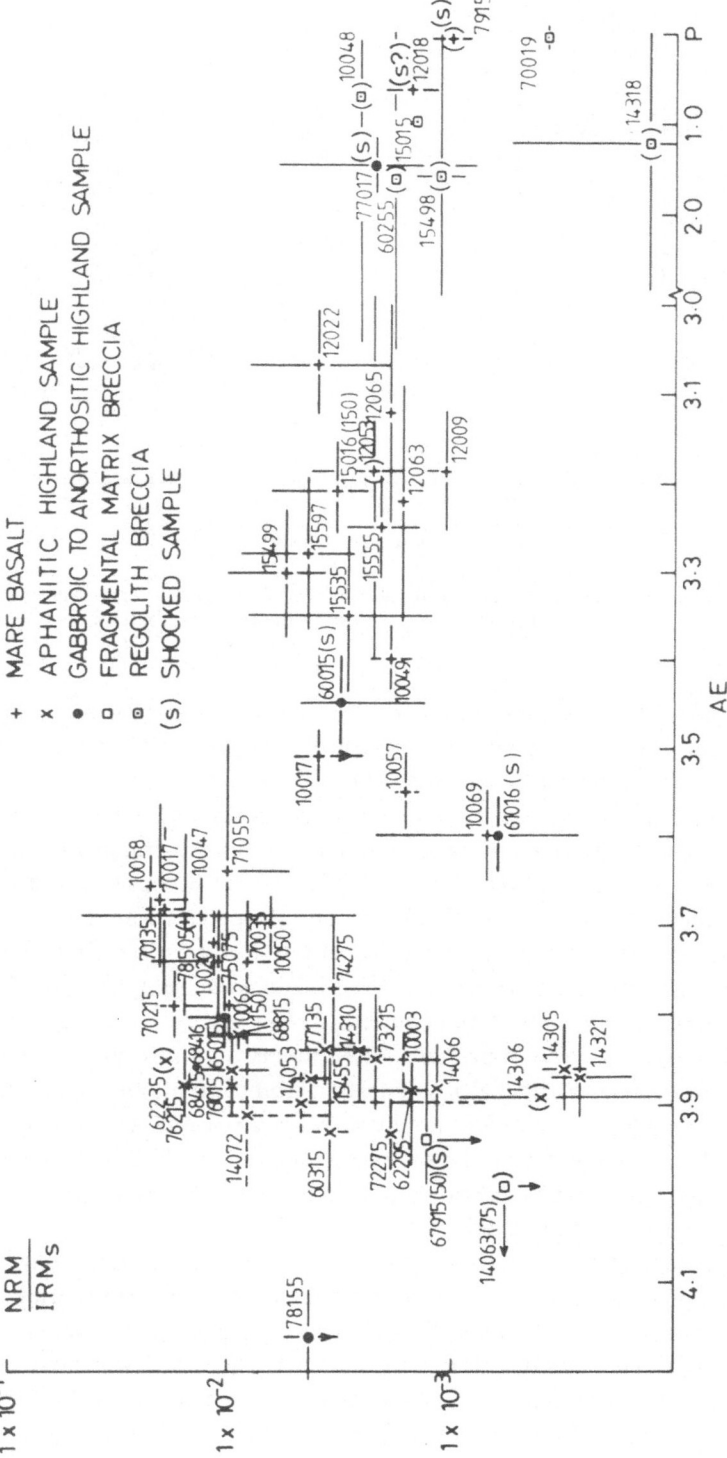

Fig. 9.   Plot of normalized remanence (NRM/IRM$_s$) against radiometric age in lunar samples. Horizontal scale is in 10^9 yr (aeons), and there is a discontinuity in the scale at 3.0 aeons (from Cisowski *et al.*, 1983). The NRM and IRM values used are those remaining after 20 mT a.f. demagnetization.

by ago. Although subject to errors along the magnetization ratio axis due to differing results in chips from the same samples and intensity differences on repeat demagnetization, and along the age axis due to published uncertainties in radiometric age determinations, the above trend is comparatively well-defined. It is also broadly consistent with the trend of Figure 8, the absolute palaeointensities based on the Thellier and ARM techniques.

The most important inference which may reasonably be drawn from Figure 9 is that the relative palaeointensities are far from randomly distributed. It is unlikely that transient magnetic fields associated with impacts would exhibit any systematic variation with time, and the likelihood of such magnetic fields being the origin of lunar sample magnetization now recedes significantly. Additional evidence against impact-generated fields arises from the lack of correlation between magnetization intensity and severity of shock in samples. An interpretation of Figure 9 is that the ancient lunar field grew rather rapidly at about 3.9 by, maintained a substantial magnitude for ~ 250 my and then decayed to a lower value for the period covered by lunar volcanism, i.e. to about 3.1 by ago.

Comparison of the (NRM/IRM$_s$) ratio with ancient field strength in samples in which the latter has been estimated by the Thellier of ARM method indicates an approximate linear relationship between the two parameters. The average value of the factor relating field strength (in μT) to (NRM/IRM$_s$) is 4700 (Cisowski et al., 1983), i.e. multiplication of ordinate values in Figure 9 by this factor gives the field strength in μT with an uncertainty of perhaps $\pm 15\%$. Thus, a magnetic field of maximum strength of the order of 100 μT is suggested by the data of Figure 9.

5. Discussion

The results of the absolute and relative lunar palaeointensity determinations have reinforced the evidence for an intrinsic lunar magnetic field and at the same time somewhat increased the difficulty of explaining it by a core dynamo process, because of the apparently high value of the field of $\sim 100\,\mu$T, existing for a significant period of lunar history. The data are also generally consistent with the present surface field strength determinations of Lin (1979), using the electron reflection method. Over the regions covered by the Apollo 15 and 16 subsatellite observations of reflection intensity, stronger surface fields, arising from crustal magnetization, were recorded over areas older than ~ 3.7 by. There is some disagreement between Lin's estimated lunar palaeointensity which would provide the observed present surface fields over older cratered uniformly magnetized surfaces and the sample-derived palaeointensity. However, Lin's estimate is not securely based, and an interesting possibility is that the lunar field underwent reversals, now well-documented in connection with the geomagnetic field, which would weaken the average crustal magnetization (Lin, 1979; Srnka and Mendenhall 1979). The most recent contribution to evidence for the ancient lunar field is Keith Runcorn's work concerning lunar polar wandering (Runcorn, 1982). This analysis derives from the mapping of small magnetic anomalies detected over the lunar surface by the Apollo 15 and 16 subsatellites, and their representation by discrete (dipolar) areas of uniform crustal magnetization (Hood *et al.*, 1978).

This elegant magnetic mapping provides indirect evidence for the direction of magnetization of some crustal regions of 10–100 km in typical dimensions, thus providing the vital information on NRM directions which is not available from the magnetic measurements on lunar samples. Runcorn (1978, 1980) treated these derived NRM directions using the profitable procedure employed in terrestrial palaeomagnetism, namely the determination, on the assumption of an ancient dipolar magnetizing field, of the axis of the field, and the position of the ancient magnetic pole of the Moon. Dynamo theory indicates that, because of the likely dominance of the Coriolis force in the lunar core, the average dipole field axis and rotation axis of the Moon will coincide: thus the directional information derived from the modelled anomaly dipoles potentially provides information on the Moon's ancient rotation axis. Runcorn (1980, 1982) showed that there was some grouping of ancient pole positions derived from selected areas of the Moon of different ages, and that the mean pole position for each age range was different and could be explained by 'polar wandering', or the variation with time of the rotation axis of the Moon with respect to the crust. The cause of lunar polar wandering is most reasonably associated with changes in the axes of maximum and minimum moment of inertia of the Moon, and consequent readjustment of the rotation axis, due to meteoritic excavation and subsequent flooding of the circular maria.

An interesting consequence of an early dynamo dipolar field, relevant to the present lunar dipole moment, was first pointed out by Runcorn (1975). He showed that the external field of a spherical shell magnetized by an internal dipole field is zero (neglecting permeability effects), when the dipole is removed. This is relevant to the Moon, the large-scale melting of the outer regions, producing the magma which filled the basins, resulting in a thermoremanent magnetization in an outer shell acquired in the lunar field on subsequent cooling. In a more rigorous analysis including the effects of finite magnetic permeability of different regions of the Moon, Stephenson (1976) showed that the external fields, and corresponding apparent dipole moment, is nonzero but still within the upper limit deduced by Russell *et al.* (1974) of 1.3×10^{15} A m^2. This result is inconsistent with a Moon originally uniformly magnetized in an external field.

With the accumulating evidence for an ancient, dynamo-generated lunar magnetic field, it is pertinent to enquire into two important aspects of such a field, namely the existence of a lunar core in which the field was generated, and whether, given the presence of a suitable core, field generation is theoretically feasible.

There is, at present, no firm evidence for the presence of a heavy, electrically-conducting iron (or iron-sulphur) core in the Moon. However, the relevant observations do allow the presence of such a core. The lunar moment of inertia factor (the ratio of the polar moment of inertia to the product of the lunar mass times the mean equatorial radius), a measure of central condensation, is now well-established at 0.3905 ± 0.0023 (Ferrari *et al.*, 1980), permitting an iron core of ~ 300–360 km radius, according to density. The seismic evidence from the one far side impact is ambiguous but the overall seismic data allow a core of up to $\sim$ 360 km radius (Latham *et al.*, 1978; Goins *et al.*, 1979). Electrical conductivity data for the lunar interior has been obtained by analysis of the Moon's response to the solar wind magnetic field, using either the Apollo surface magnetometers and the Explorer 35 satellite (Wiskerchen and Sonett, 1977; Hood *et al.*, 1982) or the Apollo 15 and 16 subsatellite magnetometers (Russell *et al.*, 1981). The observations do not place strong constraints on the conductivity of the deep lunar interior, but are consistent with a 300–400 km core of conductivity appropriate to iron. Prior to the Apollo missions, Runcorn (1967) had proposed the presence of a small lunar core to provide a possible convection pattern in the lunar interior to account for the non-hydrostatic shape of the Moon.

If present, it appears that the maximum size of a lunar iron core is $\sim$ 400 km, i.e. between a fifth and a quarter of the Moon's radius. Thus, a surface dipole magnetic field of $\sim$ 100 µT, as indicated by the palaeointensity results outlined earlier, implies a field on the core surface of $\sim$ 10 mT. The combination of small core, slow rotation and apparently strong field has led to reluctance on the part of some lunar scientists to accept the possibility of a lunar dynamo. Except for some general principles (e.g. the condition that the magnetic Reynolds number of the core must be $> \sim$ 10 for dynamo action), current dynamo theories do not allow field generation to be predicted for a given core size, conductivity, angular velocity, etc. Using as a model the type of dynamo believed to generate the geomagnetic field, Levy (1972) scaled the dynamo for a lunar core and found that an unacceptably rapid lunar rotation rate would be required. An assumption made was that the geomagnetic dynamo is only operating marginally, which might be disputed by other theoreticians. Russell and Goldstein (1976) use a scaling law based on a precessionally-driven dynamo and find a maximum surface field of only $\sim$ 300 nT is attainable, and the analysis of Anderson (1983), based on Busse's (1978) theory of planetary magnetism, suggests that a 20-fold increase of rotation rate combined with a lunar core radius $>$ 750 km is required to produce a field of $\sim$ 2000 nT.

Another problem associated with an ancient lunar dynamo is the nature of the energy source which drove it. This problem is still unresolved. Possible candidate sources are radioactivity, decay of the Moon's rotation rate, partial solidification of the core and decay of super-heavy elements (Runcorn, 1978). A connected problem is the heat source which originally caused interior melting and formation of the core.

It is apparent that at the present time lunar magnetism research has reached an intriguing stage. As the evidence accumulates for an ancient global lunar magnetic field of dynamo origin, the feasibility of dynamo action in a lunar core remains controversial. The controversy is further fuelled by lack of positive evidence for the existence of the core and the requirement of a strong lunar surface field of the order of 100 µT for a period of 200–300 my around 3.7 by ago. However, in view of our lack of knowledge of the dynamics, structure and internal properties of the Moon at this stage of its history, the difficulties of applying dynamo theory to planetary cores with poorly-defined properties, and the evidence of the importance of the dynamo process elsewhere in the Solar system, as shown by the magnetic fields of Earth, Jupiter, Saturn, and possibly Mercury, the existence of an ancient lunar dynamo is a valid interpretation of all the relevant lunar magnetic data.

Research is continuing and hopefully more data will come to hand to constrain more rigorously the characteristics of the lunar magnetic field. Further absolute palaeointensity determinations are being carried out to provide more evidence of field strength, particularly in the period 3.5–3.0 by, and evidence for a significant field between 3.0 by and the present is also being investigated. If the proposed Lunar Polar Orbiter (POLO) mission of the European Space Agency, vigorously promoted by Keith Runcorn, comes to fruition additional valuable magnetic anomaly data covering the whole lunar surface will become available, much increasing our knowledge concerning directions of crustal magnetization and providing more data with which to study lunar polar wandering.

The results of the investigations of lunar magnetism and palaeomagnetism have fully justified the research effort expended, and the data obtained have a much more far-reaching importance than merely establishing the presence of an ancient lunar magnetic field. The likely origin of the field provides vital evidence for the structure, energy sources and thermal history of the Moon. Looking to the future, it will be of exceptional interest to study the magnetic properties of samples returned from Mars, and perhaps ultimately from Venus and other bodies in the Solar System, such as the asteroids. The results of the lunar investigation will surely ensure that magnetic studies on returned samples from these bodies receive high priority.

References

Anderson, K. A.: 1983, 'Magnetic Dipole Moment, Estimates for an Ancient Lunar Dynamo', *Proc. 13th Lun. Plan. Sci. Conf.*, *J. Geophys. Res.* **88**, A588–A590.

Busse, F. H.: 1978, 'Magnetohydrodynamics of the Earth's Dynamo', *Ann. Rev. Fluid Mech.* **10**, 435–462.

Cisowski, S. M., Fuller, M., Rose, M. F., and Wasilewski, P. J.: 1973, 'Magnetic Effects of Explosive Shocking of Lunar Soil', *Proc. 4th Lun. Sci. Conf.* **3**, 3003–3017.

Cisowski, S. M., Dunn, J. R., Fuller, M., Rose, M. F., and Wasilewski, P. J.: 1974, 'Impact Processes and Lunar Magnetism', *Proc. 5th Lun. Sci. Conf.* **3**, 2841–2858.

Cisowski, S. M., Hale, C., and Fuller, M.: 1977, 'On the Intensity of Ancient Lunar Fields', *Proc. 8th Lun. Sci. Conf.* **1**, 725–750.

Cisowski, S. M., Collinson, D. W., Runcorn, S. K., and Stephenson, A.: 1983, 'A Review of Lunar Paleointensity Data and Implications for the Origin of Lunar Magnetism', *Proc. 13th Lun. Plan. Sci. Conf.*, *J. Geophys. Res.* **88**, A691–A704.

Collinson, D. W.: 1982, *Methods in Rock Magnetism and Palaeomagnetism: Techniques and Instrumentation*, Chapman and Hall, London.

Collinson, D. W. and Stephenson, A.: 1977, 'Paleointensity Determinations of Lunar Samples', *Phys. Earth Plan. Int.* **13**, 380–385.

Collinson, D. W., Runcorn, S. K., Stephenson, A., and Manson, A. J.: 1972, 'Magnetic Properties of Apollo 14 Rocks and Fines', *Proc. 3rd Lun. Sci. Conf.* **3**, 2343–2361.

Collinson, D. W., Stephenson, A., and Runcorn, S. K.: 1973, 'Magnetic Properties of Apollo 15 and 16 Rocks', *Proc. 4th Lun. Sci. Conf.* **3**, 2963–2976.

Daily, W. D. and Dyal, P.: 1979, 'Theories for the Origin of Lunar Magnetism', *Phys. Earth Plan. Int.* **20**, 255–270.

Ferrari, A. J., Sinclair, W. S., Sjogren, W. L., Williams, J. G., and Yoder, C. F.: 1980, 'Geophysical Parameters of the Earth-Moon System', *J. Geophys. Res.* **85**, 3939–3951.

Fuller, M.: 1974, 'Lunar Magnetism', *Rev. Geophys. Space. Phys.* **12**, 23–71.

Goins, N. R., Toksöz, M. N., and Dainty, A. M.: 1979, 'The Lunar Interior: A Summary Report', *Proc. 10th Lun. Plan. Sci. Conf.* **3**, 2403–2420.

Gose, W. A., Strangway, D. W., and Pearce, G. W.: 1973, 'A Determination of the Intensity of the Ancient Lunar Magnetic Field', *The Moon*, **7**, 196–201.

Hoffman, K. A.: 1979, 'Combining Paleointensity Methods: A Dual-Valued Determination on Lunar Sample 10 017.135', *Phys. Earth Plan. Int.* **20**, 317–323.

Hood, L. L., Russell, C. T., and Coleman, P. J.: 1978: 'The Magnetization of the Lunar Crust as Deduced from Orbital Surveys', *Proc. 9th Lun. Plan. Sci. Conf.* **3**, 3057–3078.

Hood, L. L., Herbert, G., and Sonett, C. P.: 1982, 'The Deep Lunar Electrical Conductivity Profile: Structural and Thermal Inferences', *J. Geophys. Res.* **87**, 5311–5326.

Latham, G. V., Dorman, H. J., Horvath, P., Ibrahim, A. K., Koyama, J., and Nakamura, Y.: 1978, 'Passive Seismic Experiment: A Summary of Current Status', *Proc. 9th Lun. Plan. Sci. Conf.* **3**, 3609–3614.

Levy, E. H.: 1972, 'Magnetic Dynamo in the Moon: A Comparison with Earth', *Science* **178**, 52–53.

Lin, R. P.: 1979, 'Constraints on the Origins of Lunar Magnetism from Electron Reflection Measurements of Surface Magnetic Fields', *Phys. Earth Plan. Int.* **20**, 271–280.

Parkin, C. W., Dyal, P., and Daily, W. D.: 1973, 'Iron Abundance in the Moon from Magnetometer Measurements', *Proc. 4th Lun. Sci. Conf.* **3**, 2947–2962.

Pearce, G. W., Gose, W. A., and Strangway, D. W.: 1973, 'Magnetic Studies on Apollo 15 and 16 Lunar Samples', *Proc. 4th Lun. Sci. Conf.* **3**, 3045–3076.

Pearce, G. W., Hoye, G. S., Strangway, D. W., Walker, B. M., and Taylor, L. A.: 1976, 'Some Complexities in the Determinations of Lunar Paleointensities', *Proc. 7th Lun. Sci. Conf.* **3**, 3271–3298.

Runcorn, S. K.: 1967, 'Convection in the Moon and the Existence of a Lunar Core', *Proc. Roy. Soc.* A. **296**, 270–284.

Runcorn, S. K.: 1975, 'An Ancient Lunar Magnetic Dipole Field', *Nature* **253**, 701–703.

Runcorn, S. K.: 1978a, 'The Origin of Lunar Palaeomagnetism', *Nature* **275**, 430–432.

Runcorn, S. K.: 1978b, 'The Ancient Lunar Core Dynamo', *Science* **199**, 771–773.

Runcorn, S. K.: 1980, 'Lunar Polar Wandering', *Proc. 11th Lun. Plan. Sci. Conf.*, **3**, 1867–1878.

Runcorn, S. K.: 1982, 'Primeval Displacements of the Lunar Pole', *Phys. Earth Plan. Int.* **29**, 135–147.

Runcorn, S. K. and Urey, H. C.: 1973, 'A New Theory of Lunar Magnetism', *Science* **180**, 636–638.

Russell, C. T. and Goldstein, B. E.: 1976, 'The Geomagnetic Dynamos of the Moon and Venus: Comparisons with a Recent Scaling Law', *Proc. 7th Lun. Plan. Sci. Conf.* **3**, 3343–3355.

Russell, C. T., Coleman, P. J., Lichtenstein, B. R., and Schubert, G.: 1974, 'The Permanent and Induced Magnetic Dipole Moment of the Moon', *Proc. 5th Lun. Sci. Conf.* **3**, 2747–2760.

Russell, C. T., Coleman, P. J., and Goldstein, B. E.: 1981, 'Measurements of the Lunar Induced Magnetic Moment in the Geomagnetic Tail: Evidence for a Lunar Core?', *Proc. 12th Lun. Plan. Sci. Conf.* **2**, 831–836.

Srnka, L. J.: 1977, 'Spontaneous Magnetic Field Generation in Hypervelocity Impacts', *Proc. 8th Lun. Plan. Sci. Conf.* **1**, 785–792.

Srnka, L. J. and Mendenhall, M. H.: 1979, 'Models of an Early Lunar Dynamo', *Proc. 10th Lun. Plan. Sci. Conf.* **3**, 2343–2356.

Stamper, J. A., Papadopoulos, K., Sudan, R. N., Dean, S. O., McLean, E. A., and Dawson, J. M.: 1971, 'Spontaneous Magnetic Fields in Laser-Produced Plasmas', *Phys. Rev. Lett.* **46**, 1012–1015.

Stephenson, A.: 1976, 'The residual Permanent Magnetic Dipole Moment of the Moon', *The Moon* **15**, 67–81.

Stephenson, A. and Collinson, D. W.: 1974, 'Lunar Magnetic Field Palaeointensities Determined by an Anhysteretic Magnetization Method', *Earth Plan. Sci. Lett.* **23**, 220–228.

Stephenson, A. and Collinson, D. W.: 1975, 'On Changes in the Intensity of the Ancient Lunar Magnetic Field', *Proc. 6th Lun. Sci. Conf.* **3**, 3049–3062.

Stephenson, A., Runcorn, S. K., and Collinson, D. W.: 1977, 'Paleointensity Estimates from Lunar Samples 10017 and 10020', *Proc. 8th Lun. Sci. Conf.* **1**, 679–687.

Sugiura, N. and Strangway, D. W.: 1983, 'Magnetic Paleointensity Determination on Lunar Sample 62 235', *Proc. 13th Lun. Plan. Sci. Conf.*, *J. Geophys. Res.* **88**, A684–A690.

Thellier, E. and Thellier, O.: 1959, 'Sur l'intensité du champ magnétique terrestre, dans le passé historique et géologique', *Ann Geophys.* **15**, 285–376.

Tidman, D. A.: 1974, 'Strong Magnetic Fields Produced by Compositional Discontinuities in Laser Produced Plasmas', *Phys. Rev. Lett.* **32**, 1179–1181.

Wiskerchen, M. J. and Sonett, C. P.: 1977, 'A Lunar Metal Core?', *Proc. 8th Lun. Sci. Conf.* **1**, 515–536.

ON THE GRAIN SIZE DEPENDENCE OF THE BEHAVIOUR OF
FINE MAGNETIC PARTICLES IN ROCKS

M. FULLER

Department of Geological Sciences, University of California, Santa Barbara, CA 93106, U.S.A.

Abstract. The grain size dependence of the ratio of saturation remanent magnetization to saturation magnetization $(J_R : J_S)$, weak field susceptibility (X_0), thermoremanent magnetization (TRM) and its stability against AF demagnetization are interpreted in terms of nucleation theory. It is concluded that each of these parameters exhibits grain size dependence due to two effects. The first is the increasing difficulty with which domain walls are neucleated as grain size decreases. The second is an intrinsic grain size dependence of the parameters in multidomain particles.

1. Introduction

The success of paleomagnetism is now undisputed. The revolution in geological thinking, which brought about the modern mobilist view of the earth's crust, was to a large extent prompted by paleomagnetism. Subsequently paleomagnetism has made contributions to stratigraphy and has entered a new phase in its tectonic applications by elucidating the details of phenomena at plate boundaries. However, when Professor Runcorn began his work the situation was very different. Paleomagnetists were subjected to ridicule which is regrettably all too often reserved for those who introduce new thinking into a science. Professor Runcorn was of course a leader of the small band of workers, whose enthusiasm and energy, as well as their good scientific judgment, withstood the criticism and produced the key data to confirm the then somewhat discarded theory of continental drift. As geologists, we owe him a great debt and it is an honour to be invited to contribute to this volume dedicated to him.

It has frequently been noted that despite the obvious success of paleomagnetism, the theoretical basis of rock magnetism remains somewhat insecure. It therefore seems appropriate to review the development of rock magnetism and consider the major problems. In particular, the paper will be concerned with the nature of the domain state of the fine magnetic particles, which carry the paleomagnetic record, and its influence on their magnetic behaviour.

2. Some Historical Perspectives of Rock Magnetism

One of the problems which faced the early paleomagnetists was the unfamiliarity of geologists and geophysicists with magnetism, so that there was an air of magic about the results, which invited scepticism. A good deal of the sceptical criticism was uninformed, e.g., Jeffreys (1959). Nevertheless, the aspects of magnetic theory most relevant to paleomagnetism had not been matters of importance to those who had worked on the physics of magnetic materials. They had been interested in the fundamental forms of magnetic order and the interpretation of hysteresis. On the other

Geophysical Surveys 7 (1984) 75-87. 0046-5763/84/0071-0075$01.95.

hand, paleomagnetists were concerned with how sedimentary and igneous rocks acquired their Natural Remanent Magnetization (NRM) in the weak geomagnetic field and how this record was preserved over the geological aeons. Néel (1955), noted that there were a number of aspects of the magnetism of rocks which made the relevant physics different from that commonly encountered in studies of magnetic materials.

In two fundamental papers, Néel (1949, 1955) presented what has become the cornerstone of rock magnetism. He treated remanent magnetism as a typical relaxation phenomenon governed by the familiar equation

$$M(t) = M(0) \; e^{-t/\tau}$$

where $M(0)$ is the initial magnetization and $M(t)$ the magnetization after a time t and τ is a relaxation time defined as

$$\frac{1}{\tau} = f \; e^{-K_u V/kT}$$

where K_u is the anisotropy energy, V the particle volume, k Boltzmann's constant and T absolute temperature. In his model of single domain TRM, Néel developed an intuitive theory to calculate the frequency factor f. This aspect of his work was later criticized by Brown (1959) who derived f from more generalized theory. However, as Brown acknowledged the final estimates are similar.

In his theory of TRM, Néel assumed randomly oriented, uniaxial particles. The magnetization is therefore constrained, in the absence of other energy terms, to lie along the easy axis, in either of the two possible orientations. He then noted that the energy levels are split by the magnetostatic energy in the external field. He next showed that the relaxation times for switching from parallel to the field to antiparallel would be greater than for switching in the opposite sense. This then gives a statistical bias to the population in the field direction. As the temperature falls, this distribution is frozen in as TRM. The treatment is analogous to spin 1/2 paramagnetism and gives the same tanh field dependence. The model is qualitatively successful in explaining many aspects of TRM, although it saturates too quickly with H, as noted by Day (1977) in a comprehensive review of TRM.

Néel's work gave a physical basis for paleomagnetism, but by the late 1950's and early 1960's, there was a growing recognition that the magnetism of rocks might be a complicated and messy business. Moreover, these complexities could involve systematic distortions of the paleomagnetic record. Graham et al. (1957) drew attention to the possible effects of magnetostriction. Others were concerned about the effects of magnetic anisotropy (Uyeda et al., 1963). It appears in retrospect, that these results were accepted somewhat uncritically and given too much weight by some, who found continental drift unpalatable.

It was however clear in the early 1960's that the theory of rock magnetism was marred by one fundamental paradox. Néel's single domain theory was undeniably successful. Yet, the grain size of the magnetic particles in rocks was far too large for them to be single domain according to standard theory (e.g., Kittel, 1949). An associated problem, which emerged as rock magnetism began to acquire more of an observational data base, was the grain size dependent behaviour exhibited by particles of a few microns to a few tens of microns. The term pseudosingle domain (PSD) behaviour was coined by Stacey (1963) to describe the phenomena. A number of speculations emerged to explain the PSD behaviour. Essentially, they consisted of ascribing unusual characteristics to the walls in such particles. For example, Stacey (1963) suggested that there was anomalous Barkhausen discreteness. Stacey and Banerjee (1974) considered the possibility that the PSD moments were carried by small surface domains. Dunlop (1977) appealed to wall moments.

It now appears that at least part of the explanation of PSD behaviour lies in the increasing difficulty with which domain walls are nucleated by particles, as their grain size decreases (Halgedahl and Fuller, 1980, 1983). The explanation was cast in the form of a simple model of the grain size dependence of the ratio of saturation remanent magnetization to saturation remanence ($J_R : J_S$). The idea, if correct, presumably has broader significance and so in this paper an attempt is made to generalize the approach. Unfortunately, the critical data required to test the models are not presently available. We therefore have to make do with unsatisfactory and preliminary tests.

3. Domain Wall Nucleation

Domain wall nucleation has long been recognized as an important aspect of magnetic behaviour (e.g., De Blois and Bean, 1959; Brown, 1963; and Becker, 1969), it has been largely, though not completely (Levi and Merrill, 1978), ignored in rock magnetism until recently. It has now been demonstrated that a substantial fraction of particles up to tens of microns in diameter may remain saturated at remanence in a hysteresis loop and only nucleate walls when a reverse field is applied (Halgedahl and Fuller, 1980, 1983). Thus the particles remain single domain in saturation remanent magnetization (J_R or IRM_S) but magnetization reversal is achieved by backfield nucleation followed by wall motion in the usual sense. In an ensemble of particles with distributed grain size, the contribution of such particles to IRM_S will be large. It will be large compared with true single domain particles because the volume of individual metastable single domain particles is so much greater than of true single domain particles. It will be large compared with multidomain particles because the remanent moment per unit volume will be much greater due to their saturated state. The moment of individual particles is the product of their volume and the saturation magnetization per unit volume, giving for a spherical particle of magnetite, with a radius of one micron (μm) it will be $2 \times 10^{-9} \, J \, G \, cm^3$ ($2 \times 10^{-12} \, A \, m^2$). Since the moment varies with the volume, or the cube of the radius, a similar particle with a 10 μm radius will have a moment as large as $2 \times 10^{-6} \, J \, G \, cm^3$. Such a moment is readily measured with modern magnetometers and is larger than the natural remanent magnetization (NRM) moment of many of the standard samples used in paleomagnetism!

The physical basis of nucleation was established theoretically by Brown (1963), who demonstrated that the internal field must be strong enough to rotate spins against the anisotropy field before walls can form. An important extension of this idea came from Becker (1971, 1973, and 1976), who pointed out that the nucleated wall must also be able to escape from the nucleation site, which can act as a trap and pin the tiny nucleus of the wall. It is therefore useful to distinguish two stages of nucleation, true nucleation, or the initial creation of the wall, and secondary nucleation, or the initial unpinning of the trapped nucleus. In a spherical particle of magnetite the demagnetizing field will be $4/3 \pi$ Ms ~ 2000 Oe (~ 25 Am^{-1}) and thus the observed pinning of the nucleus is against a very substantial field. It should be recognized that it is the nucleus of the reversed domain which is trapped at the surface nucleating site, so that it is not correct to regard the magnetic moment of pseudosingle domain particles as residing in surface pinned domains, as was suggested earlier by Stacey and Banerjee (1974).

Difficulties in true nucleation and the escape of the wall from the nucleation site can account for the existence of particles in a metastable single domain state in a rock carrying remanent magnetization. These particles may either reverse magnetization by nucleation of walls in small backfields followed by normal wall motion, or they may require so large a backfield to nucleate the wall that, once formed and initially unpinned, the wall traverses the particle almost instantaneously. Difficulties in wall nucleation may also explain why those particles, which are multidomain, contain far fewer domains than Kittel (1949) theory predicts, an observation common to domain studies of fine particles in rocks, whether they be by Bitter pattern methods (Bogdanov and Vlasov, 1966; Soffel, 1977; Halgedahl and Fuller, 1983) or by electron microscopy (Smith, 1980).

4. Grain Size Dependence of $J_R : J_S$

PSD behaviour is clearly manifested in the grain size dependence of hysteresis characteristics. From Stoner–Wohlfarth theory, the ratio of saturation remanence to saturation magnetization ($J_R' : J_S$) for randomly oriented single domain particles with uniaxial anisotropy is 0.5. Values for assemblages of particles with restricted grain size ranges show a grain size dependence of the ratio $J_R : J_S$ between the single domain and multidomain values. Other properties of such sized fractions also vary in a compatible manner [Table I, (from Day et al., 1977)].

A very simple nucleation model of the grain size dependence of the ratio $J_R : J_S$ was given in Halgedahl and Fuller (1980). Nucleation centers were assumed to be randomly distributed through an ensemble of particles, with successful nucleation at a site in zero field treated as a very rare event. This then gives a grain size dependent probability that a particle will contain a particular number (w) of domain walls. This probability $P(w)$ is obtained from Poisson statistics to be

$$P(w) = \lambda^w \frac{e^{-\lambda}}{w!}$$

where λ is the average number of walls.

To obtain the value of $J_R : J_S$ for the assemblage, one assumes that only those particles, which have not nucleated a wall, contribute significantly to the saturation remanent magnetization (J_R). Knowing the value of $J_R : J_S$ for a single domain particle gives the ratio for the PSD ensemble according to the relation

$$\frac{J_R}{J_S} = e^{-\lambda} \frac{J_R}{J_S} (\text{SD})$$

in which $e^{-\lambda}$ acts as a weighting factor of the SD value. λ can be obtained observationally, or it can be calculated from the theory of Kittel (1949). In the latter case it is sensitive to the choice of the critical size for single domain behaviour. In the published model, that size was taken to be $1.4\,\mu m$. The model was compared with data from Day *et al.* (1977) and Co--Fe alloy particles (Kneller and Luborsky, 1963) and was shown to be reasonably successful in the explaining the grain size dependence of $J_R : J_S$ in those two very different magnetic materials.

The initial model is slightly modified here and extended to a more general role in explaining grain size dependence of PSD behaviour. Zero multidomain moment was assumed (Figure 1a), so that the predicted value of $J_R : J_S$ for the ensemble continues to fall below the multidomain value at large grain sizes. This can be rectified by adding a term to represent the increasing effect of the particles which have nucleated a wall, as the grain size of the ensemble increases. This probability is obtained from the condition that the sum of the probabilities that a particle has and has not nucleated a wall must be 1. Thus

$$P(w = 0) + P(w > 0) = 1$$
$$P(w > 0) = 1 - e^{-\lambda}.$$

The original expression for $J_R : J_S$ for the ensemble thus becomes modified to

$$\frac{J_R}{J_S} = e^{-\lambda} \frac{J_R}{J_S} (\text{SD}) + (1 - e^{-\lambda}) \frac{J_R}{J_S} (\text{MD}).$$

For large λ, this gives the multidomain value and for $\lambda = 0$, the single domain value obtains.

Figure 1b illustrates the fit of the modified model to the $J_R : J_S$ data of Day *et al.* (1977). The same value of λ derived from Kittel theory is used which assumes a single domain critical size of $1.4\,\mu m$. The model fits the data better, as one would expect from the addition of another term, but it still underestimates the value of $J_R : J_S$.

On reexamination of the data by Day *et al.* (1977) a value of $0.8\,\mu m$ has been chosen as the critical single domain size. Note that although $J_R : J_S$ is indeed 0.5 for $1.4\,\mu m$, there is a continued increase in H_C and decrease in susceptibility and $H_{RC} : H_C$ down to $0.8\,\mu m$. Hence the maximum single domain size should be $0.8\,\mu m$. When the modified model is plotted against the data, using the value of λ, which comes from the choice of

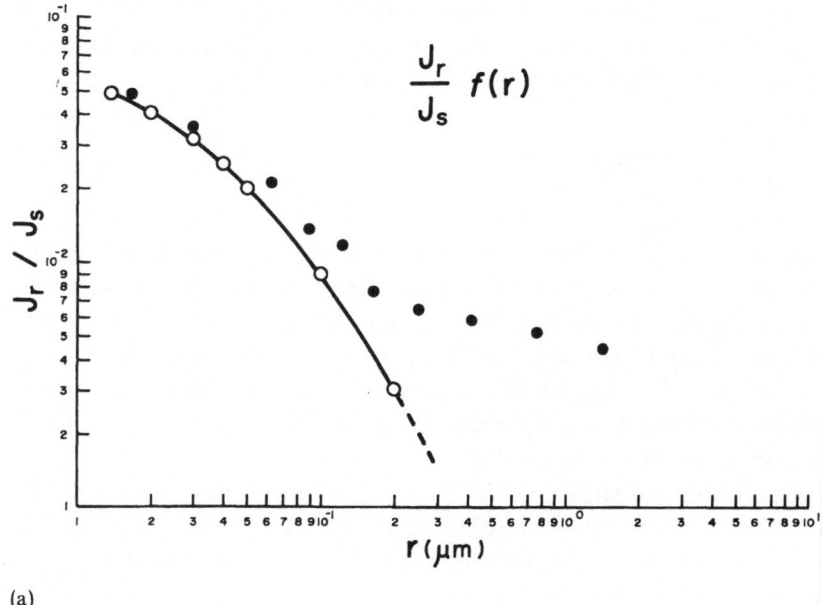

(a)

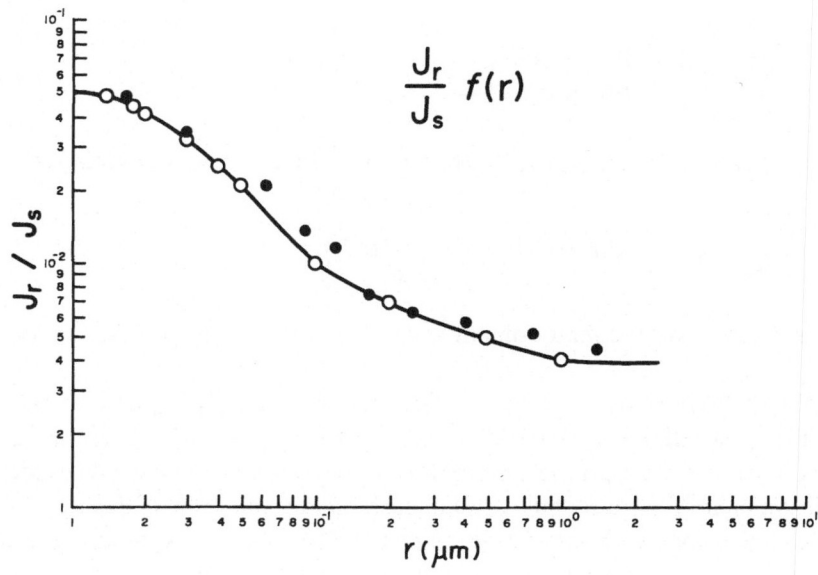

(b)

Fig. 1   (a). $J_R : J_S f(r)$ – S.D. model; (b). $J_R : J_S f(r)$ – S.D. + M.D. model.

0.8 μm for the critical size, the fit is not so good as before. The model now systematically underestimates $J_R : J_S$.

There are a number of difficulties in applying this simple model to the $J_R : J_S$ data. The first problem lies in the choice of the value of λ. As we noted above, λ is not available at present from direct observations. The Kittel value is not for J_R and, moreover, is sensitive to the choice of the single domain critical size. It is therefore important to observe λ. Unfortunately, it is hard to get good domain patterns on 0.6 X titanomagnetites, so it may be some time before the necessary counts can be done. A second obvious weakness of the model is that it assumes that once a wall is nucleated in a particle, the particle achieves the characteristic MD value of $J_R : J_S$. This implies that the efficiency of self demagnetization in multidomain particles is independent of grain size, which is unlikely. A large particle, with many walls, has a higher probability of achieving more efficient self demagnetization than does a small particle with a single wall. Hence, there is likely to be a grain size dependence of the MD contribution to the saturation remanent magnetization (J_R).

It appears that the nucleation of domain walls plays a critical role in explaining the grain size dependence of the ratio of $J_R : J_S$. However, a probable additional effect is the intrinsic grain size dependence of the multidomain value of $J_R : J_S$.

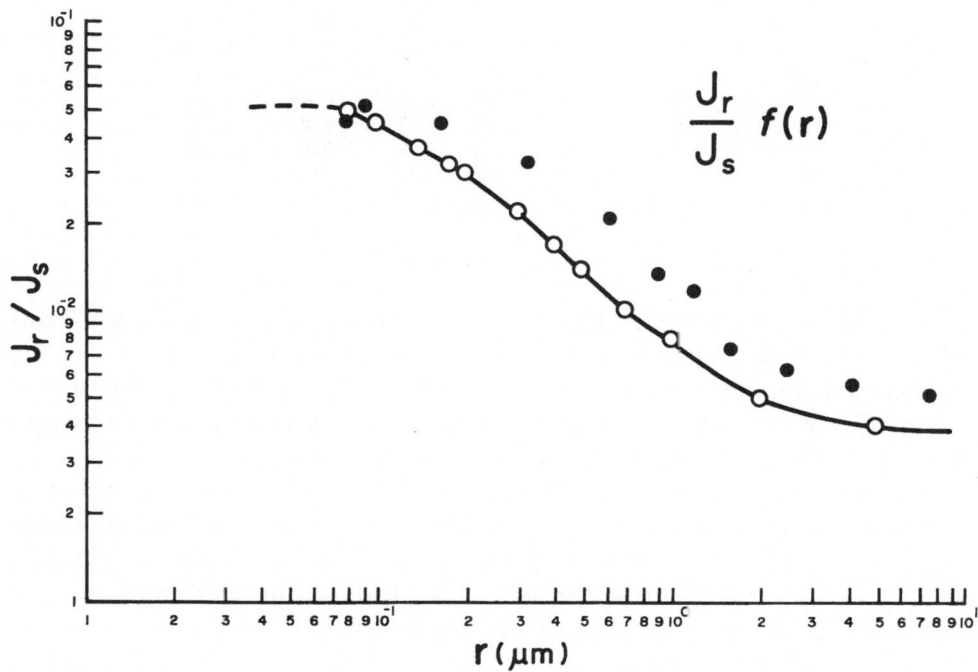

Fig. 2. $J_R : J_S f(r)$ – S.D. + M.D. model.

5. Grain Size Dependence of Weak Field Susceptibility

The initial or weak field susceptibility of PSD particles is grain size dependent (Table I). The weak field susceptibility of multidomain particles is about an order of magnitude larger than that of single domain particles. In an ensemble of PSD particles, it is therefore those particles, which have nucleated a wall, that make the dominant contribution to susceptibility. This reflects the ease with which domain walls can be moved by magnetic fields compared with the difficulty of achieving magnetization rotation.

TABLE I

Grain size dependence of magnetic properties of $\chi = 0.6$, titanomagnetite
(after Day *et al.*, 1977)

Size (μm)	Standard deviation	H_C (Oe)	H_{RC} (Oe)	H_{RC}/H_C	J_{RS}/J_S	χ_0 (G/Oe)
$\chi = 0.6$:						
0.8	–	1584	2130	1.34	0.473	0.018
0.94	–	1115	1815	1.63	0.511	0.030
1.7	1.0	581	1115	1.94	0.459	0.048
3.3	1.3	280	625	2.23	0.339	0.074
6.4	2.4	132	330	2.50	0.210	0.105
9.0	3.1	92	260	2.83	0.131	0.118
12.0	3.7	82	235	2.82	0.117	0.176
16.0	5.4	73	201	2.75	0.075	0.217
25.5	7.9	61	185	3.03	0.063	0.237
41	11.2	52	175	3.37	0.058	0.258
78	19	46	160	3.48	0.051	0.266
140	24	39	148	3.79	0.043	0.268

To apply the nucleation model, the probability of there being at least one wall in the particle is used as a weighting factor for the multidomain susceptibility value. The smaller single domain contribution is then weighted with the probability of nucleation failure. Figure 3 illustrates the results of applying this model to explain the susceptibility data. The model clearly overestimates susceptibility seriously.

Additional evidence of the overestimation of susceptibility by the nucleation approach comes from studies on pyrrhotite. The weak field susceptibility of pyrrhotite falls to about 30% of its multidomain value with a grain size decrease to 10 μm (Fuller, 1961). The average number of walls in pyrrhotite of 10 μm size has been observed to be 4 (Halgedahl and Fuller, 1983). Hence, the probability of failure to nucleate a wall at this grain size is 0.02. Thus the presence of such metastable single domain particles should not be important in an ensemble with an average grain size of 10 μm and there must be some other effect present.

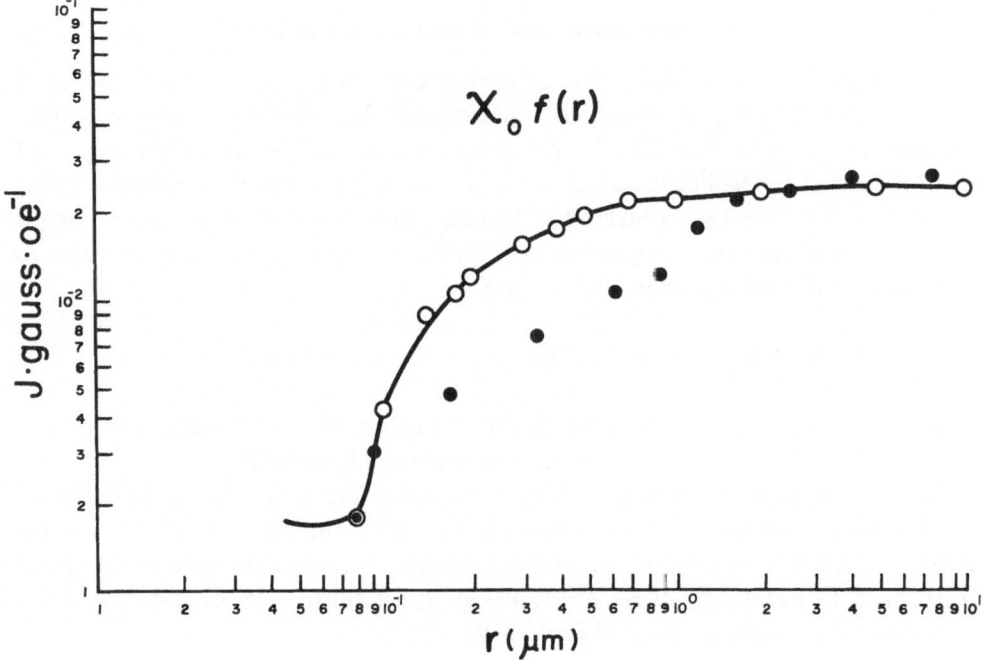

Fig. 3. $X_0 f(r)$ – S.D. + M.D. model.

The standard expression for the contribution of 180° walls to initial, or weak field, susceptibility is

$$X_0 = \frac{4 J_s^2 \cos^2 \varphi}{B} S$$

where S is the wall area per unit volume, φ the angle between the axis of domain magnetization and the applied field H and B a measure of the barriers to wall motion. B is the coefficient of the X^2 term, which gives the wall energy as a function of position. A positive first derivative represents a barrier to wall motion. If either B or S vary with grain size, then there will be a size dependence of weak field susceptibility in multidomain grains. It is frequently assumed that in large particle with numerous domain walls, there is a higher probability for sections of walls to be faced by small barriers to wall motion and so be able to respond to weak fields.

The decrease of susceptibility with decreasing grain size appears to be due to two effects. First, there is an increasing intrinsic grain size dependence of the susceptibility of multidomain grains. Second, the increasing number of particles which failed to nucleate a wall as the grain size of the ensemble decrease, brings about a further decrease in susceptibility, until the single domain value is reached at which the probability of stable walls is zero.

6. Thermoremanent Magnetization (TRM)

Although the interpretation of PSD behaviour expressed in the $J_R : J_S$ ratio and in weak field susceptibility are of interest to paleomagnetists, the more directly important phenomenon is TRM. That the ideas discussed above are relevant to TRM is suggested by the grain size dependence which all three exhibit. Moreover, metastable single domain grains have been observed in igneous rocks, interpreted to be carrying a primary thermoremanent magnetization (TRM). It should therefore be possible to express the TRM of an ensemble of particles as

$$J(TRM) = e^{-\lambda'} J(TRM\,SD) + (1 - e^{-\lambda'}) J(TRM\,MD)$$

The value λ' is the value appropriate after weak field cooling of the material and is not the same as that in the state of saturation remanence (J_R or IRM_S).

Once again we face the situation that the appropriate value of λ' is available for pyrrhotite, but not for the titanomagnetites for which the TRM data are available (Day, 1977). It is however known that the probability of nucleation failure is decreased by the weak field cooling process. The values for pyrrhotite for $\lambda\,IRM_S$ and $\lambda\,NRM$ in a basalt are given in Halgedahl and Fuller (1983) as

$$\lambda_{IRM_s} = 0.7r^{0.5} - 1$$

$$\lambda_{NRM} = 0.79r^{0.56} - 1.$$

As an interim test in the absence of the necessary value for λ', the values for pyrrhotite are used to correct the $\lambda\,IRM_S$ value for the titanomagnetites. Values of 1.5 and 0.3 J G are used for single and multidomain TRM respectively.

Figure 4 illustrates the comparison of the model predictions and the data for TRM in a one oersted field for the sized titanomagnetites (Day, 1977). Again the model underestimates the magnitude of the TRM. Clearly with so many assumptions introduced into the model, there may be numerous explanations. One possibility is that the depression of the probability for nucleation failure may be different in the titanomagnetites from that in pyrrhotite. A second possibility is underestimation of the contribution of fine multidomain particles.

7. Alternating Field Demagnetization Characteristics of TRM in the PSD Range

The greater stability of weak field TRM compared with saturation remanence (J_R) has been known since the classic work of Rimbert (1959). In the interpretation of this observation, we follow a suggestion made earlier by Levi and Merrill (1978), which can now be confirmed observationally.

As noted in the last section, the TRM ensembles of fine particles in the PSD range is interpreted to be dominated by those few particles, which do not exhibit a wall after the

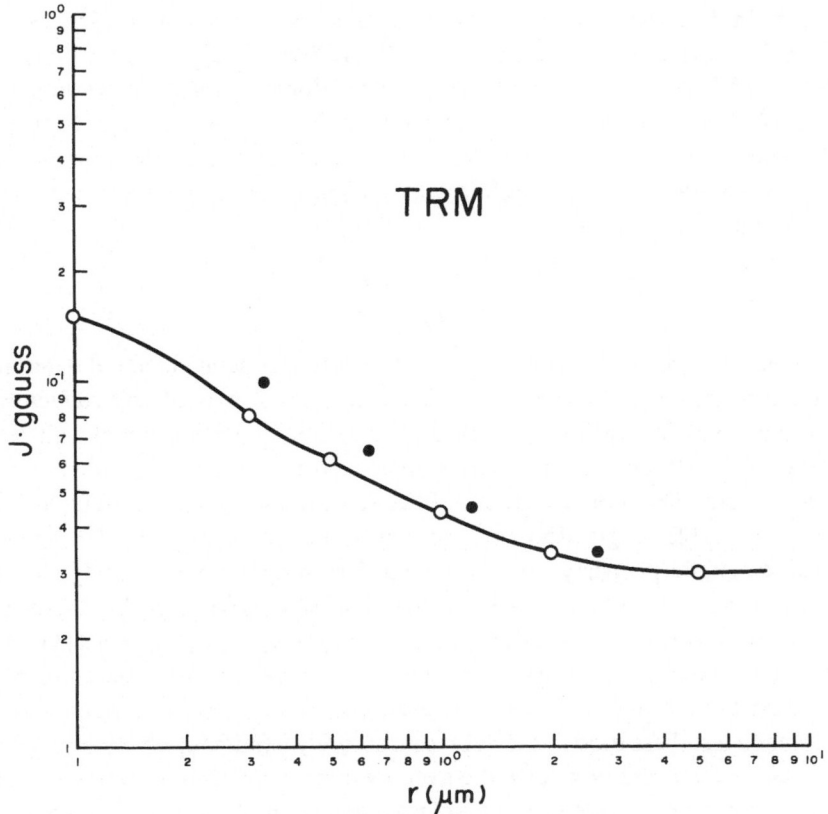

Fig. 4.   TRM$f(r)$ – S.D. + M.D. model

weak field cooling process. Now, these particles failed to exhibit a wall after the possibility of nucleation at high temperature. Since the critical pinning anisotropy surely varies with the saturation magnetization (J_S) to a higher power than does the demagnetizing energy, nucleation at high temperature should be much easier than at room temperature. Again, if a wall were nucleated and trapped at the nucleation site, it should have been able to escape more easily at high temperature than at room temperature. Thus, those particles which failed to exhibit a wall at room temperature after weak field cooling can only contain very unfavourable nucleation sites. In contrast, the saturation remanent magnetization (J_R) is carried by particles which fail to nucleate a wall at room temperature. These particles need not have such un-favourable nucleation sites. Thus AF demagnetization IRM_S will result in wall nucleation and wall motion in these particles at relatively low fields compared with the AF fields required to nucleate walls in the particles carrying TRM. The result will then be that the weak field TRM will be harder against AF demagnetization than is saturation remanent magnetization (J_R).

These AF demagnetization characteristics have long been used in paleomagnetism to distinguish single domain from multidomain material (Lowrie and Fuller, 1971). The characterisic of weak field TRM more stable than IRM_S should however be attributed to PSD ensembles, in which metastable single domain particles account for the hardness of the weak field TRM. The opposite characteristic for coarse grain multidomain material has been shown elsewhere to be due to the presence of soft sections of walls after weak field cooling, which are not present in saturation remanence (Halgedahl and Fuller, 1981).

8. Discussion

In this paper, a generalized approach to the grain size dependence of a number of magnetic properties has been suggested. The interpretation involves two factors. The first is domain wall nucleation. As grain size decreases the nucleation of wall becomes increasingly difficult, until at the single domain critical size the probability of a stable wall reaches zero. The domain states in PSD ensembles may therefore include single domain, metastable single domain and multidomain. To obtain the value of a particular magnetic property, one must therefore weight appropriately the contributions from the particles in the various states. The second factor is that there appears to be an intrinsic grain size dependence of the various magnetic properties of interest in multidomain particles. This is open to experimental investigation by measurements on individual particles, which are known from independent observations to contain walls.

Attempts to test these models are at present severely handicapped because we do not have sufficient determinations of λ the average number of domain walls for different materials as a function of size and states of magnetization. If this can be achieved, more rigorous tests of the models for the properties discussed should be possible. Other properties, such as coercive force, are amenable to a similar approach, but it requires observation of λ as a function of field, an experimentally daunting requirement.

The recognition of the importance of nucleation of domain walls may provide a resolution to the paradoxical success of Néel's single domain model in explaining the behaviour of the fine magnetic particles in rocks, which are too large to be single domain. Néel theory works well because, although the particles are indeed too large to be single domain, they are small enough that a nucleation barrier maintains many of them in a metastable single domain state. It now remains to test this suggestion rigorously.

Acknowledgements

The ideas developed in this paper have benefited from the efforts of a number of people with whom I have had the pleasure to work with in rock magnetism, most particularly W. Lowrie, V. Schmidt, R. Day, Y. Hamano, and most recently S. Halgedahl. It is a pleasure to acknowledge how much I have learned from them. This work was supported by grants from the National Science Foundation which are gratefully acknowledged.

References

Becker, J. J. 1969, 'Observations of Magnetization Reversal in Cobalt-Rare-Earth Particles', *IEEE Trans. Magn. MAG-5*, 211–214.

Becker, J. J.: 1971, 'Interpretation of Hysteresis Loops of Cobalt-Rare-Earth', *IEEE Trans. Magn. MAG-7*, 644–647.

Becker, J. J.: 1973, 'A Model for the Field Dependence of Magnetization Discontinuities in High Anisotropy Materials', *IEEE Trans. Magn. MAG-9*, 161–167.

Becker, J. J.: 1976, 'Reversal Mechanisms in Copper Modified Cobalt Rare Earths', *IEEE Trans. Magn. MAG-11*, 965–967.

Becker, R. and Kersten, M.: 1930, 'Magnetization of Nickel Wire Under Large Stress', *Z. Phys.* **64**, 660–681.

Bogdanov, A. A. and Vlasov, A. Ya.: 1966, 'The Domain Structure on Magnetite Particles', *Izv. Earth Physics* **9**, 577–581.

Bozorth, R. M. and Williams, H. J.: 1945, 'Effect of Small Stress on Magnetic Properties', *Rev. Mod. Phys.* **17**, 72–80.

Brown, W. F.: 1959, 'Relaxational Behaviour of Fine Magnetic Particles', *J. Appl. Phys.* **30**, 1305.

Brown, W. F.: 1963, *Micromagnetics*, John Wiley, New York.

Butler, R. F. and Banerjee, S. K.: 1975, 'Theoretical Single Domain Grain Size in Magnetite and Titanomagnetite', *J. Geophys. Res.* **80**, 4049–4058.

Day, R.: 1977, 'TRM and Its Variation with Grain Size', *J. Geomag. Geoelec.* **29**, 233–265.

Day, R., Fuller, M., and Schmidt, V.: 1977, 'Hysteresis Properties of Titanomagnetites: Grain Size and Compositional Dependence', *Phys. Earth Planetary Int.* **13**, 260–267.

De Blois, R. W. and Bean, C. P.: 1959, 'Nucleation of Ferromagnetic Domains in Iron Whiskers', *J. Appl. Phys.* **30**, 225–226S.

Dunlop, D. W.: 1977, 'The Hunting of the 'Psark'', *J. Geomag. Geoelec.* **29**, 293–318.

Fuller, M.: 1961, Ph. D. Thesis, University of Cambridge.

Graham, J. W., Buddington, A. F., and Balsey, J. R.: 1957, 'Stress Induced Magnetization of some Rocks with Analyzed Magnetic Minerals', *J. Geophys. Res.* **62**, 465–482.

Halgedahl, S. and Fuller, M.: 1980, 'Magnetic Domain Observations of Nucleation Processes in Fine Particles of Intermediate Titanomagnetites', *Nature* **288**, 6, 70–72.

Halgedahl, S. and Fuller, M.: 1981, 'The Dependence of Magnetic Domain Structure upon Magnetization State in Polycrystalline Pyrrhotite', *Phys. Earth Planetary Int.* **26**, 93–97.

Halgedahl, S. and Fuller, M.: 1983, 'The Dependence of Magnetic Domain Structure upon Magnetization State with Emphasis upon Nucleation as a Mechanism for Pseudosingle Domain Behaviour', *J. Geophys. Res.* **88**, B8, 6505–6522.

Jeffreys, W.: 1959, *The Earth*, Cambridge Univ. Press.

Kittel, C.: 1949, 'Physical Theory of Ferromagnetic Domains', *Rev. Mod. Phys.* **21**, 541–583.

Kneller, E. F. and Luborsky, F. E.: 1963, 'Particle Size Dependence of Coercivity and Remanence of Single Domain Particles', *J. Appl. Phys.* **34**, 656.

Levi, S. and Merrill, R.: 1978, 'Properties of Single Domain, Pseudo-Single Domain and Multidomain Magnetite', *J. Geophys. Res.* **83**, B1, 309–323.

Lowrie, W. and Fuller, M.: 1971, 'On the Alternating Field Demagnetization Characteristics of Multidomain Magnetization in Magnetite', *J. Geophys. Res.* **76**, 6339–6349.

Nagata, T.: 1970, 'Basic Properties of Rocks under the Effect of Stress', *Tectonophysics* **9**, 167–195.

Néel, L.: 1949, 'Théorie du traînage magnétique en grains fins avec application aux terres cuites, *Ann. Géophys.* **5**, 99.

Néel, L.: 1955, 'Some Theoretical Aspects of Rock Magnetism', *Adv. Phys.* **4**, 191–243.

Rimbert, F.: 1959, 'Contribution à l'étude de l'action de champs alternatif sur les aimantation rémanentes des roches', *Appl. Geophys., Rev. Int. Fr. Pétrol.* **14**, 17–54 and 123–134

Smith, P.: 1980, 'The Application of Lorentz Electron Microscopy to the Study of Rock Magnetism', *Inst. Phys. Conf. Ser.* **52**.

Soffel, H.: 1977, 'Pseudo-Single-Domain Effects and Single-Domain-Multidomain Transition in Natural Pyrrhotite Deduced from Domain Structure Observations', *J. Geophys Res.* **42**, 351–359.

Stacey, F. D.: 1963, 'The Physical Theory of Rock Magnetism', *Advances Phys.* **12**, 46–133.

Stacey, F. D. and Banerjee, S. K.: *The Pysical Principles of Rock Magnetism*, Elsevier, New York, 1974.

Uyeda, S., Fuller, M., Belshé, J., and Girdler, R.: 1963, 'Anisotropy of Magnetic Susceptibility of Rocks and Minerals', *J. Geophys. Res.* **68**, 1, 279–291.

S. K. RUNCORN'S COMMENTARY

In 1965 I was appointed Chairman of the Moon and Planets Subcommittee of a newly constituted Science Research Council Space Board under Sir Harrie Massey. On returning from the first meeting – a very dull one of routine business generated by the S.R.C. bureaucracy – I reflected how stupid I had been to accept this task.

The following day I received a circular from NASA in a pile of the many papers which a head of a department in a British University receives from the University administration. Fortunately I read it and found that it invited foreign nationals to apply to study the rocks to be returned from the Moon in the many landings being then planned as the Apollo project. I at once rang the secretary of the committee and asked that all physics, chemistry, astronomy and geology departments in British Universities be circulated with this information and asked to submit their proposals to us. Typically the plan had to be at NASA in about 2 weeks' time and there followed a hectic series of meetings with would-be investigators. The committee was small and Dr John Kerridge (then at Birkbeck College), Dr Gilbert Fielder and Professor Jim Ring and I worked hard at the matter, and when the plan was ready, 17 British laboratories had applied and with the applications went a flow diagram, in which the minute sample envisaged was to go through a sequence of tests from the non-destructive onwards until finally the sample was to be returned to NASA – another stipulation – in the form of a solution! Naturally my colleagues David Collinson, Alan Stephenson, and I wanted to do rock magnetism but I do not think we dared to suggest that we might measure the remanent magnetization – we would certainly have invited rejection, for it was already known that the Moon had no magnetic field – so we proposed to use rock magnetism techniques to study the opaque minerals. Magnetism was to be a handmaiden to the petrologists – geologists were quickly coming to the fore in the planning.

This initiation of the committee and its enthusiastic work – Kerridge and Fielder drew up the famous flow diagram – bore fruit later when, after the Apollo landings, British investigators took an important role.

We in Newcastle got some lunar soil to study – material appropriate to our proposal, but we thought it worthwhile to look for remanent magnetism and of course I knew that Professor Bastin had received a beautifully cut piece of basalt – rather like a small bar magnet for study of the thermal conductivity, and he kindly lent it to us. Its remanent magnetization had the properties suggesting that it had been acquired in a lunar field and we suggested in our first paper that it possibly was evidence of the former existence of a lunar magnetic field – a good example of Mark Twain's observation that the great thing about science is that one gets such a great return from such a small investment of fact. Other groups, especially those of Mike Fuller and David Strangway, discovered remanent magnetization in the basalts and breccia and the subject took off! One of the great pleasures of this work has been to see how the pieces of the jigsaw have fitted together until now there is wide agreement that the Moon did possess a strong magnetic field from at least 4.2 G yr to 3.2 G yr, that this was generated by a core dynamo in a small iron core and driven by heat sources which remain obscure. Another pleasure of the subject has been to see how each worker has been responsible for different advances. David Strangway early showed empirical evidence for the magnetic stability and Mike Fuller made among other contributions, advances to our understanding of the physical processes of magnetization that are contributions to terrestrial studies as well. My colleagues David Collinson and Alan Stephenson have made advances in palaeointensity studies, notably the ARM method, which will have wider applications than to the lunar rocks.

Finally it has been very satisfying to find that just as palaeomagnetists showed the geologists and other geophysicists against their strong initial prejudices to the contrary, that the subject was relevant to such problems as continental drift v tectonics, so we are now seeing the relevance of palaeomagnetism to the early evolution of the Moon and the origin of the Earth/Moon system.

THE GEOMAGNETIC DYNAMO – ELEMENTARY ENERGETICS
AND THERMODYNAMICS

F. J. LOWES

School of Physics, The University, Newcastle upon Tyne. NE1 7RU, U.K.

Abstract. This paper is a non-mathematical review, summarising the work in this field.

Estimates are made of the power needed to maintain the electric currents which give the main geomagnetic field. The observed surface field needs at least 2×10^8 W, but unobservable fields may need much more; a toroidal field of peak value 10 or 50 nT would need $\sim 10^{10}$ or $2.5 \times 10^{1-}$ W.

Ways of obtaining this power from the Earth's rotation, particularly through precession, are considered and rejected.

Thermal power sources have the disadvantages that there is inherent thermodynamic inefficiency in driving the dynamo, and that a significant fraction of the heat input will be carried away by conduction rather than convection. Radioactivity will only be important if there is a substantial amount of potassium in the core. If this is not the case the core might be cooling; cooling at 20 K per 10^9 yr would release specific heat at a rate of $\sim 10^{12}$ W. If the cooling causes the inner core to grow by freezing from the liquid core, then an additional $\sim 10^{12}$ W would be released from the latent heat of freezing. These heat fluxes might support a dynamo having a small toroidal field.

If, as seems likely, the solid inner core is significantly denser than the liquid, such cooling would also release $\sim 0.6 \times 10^{12}$ W of gravitational energy, giving compositional convection which would drive the dynamo very efficiently and give a large toroidal field.

1. Introduction

We now believe that the main geomagnetic field is due to electric currents flowing in the Earth's conducting liquid core, these currents being maintained against ohmic dissipation by a self-exciting dynamo mechanism in which the necessary e.m.f.s. are induced by the motion of the conducting liquid through the magnetic field. The details of the mechanism (e.g. what sorts of motion are needed) are very complicated , and are studied mainly by applied mathematicians; unfortunately very few results of direct geophysical relevance have yet been produced.

However there is one aspect of the problem, more concerned with the global (geo)physics than detailed magnetohydrodynamics, which is, at least in part, becoming better understood. This is the energetics of the system, the discussion of how much power is required to drive the dynamo and what geophysically plausible sources could produce it.

The dynamo is driven by motions of the liquid core. In what follows I am not concerned with whether or not the motions produced by any particular mechanism are suitable to give self-excitation, but only with the power transferred. (There is now a general feeling that almost any sufficiently vigorous stirring will act as a dynamo.) The suggested direct or indirect causes of the motions are departures of the gravitational force density ρg from its steady equilibrium value; these departures are very small so we can think in terms of forces of the type $\rho \Delta g$ and $g \Delta \rho$.

The $\rho \Delta g$ forces are time varying and come from the varying positions of the Moon,

Geophysical Surveys 7 (1984) 91-105. 0046-5763/84/0071-0091$02.25.

and to a small extent the Sun – tides, tidal friction, precession – but these act only indirectly and are now generally thought to have no significant effect in producing quasi steady motions. If any power is transferred to such motions it comes from the energy of the Earth's rotation.

Our main consideration therefore is with the convective motions given by $g\Delta\rho$ forces, where $\Delta\rho$ is the local departure from the equilibrium value, the departure being due to either thermal expansion or to compositional differentiation. Most of us are much more familiar with the former than with the latter, so I will explain a little more. To anticipate, the suggestion is that the inner core is still being formed by freezing of the liquid outer core; because of the particular conditions this leaves behind in the liquid a lighter fraction, giving us our $\Delta\rho$.

Whatever the origin of the $\Delta\rho$ it will not result in gross motion unless the $\Delta\rho$ is of the right sign (we need lighter material at the bottom or heavier material at the top) and the power input exceeds an appropriate minimum. If $\Delta\rho$ is produced by heating there will be instability, leading to thermal convection, only if the magnitude of the resultant temperature gradient exceeds a critical value, the adiabatic gradient. Only if thermal diffusion, i.e. thermal conduction, down the adiabatic gradient is insufficient to remove the input heat will there be convection. Similarly when the $\Delta\rho$ is due to change of concentration of some constituent there is the analogous minimum concentration gradient below which there will not be compositional (mechanical) convection. Again, there will be (mechanical) convection only if (mechanical) diffusion (mechanical 'conduction') down this critical gradient is insufficient to cope with the rate of separation. Just as a thermally unstable system will convect even if it is stable against mechanical convection (the possibility of the latter is not usually considered) so also will a mechanically unstable system convect even if it is stable against thermal convection; in each case the other mechanism can help (or hinder) instability.

Whatever the origin of the $\Delta\rho$ the immediate force acting on the fluid is gravitational, but if the $\Delta\rho$ is due to heating then it is this heating which is the ultimate power source of power for the core motions, while if the $\Delta\rho$ is due to differentiation it is the resultant loss of gravitational potential energy which is the basic power source.

In this paper I will try to present, in simple non-mathematical terms, the main results of investigations of dynamo energetics, making clear what assumptions about physics, and the properties of materials, are still involved. I will be concerned with principles rather than detail, and so will refer only briefly to the actual problem of estimating the numerical values of the relevant properties.

I start in Section 2 by summarising 'reasonable' values of the important properties of the core. Then in Section 3 I discuss how we can estimate how much power is actually needed to drive the dynamo; it turns out that we can in fact produce a reasonable estimate of the *minimum* power required, i.e. that needed in the most favourable situation.

In Section 4 I discuss, and dismiss, the possibility of diverting some of the rotational energy of the Earth to the dynamo via the tides, tidal friction, and precession.

Possible thermal power sources (radioactivity, specific heat, latent heat of freezing) and the complications of the innate inefficiency of heat engines and of the thermal conduction 'bypass' are discussed in Section 5, and the gravitational power source, the favourite, in Section 6. I conclude with a brief summary in Section 7.

2. Properties of the Earth's Core

It is *very* difficult to make measurements at the pressure and temperature of the core, so in many cases what is involved is a large extrapolation, involving the imperfectly understood behaviour of a material whose composition is uncertain! Table III of Loper and Roberts (1983) gives a good summary of the ranges of values suggested. Unfortunately several of the papers I will quote have numerical mistakes and/or do not use a consistent set of values. I will use a set of 'middle of the road' values, but because I am relying on the numerical work of other authors I do not guarantee complete consistency.

It is generally agreed that the core material is essentially iron (or iron/nickel with the nickel having negligible effect on its properties) 'alloyed' with (10–20)% of some lighter element such as silicon or sulphur. It is therefore a comparatively good electrical conductor, and various estimates of its conductivity seem to be converging, conveniently, on the conventional value of $3 \times 10^5 \, \Omega^{-1} \, \mathrm{m}^{-1}$ first suggested by Bullard (1949) as an (incorrect!) average of his and Elsasser's 'guess timates'. The iron is well above its Curie temperature, and it is safe to take $\mu = \mu_0$ in all the regions of interest.

The absolute temperature of the core is not well known, and even more uncertain is its adiabatic gradient, the radial temperature gradient below which there will be no thermal convention; for illustration I will assume the temperature profile used by Gubbins and Masters (1979) in which the core temperature varies from 3300 K at the outside ($r = 3500$ km) to 4100 K at the inner core boundary ($r = 1200$ km), giving a radial temperature difference $\Delta T = 800$ K.

For metallic conductors in laboratory situations, where thermal conduction is essentially also purely electronic, the thermal conductivity is related to the electrical conductivity by the Wiedemann-Franz law; assuming that this holds also in the core gives a thermal conductivity of about $35 \, \mathrm{W} \, \mathrm{m}^{-1} \, \mathrm{K}^{-1}$.

There is surprisingly good agreement about the specific heat capacity of the liquid, $700 \, \mathrm{J} \, \mathrm{kg}^{-1} \, \mathrm{K}^{-1}$, and the specific latent heat capacity of freezing of the liquid, $1 \times 10^6 \, \mathrm{J} \, \mathrm{kg}^{-1}$.

Although estimates of the viscosity of the core cover a wide range, they are all so low that viscous effects can be ignored in the body of the core.

In papers on dynamo theory the solid inner core is usually ignored unless its presence helps the theory; in the content of energy sources its presence might well be useful! It is generally assumed to have frozen out of the liquid core when (or as) the temperature dropped. The present estimate (Masters, 1979) of the density jump $\delta\rho$ from liquid to solid is $0.7 \times 10^3 \, \mathrm{kg} \, \mathrm{m}^{-3}$ with an uncertainty of $\pm(0.3 - 0.4) \times 10^3 \, \mathrm{kg} \, \mathrm{m}^{-3}$. This large value of the jump can only be reasonably explained if there is a difference in composition, the solid being significantly less dilute than the liquid, presumably because the liquid is more metal-rich than the eutectic composition. In fact this $\delta\rho$ is consistent with the inner core being pure ion. (This $\delta\rho$ is not quite the same as the resultant $\Delta\rho$ produced in the remaining fluid, but the difference is usually ignored.)

The compositional convection analogy to the adiabatic gradient is discussed by Loper and Roberts (1983).

3. Power needed to Maintain the Dynamo

To drive a dynamo requires power from outside, but the problem of calculating the power required for the terrestrial dynamo from the forces involved is formidable. Fortunately, in the steady state all the power input to the dynamo through the motions is used to produce the electromotive forces which maintain the electric currents in the core, so we can equate the input mechanical power to the resultant Joule heating produced by these currents.

If we have a current I flowing through a resistance R then the power dissipated as Joule heating is I^2R; if we work in term of conductance $S = 1/R$ then the power is I^2/S. The Earth's core is a more complicated 'resistance', but if at any point the current density is $\mathbf{j}(r, \theta, \phi)\,\mathrm{A\,m^{-2}}$, and the electrical conductivity is $\sigma(r, \theta, \phi)$, then the total Joule heating is

$$P = \int (j^2/\sigma)\mathrm{d}v, \tag{1}$$

where the integral is over the core. (Some currents leak into the mantle, but their contribution is probably insignificant.) It is unlikely that σ varies much throughout the core, so it is usual to assume that it is a constant and to take it outside the integral, giving

$$P = (1/\sigma)\int j^2(r, \theta, \phi)\,\mathrm{d}v. \tag{2}$$

It is this current $\mathbf{j}$ which gives us the observed (and some unobserved?) magnetic field $\mathbf{B}$; how much can we deduce about $\mathbf{j}$ from our knowledge of $\mathbf{B}$? The dipole type of $\mathbf{B}$ which we observe outside the core is called 'poloidal', and in general has radial components. It is given by electric currents in the core which have no radial components, and which are called 'toroidal'. Conversely, if there were poloidal currents in the core they will give toroidal magnetic fields which are confined to the core. It is convenient to express the $\mathbf{j}(r, \theta, \lambda)$ of each type, toroidal or poloidal, as a series of terms each of which varies with latitude and longitude as a particular surface harmonic such as $P_n^m(\cos\theta)\cos m\phi$, but with an arbitrary variation with radius; each current term is associately uniquely with the appropriate, poloidal or toroidal, magnetic field term. Because of the orthogonality of the harmonics, each current/field term contributes independently to the integral (2) over the spherical core, but without knowing the radial variation we cannot actually perform the integration. (Exactly the same problem arises in the equivalent calculation of input power by dividing the total stored magnetic energy $\int (B^2/2\mu_0)\,\mathrm{d}v$ for each term by the appropriate decay time.)

However, Parker (1972) showed that for currents which produce a given magnitude dipole field the integral (2) was a minimum for a particular radial variation (j proportional to radius in this case). Extending this method to the other, non-dipole, observed poloidal field (and including a small addition for higher harmonics given by extrapolating the spatial power spectrum of the core field (Lowes, 1974)) gives the results of the first two lines of Table I. (So the field which makes compass navigation possible could be maintained by a quite small modern power station dynamo!)

The third line of Table I indicates the problem that, partly because of geometrical attentuation and partly because of masking by the crustal field, we will never be certain that there are not significant small scale poloidal magnetic fields in the core.

More importantly, we have no observational evidence for or against the existence of toroidal fields inside the core; all attempts to provide an upper limit to the magnitude of a toroidal field by its possible effect on seismic waves in the core (e.g. Knopoff, 1955; Crossley and Smylie, 1975) have not been useful.

Quite large toroidal fields would be produced by the shearing of the dipole lines of force if there were significant differential rotation in the core. One theoretical limit that has been suggested is to assume that all the Coriolis force density $2\rho\boldsymbol{\omega} \times \mathbf{u}$ is balanced by the Lorenz force density $\mathbf{j} \times \mathbf{B}$ of a toroidal field, the so-called magnetostrophic balance. In the absence of any other information it is conveniental to take a 'typical' value of the fluid velocity u to be $4 \times 10^{-4}\,\mathrm{m\,s}^{-1}$, i.e. that corresponding to a westward drift of $0.2°\,\mathrm{yr}^{-1}$ at the core surface; this gives a toroidal field of about 50 mT (and what dynamo theorists call a 'strong field' or '$\alpha\omega$' dynamo). However Busse (1973) argues that much of the Coriolis force may be balanced by pressure force (as it would have to be in a non-conducting fluid), and that the toroidal field would be of the same magnitude as the poloidal field (a 'weak field' or 'α^2' dynamo); there is also evidence that the Lorenz force balance does not hold in the solar convection zone, and in actively convecting stars. There is considerable controversy as to the likely magnitude of the toroidal field; for illustrative purposes only a 'typical' value of 10 mT (100 G) is used to give the fourth line of Table 1.

It is clear that the actual power requirement of the dynamo will be dominated by the

TABLE I

Power needed to maintain various magnetic fields in the Earth

	P	B_{rms}	B_{peak}
Dipole	$\geqslant 1.0 \times 10^8\,\mathrm{W}$	0.27 mT	1 mT
Non-dipole	$\geqslant 2.2 \times 10^8$	0.27	
Small-scale poloidal	?	?	?
Toroidal	$\sim 10^{10}$	0	say 10

P = the power dissipated (for the optimum current distribution).
B_{rms} = the r.m.s. field at the core surface. (1 mT = 10 G).
B_{peak} = the corresponding maximum field inside the core.
(It is assumed that $\sigma = 3 \times 10^5\,\Omega^{-1}\mathrm{m}^{-1}$).

(invisible and unknown) toroidal field if that has significant magnitude.

On the assumption that the Earth is not heating up, any power dissipated in the core by the dynamo must presumably flow out through the Earth's surface. (This is an over-simplification, but is adequate for now.) This surface heat flow is about 4×10^{13} W (Sclater *et al.*, 1980), but at least 3×10^{13} W of this is thought to come from radioactivity in the crust and mantle, so *at most* 1×10^{13} W is coming from the core.

The gap between 10^8 W (minimum for dipole field) and 10^{13} W (maximum allowable heat flux from core), seems comfortably large even by geophysical standards. However, the lower limit is almost certainly too low (e.g. a 50 mT toroidal field would dissipate $> 2.5 \times 10^{11}$ W), and as we shall see in Section 5 inevitable inefficiencies in effect reduce the upper limit of power available to the dynamo, so there is indeed a problem.

4. Rotational Power Sources

These are sources for which any power input would come from the Earth's rotational kinetic energy. They act via $\rho \Delta g$ type forces, where the Δg is due to the Moon (and to a much smaller extent the Sun).

The gravity gradient given by the Moon produces not only the visible tides in the oceans, but also body tides of the core and mantle; the core/mantle interface will have a radial oscillation of amplitude 5 to 10 cm. Bullard (1949) in effect rejected this mechanism with a very brief qualitative discussion. Elsasser (1950) discussed it in a little more detail, and argued that the known non-linearities were insufficient to convert such semi-diurnal oscillations (as seen by an observer rotating with the core) to significant quasi-steady (as seen by the observer) motions of the core fluid.

Because of the lag of the ocean tides there is a torque which slows the Earth's rotation, and the resulting rate of loss of kinetic energy by the core is large, 10^{11} W. In his first papers Elsasser (1947) argued that as most of the dissipation is in the oceans this would leave the core rotating too fast, and that the resulting torque could somehow transfer energy to large scale motions in the core. However, Bullard disproved this also in his 1949 paper, showing that, because of conservation of momentum, such a differential rotation would transfer only a trivial amount of energy to other motions.

The spin axis of the oblate Earth is tilted at $23\frac{1}{2}°$ to the perpendicular to the ecliptic, the plane of the Earth's orbit about the Sun. The Moon is (almost) in this plane, and its gravitational torque on the equatorial bulge causes the spin axis to precess in a cone of semi-angle $23\frac{1}{2}°$ with a period of 26 000 yr. The suggestion that this might power the dynamo was first made by Bullard (1949); the idea is that, because of their different ratio of axial to equatorial moments of inertia, the core and mantle would have different precession rates, so again there is the tendency to produce relative motion. Bullard wanted more information about the applicability of a theorem by Poincaré, so did not commit himself. But Elsasser (1950) pointed out, in effect, that what was involved was diurnally varying shear stresses at the core/mantle interface, and dismissed precession as a source of dynamo power using the same argument as for tides – lack of non-linearity to convert short period to quasi-steady forces.

Although since these early rejections there has been no serious further discussion of tides and tidal friction as power sources, this has not been the case for precession!

Malkus (1963, 1968), in two unrelated and difficult papers, produced superficially convincing arguments that precession *could* power the dynamo. For a clear discussion of the effects involved, and a summary of the literature, see Rochester *et al.* (1975) and Loper (1975), where many mistakes are pointed out. It seems clear that the particular mechanisms which appeared to be suggested by Malkus would not be effective, and it is likely that any transfer of power would be very inefficient. Gubbins and Masters (1979) have pointed out that Rochester *et al.* and Loper treated rigorously only a *laminar* magneto-hydrodynamic boundary layer, and it has not yet been *proved* that a turbulent layer would not be more efficient, but at present the general feeling is that precession is not a major power source. The fact that other planetary fields exist in the absence of precession certainly makes it an unattractive source.

Given the complexity of the theory, can any limit be derived from astronomical observations? Precessional power dissipated in the core would lead to a reduction of the $23\frac{1}{2}°$ obliquity. At present this is reducing at $47'' \, \mathrm{cy}^{-1}$ owing mainly to perturbation of the Earth's orbital motion by the other planets. Over the last twenty years the discrepancy between the observed figure and that predicted by orbital theory has progressively reduced from $0''.3 \, \mathrm{cy}^{-1}$ to the $0''.01 \pm 0''.05 \, \mathrm{cy}^{-1}$ of Duncombe and van Flandern (1976). The discrepancy has the right sign, but even the figure 0.01 corresponds to $10^{12} \, \mathrm{W}$ dissipation, so unfortunately there is no useful limitation.

5. Thermal Power Sources

The most obvious source, and one assumed for a long time by most dynamo workers, is radioactivity. However, most geochemists now think that almost all of the uranium and thorium will have migrated to the mantle and crust as the core differentiated, leaving only trace amounts in the core.

One current suggestion is that significant amounts of potassium-40 may have been carried into the core with the sulphur, if in fact that is the diluent.

There appears to be a depletion of potassium in the mantle and crust, and putting it in the core would be one solution, but there is considerable debate among geochemists, with estimates of radioactive heating varying from 0 to $10^{13} \, \mathrm{W}$!

Present calculations suggest (Verhoogen, 1980) that if there is indeed a global depletion of potasium then the Earth as a whole will be cooling. Various workers (e.g. McKenzie and Weiss, 1975) do in fact argue that there is evidence that the mantle has cooled through 200 K in 3×10^9 yr. On this time scale, if the mantle has cooled then so has the core. In this context, the suggestion that the Earth is cooling slowly, releasing both specific heat from the whole core and latent heat of solidification at the (slowly growing) inner core boundary, was first made by Verhoogen (1961). Using not unreasonable, but by no means definite, values for specific and latent heat capacities, and for the melting point/pressure curve, Loper and Roberts (1983) suggest that a cooling of the whole core by 80 K over 4×10^9 yr would build up the inner core to its

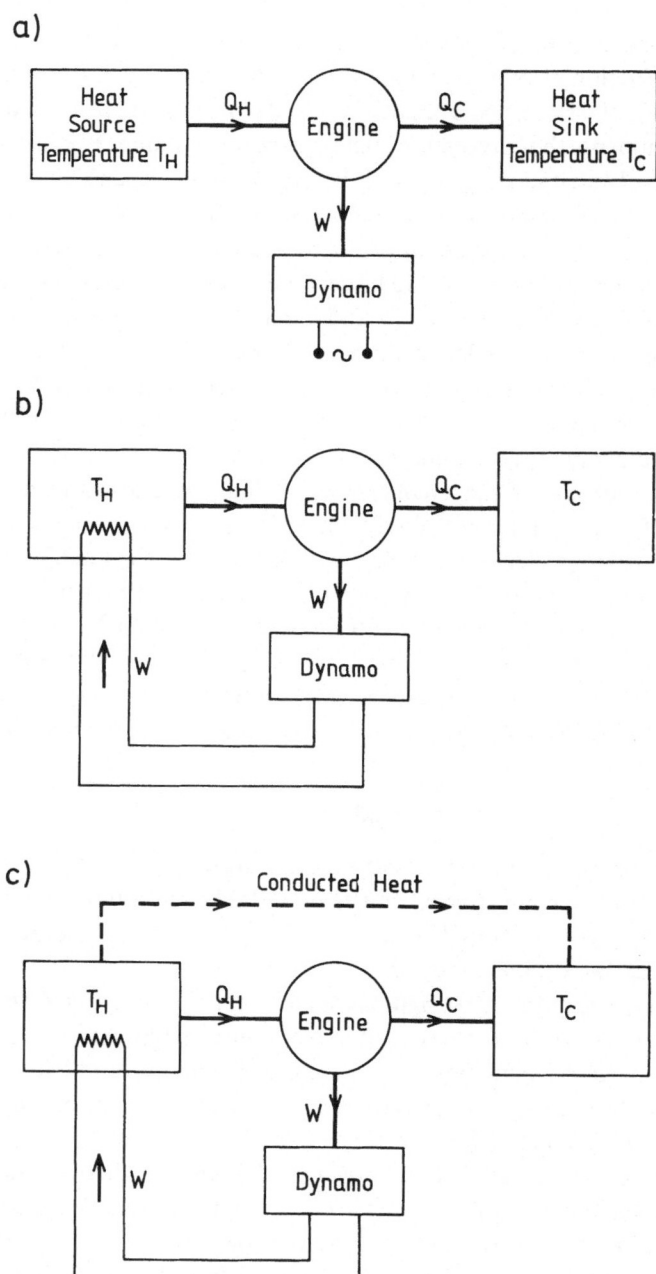

Fig. 1. Successive approximations to the thermally driven dynamo. (a) Schematic diagram of a surface power station. (b) The electrical output of the dynamo produces Joule heating in the core. (c) The dynamo is partially bypassed by conduction.

The heat engine (turbine) is fed with power from the heat source (boiler); most of it is rejected to the heat sink (condenser) but some is used to drive the dynamo; the electrical output of the dynamo is used elsewhere.

present size of 1200 km radius, and produce an average 0.75×10^{12} W from the cooling and 1.0×10^{12} W from the freezing. We do not know definitely if the Earth is in fact cooling, but a heat source of the order of 10^{12} W seems not implausible.

If a fluid is gradually heated from below then at first the temperature gradient will simply increase until the heat is conducted away. However, when the conduction gradient becomes steeper than the adiabatic gradient the fluid becomes unstable, and thermal convection starts; this convection 'stirs' the fluid, maintaining the gradient at just above the adiabatic value. These convective motions are what drive our dynamo, so we have a heat engine.

In physics texts heat engines are schematicised as in Figure 1a, and if heat Q_H is supplied at temperature T_H (H for Hot), and Q_C leaves the engine at T_C (C for Cold), we know that the mechanical work $W = (Q_H - Q_C)$ produced by the engine is limited and that

$$\frac{W}{Q_H} \leqslant \frac{T_H - T_C}{T_H} = \frac{\Delta T}{T_H}. \tag{3}$$

However this simple theory is for a system with *separate* heat source, engine, dynamo, and heat sink. The Earth's core is *not* so simple, and sorting out the complications has taken over 30 yr; as in many aspects of dynamo theory, almost every significant paper has had mistakes which have had to be corrected by a later author, who himself made mistakes...!

In his very first paper on the dynamo Elsasser (1947) used Equation (3), but he took for his ΔT the small *horizontal* difference between the hot (rising) and cold (falling) fluid. This was soon corrected by Bullard (1949) who pointed out that the correct ΔT is the, much larger, radial difference from bottom to top. With modern estimates of temperatures this leads to *maximum* efficiencies of 10–20%, so for a dynamo power P there would have to be a convected heat flux of at least (5–10) times P.

Then, quite independently, both Hewitt *et al.* (1975) and Backus (1975) pointed out another complication; the electrical power produced by the dynamo was dissipated as Joule heating *inside* the heat source ('boiler') of the engine, and re-emerged again as part of Q_H! In fact the conventional argument based on conversation of energy told us nothing, and it was the entropy equation which gave the correct answer. It turns out that, in terms of what is 'observable', the heat flux Q_C leaving the core, the maximum power W diverted through the dynamo is limited by

$$\frac{W}{Q_C} \leqslant \frac{T_H - T_C}{T_C} = \frac{\Delta T}{T_C}. \tag{4}$$

So for $T_H > 2T_C$ it was possible in this system to have 'efficiencies' greater than 100%! However, in the Earth's core T_C is not much smaller than T_H, so the 10–20% efficiency quoted above was not greatly altered. (For a conventional system Equation (3) is algebraically equivalent to (4), so (4) could have been obtained by a very simple-minded, if incorrect, argument!)

From the early days it was assumed that, as with real engines, the actual efficiency would be considerably less than the theoretical maximum. Metchmik *et al.* (1974) argued, to the contrary, that all the theoretically available power would in fact be transferred to those dissipative processes which constrained the fluid motions, in practice Joule heating, but Verhoogen (1980) has shown that this argument was invalid. However Hewitt *et al.* (1975) showed that, provided the Boussinesq approximation is valid, then the result of Metchnik et al is in fact true. (This is for the ideal case where all the heat is input at the highest temperature, e.g. from latent heat at the inner core/outer core boundary; if the heat input is distributed uniformly throughout the core, e.g. from radioactivity, they showed that the efficiency is reduced to 2/5 of $\Delta T/T_C$.) Unfortunately the Boussinesq approximation is valid only if the thickness of the convecting layer is much smaller than the temperature scale height, whereas for the core their ratio is about 1 : 3, so the result is still in doubt.

Another complication that tended to be ignored by early workers was that, quite apart from the heat transported by convection, there would also be the heat transported by conduction; power stations do not have the boiler inside the cooling tower! For my figures for thermal conductivity and average gradient this amounts to 3×10^{11} W, a major contribution. (Figure 1c is an oversimplification. As both the adiabatic gradient and the area increase outwards the conducted flux will increase with radius; there may in fact be quite a thick thermally stable layer at the outside where there is no convection.)

All the figures I have quoted so far have had to assume values for electrical and thermal conductivities, σ and k, and the radial temperature difference ΔT (or equivalently the adiabatic temperature gradient). The values used have been realistic, but are uncertain by at least a factor of two, and some authors would say by much more. Gubbins (1976) showed how some of this uncertainty could be removed. He pointed out that, for a given current distribution, Joule heating is proportional to $1/\sigma$, and is produced at an efficiency proportional to $1/\Delta T$, so that the convective heat flux needed to drive the dynamo is proportional to $1/\sigma\Delta T$; on the other hand as k is proportional to σ (Wiedemann–Franz law) the conducted heat flux is proportional to $\sigma\Delta T$. So, taking the product $\sigma\Delta T$ as a new variable, he produced Figure 2.

Clearly the sum of convected plus conducted heat flux is a minimum when the two are equal at point A; this gives an *absolute minimum* of 5×10^{10} W (for the dipole and non-dipole fields) *whatever* the separate values of σ and ΔT. (Gubbins forgot that both fluxes were present at the minimum.) This figure assumes the most efficient system, with all the heat input at the highest temperature, e.g. from latent heat released at the inner core surface. If the heat input is distributed throughout the core the system is less efficient, and the line BA is higher, while the line AC is also raised because for a given ΔT the gradient becomes larger at the outside; for uniform heat distribution the minimum flux is increased by a factor of about 2.5. Similarly, the presence of any toroidal field pushes BA up in proportion to the corresponding Joule heating. (The limit is when A has moved to C, and Gubbins used this to determine a limit for the magnitude of the toroidal field, but in fact the corresponding ΔT is 30 000 K, and the line BA no longer has any meaning!) In practice reasonable values of $\sigma\Delta T$ are well to the right of the

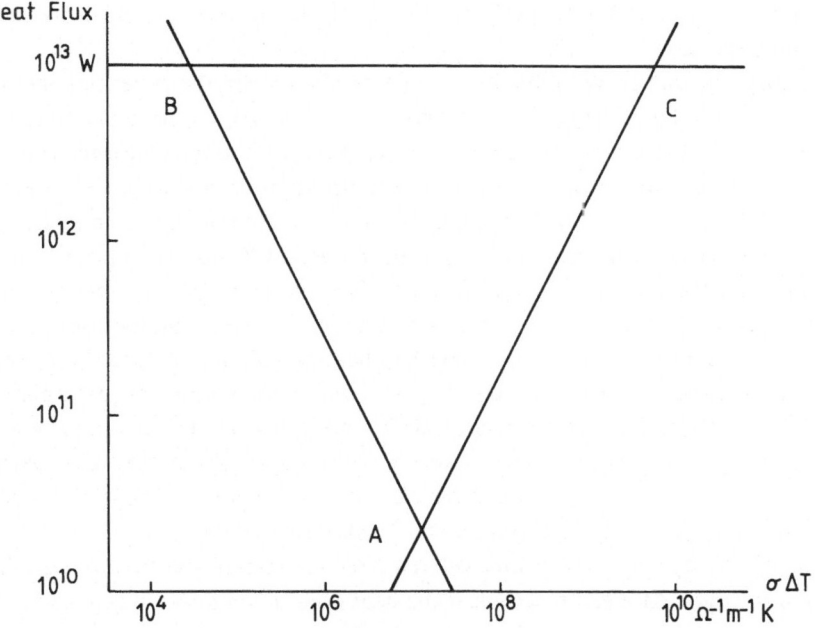

Fig. 2. Bounds on the heat flux from the Earth's core (after Gubbins, 1976). Line BA is the lower limit of the convected heat flux needed to produce the observed field. Line AC is the conducted heat flux. Line BC is an upper limit to the total heat flux from the core to the mantle.

minimum ($\sigma \Delta T = 2.4 \times 10^8 \, \Omega^{-1} \, m^{-1} \, K$ on my figures), again illustrating the important effect of conduction (in at least part of the core).

Remember that all my estimates of power requirements and of consequential convective heat flux are *under*-estimates; although it is not impossible to choose figures so as to give a total of less than 10^{13} W it is quite difficult, especially if the toroidal field is large. Although a thermal power source giving thermal convection is conceptually simple, the inevitable thermodynamic inefficiency, together with the inevitable additional thermal conduction, makes it unattractive.

6. Gravitational Power Sources

Finally I discuss the effects of $g\Delta\sigma$ type forces in which the $\Delta\rho$ comes from compositional differentiation rather than thermal expansion. Although Urey (1951) suggested iron settling out from the mantle on to the inner core, the only serious suggestion at present is production of a less dense fluid component at the surface of a growing inner core, due to the freezing out of a more dense solid. This possibility was suggested by Braginski (1963) in his first dynamo paper, just because the very large toroidal fields his dynamo needed could not be provided by thermal convection. However, the idea was largely ignored until the thermodynamics of the system was considered by Gubbins (1977) and Loper (1978), with more detail (and corrections) in succeeding papers.

Essentially the energy available is the difference between the potential energy of a mass $-\Delta m$ produced at a 'height' $-h$ in a gravitational field g, i.e. $\Delta m\,g\,h$, and that when Δm is uniformly distributed throughout the core. Because this gravitational energy stirs the liquid directly, rather than via a heat engine, there is no basic thermodynamic inefficiency; it has in fact been shown that (virtually) all of it is used to do work on the fluid (and hence drive the dynamo). Also, mechanical diffusion is so poor (compared with thermal diffusion) that only a trivial part of the compositional difference produced at the inner core bypasses the mechanical engine by diffusive 'conduction'.

Of course the thermal sources considered earlier are still present, and there are many other complications. For example the core as a whole may contract (or expand) as a result of the formation of the inner core, with consequential changes in gravitational and strain energy and the change in pressure will give adiabatic heating. One particular complication only recently sorted out is that some of the gravitational power may be used to drive a rather large heat pump, and I will now explain this.

Stirring a fluid, by whatever means, will tend to maintain the temperature gradient at the adiabatic value. At a given radius, if the heat conducted down this gradient is less than the total heat flux needed at that radius (i.e. all the heat, thermal plus dissipation of gravitational forces, generated inside) then the gradient will rise sufficiently (in practice only very slighty) above adiabatic so the additional flux is carried by thermal convection; the gravitational power available for the dynamo will be (inefficiently) augmented by this thermal convection. On the other hand, if the heat conduction down the adiabatic is greater than needed then the gradient will fall to very slightly less than adiabatic, and the core will be thermally stably stratified. However the core is still being stirred by the mechanical convection, but now some of the energy of the motions is used to 'convect' the excess heat downwards *against* the temperature gradient; the mechanical convection is acting as a heat pump and some of the gravitational power is diverted to this.

However, in a very clear paper which is a good guide to the literature, Loper and Roberts (1983) conclude that for any reasonable value of density jump the thermal contribution, positive or negative, will be small compared with the gravitational contribution to the dynamo power. They also conclude that the other known complications will have only minor effects.

The numerical results of the particular model assumed by Loper and Roberts (1983) are summarised in Table II; other authors use other models and obtain different numbers, but those of the Table are not untypical. (It must be remembered though that all these models are grossly simplified and have to use parameters which are poorly known.) Loper and Roberts estimate that 2.1×10^{12} W of their 2.4×10^{12} W total flux is transported by conduction at the outher boundary, with convection being more important at greater depth. (Once the gravitational energy has driven the dynamo and produced Joule heat it is irrelevant whether this heat is taken away by convection or conduction.)

TABLE II

The numerical results of Loper and Roberts (1983)

They assume a total cooling of $80\,K$ over 4×10^9yr, and a density jump at the inner core of $0.75 \times 10^3\,kg\,m^{-3}$: the powers are obtained by dividing total energy by total time.

Power input to core

Latent heat of freezing	$1.0 \times 10^{12}\,W$
Specific heat of outer core	0.75
Gravitational energy	0.65
Total	$2.4 \times 10^{12}\,W$

Power input to dynamo	$0.6 \times 10^{12}\,W$

As *all* of the gravitational power (in their case none is diverted to a heat pump) *has* to be dissipated as Joule heating (it can only leave the core as heat) the $0.6\,10^{12}\,W$ is not just available to the dynamo, it must be used by the dynamo. Using the figures of line 4 of Table I this means that there must be a toroidal field having B_{max} about $75\,mT$ ($750\,G$), larger than most previous suggestions.

7. Summary

To maintain the electrical currents which produce the observed geomagnetic field, it is estimated that at least $3 \times 10^8\,W$ of mechanical power needs to be diverted from the motions of the core fluid which drive the dynamo. However there is almost certainly, confined to the core and not directly observable, a toroidal magnetic field; estimates of its peak value range from about 1 to $50\,nT$ (10 to $500\,G$) which would need about 1×10^8 to $2.5 \times 10^{11}\,W$.

Present estimates of the heat flux from the core into the mantle are that it is not more than about $10^{13}\,W$.

Possible power sources for the dynamo have been divided into rotational, thermal, and gravitational.

Rotational sources involve mechanisms powered either by the kinetic energy of the daily rotation of the Earth about its spin axis (tides, tidal friction), or the kinetic energy or the $26\,000\,yr$ precession of this axis. The former have been discounted. The latter, though they have been strongly advocated by Malkus and his colleagues, suffer from the disadvantage that the actual forces involved are diurnal, and can transfer energy to quasi-steady motions of the core only very inefficiently through a dissipative boundary layer; it is now thought that any transfer will be negligible, though this has not been proved rigorously. Any transfer would affect the rate of change of obliquity of the spin axis, but at present astronomical observations are not good enough to provide a useful limit.

Thermal sources are attractive because of their conceptual simplicity. However, they have the inherent disadvantage that while it is the resulting thermal convection which drives the dynamo this is a heat engine, with its inherent thermodynamic inefficiency; also a significant fraction of the heat flux bypasses the engine because of thermal conduction through the core.

In the Earth the simplest source of heat is radioactivity. However it is now thought unlikely that there are significant amounts of uranium and thorium in the core. The current problem is whether or not an apparent depletion of potassium in the mantle and crust is due to it being in the core; what is 'missing' would provide $\sim 10^{13}$ W from the decay of potassium 40. A core heating of $10^{12} - 10^{13}$ W probably could power a dynamo. There is considerable controversy, and estimates of radioactive heating in the core range from 0 to 10^{13} W.

If radioactive heat production is much less than 10^{13} W the core is probably cooling slowly, at a rate determined mostly by processes in the mantle. Cooling the core will release specific heat throughout the core. If, as seems likely, the inner core is a solid phase in equilibrium with the liquid outer core, then core cooling will also release latent heat of freezing at the surface of the inner core as it grows. Using not unreasonable estimates of the properties involved it has been estimated that a cooling of only 80 K would enable the inner core to grow to its present size in 4×10^9 yr, producing $\sim 10^{12}$ W from specific heat and $\sim 10^{12}$ W from latent heat. Conduction will transport much of this to the outside of the core; the conducted flux will probably increase with radius to a maximum of $\sim 2 \times 10^{12}$ W at the outside; at any level the remaining flux will be convected, and could possibly drive a dynamo having only a small toroidal field, but not one with a large toroidal field.

Another consequence of freezing out of the inner core is that the rearrangement of mass considerably reduces the gravitational potential energy, so that energy is released. Whatever the composition of the solid and liquid phases there will be a small energy release because of the contraction of freezing. However present evidence is that the density jump from liquid to solid at the surface of the inner core is such that the solid must be less dilute, and hence more dense, than the fluid. Again the overall effect is to move mass towards the centre, releasing much more gravitational potential energy; a typical estimate would be a release rate of $\sim 0.6 \times 10^{12}$ W.

On this picture the growth of the inner core produces a more dilute, lighter, fluid. It appears that the instability due to this compositonal differentiation (lighter fluid underlying denser fluid) is much greater than that due to the thermal expansion given by the local heat release so that, if differentiation exists at all, it will be compositional instability, rather than thermal instability, which will control the convection. Also, in favourable situations *all* of this gravitational power *has* to be dissipated as heat, through the dynamo, before it can escape from the core. So for a given amount of power diverted through the dynamo there is considerably less heat flux leaving the core.

So, despite the formidable difficulties involved in a proper understanding of the very complicated thermodynamics, at present the 'gravitational' dynamo, driven mainly by compositional convection, is the favourite.

References

Backus, G. E.: 1975, 'Gross Thermodynamics of Heat Engines in Deep Interior of Earth', *Proc. Nat. Acad. Sci. USA* **72**, 1555–1558.

Braginskii, S. I.: 1963, 'Structure of the F Layer and Reasons for Convection in the Earth's Core', *Dokl. Akad. Nauk SSSR* **149**, 1311–1314.

Bullard, E. C.: 1949, 'The Magnetic Field within the Earth', *Proc. R. Soc. A*, **197**, 433–453.

Busse, F. H.: 1973, 'The non-linear dynamo problem', *Geophys. J. R. Astron. Soc.* **35**, 343–344.

Crossley, D. J. and Smylie, D. E.: 1975, 'Electromagnetic and Viscous Damping of Core Oscillations', *Geophys. J. R. Astron. Soc.* **42**, 1011–1033.

Duncombe, R. L. and van Flandern, T. C.: 1976, 'The Secular Variation of the Obliquity of the Ecliptic', *Astron. J.* **81**, 281–284.

Elsasser, W. M.: 1947, 'Induction Effects in Terrestrial Magnetism. Part III. Electric Modes', *Phys. Rev.* **72**, 821–833.

Elsasser, W. M.: 1950, 'Causes of Motions in the Earth's Core', *Trans. Am. Geophys. Union* **31**, 454–462.

Gubbins, D.: 1976, 'Observational Constrains on the Generation Process of the Earth's Magnetic Field', *Geophys. J. R. Astron. Soc.* **47**, 19–39.

Gubbins, D.: 1977, 'Energetics of the Earth's Core', *J. Geophys.* **43**, 453–464.

Gubbins, D. and Masters, T. G.: 1979, 'Driving Mechanisms for the Earth's Dynamo', in B. Saltzmann (ed.), *Advances in Geophysics*, Academic Press, New York, Vol. 21.

Hewitt, J. M., McKenzie, D. P., and Weiss, N. O.: 1975, 'Dissipative Heating in Convective Flows', *J. Fluid Mech.* **68**, 721–738.

Knopoff, L.: 1955, 'The Interaction between Elastic Wave Motions and a Magnetic field in Electrical Conductors', *J. Geophys. Res.* **60**, 441–456.

Loper, D. E.: 1975, 'Torque Balance and Energy Budget for the Precessionally Driven Dynamo', *Phys. Earth Planet. Int.* **11**, 43–60.

Loper, D. E.: 1978, 'The Gravitationally Powered Dynamo', *Geophys. J. R. Astron. Soc.* **54**, 389–404.

Loper, D. and Roberts, P. H.: 1983, 'Compositional Convection and the Gravitationally Powered Dynamo' in A. M. Soward (ed.), *Stellar and Planetary Magnetism*, Gordon and Breach, London, pp. 297–327.

Lowes, F. J.: 1974, 'Spatial Power Spectrum of the Main Geomagnetic Field, and Extrapolation to the Core', *Geophys. J. R. Astron. Soc.* **36**, 717–730.

McKenzie, D. P. and Weiss, N. O.: 1975, 'Speculations on the Thermal and Tectonic History of the Earth', *Geophys. J. R. Astron. Soc.* **42**, 131–174.

Malkus, W. V. R.: 1963, 'Precessional Torques as the Cause of Geomagnetism', *J. Geophys. Res.* **68**, 2871–2886.

Malkus, W. V. R.: 1968, 'Precession of the Earth as the Cause of Geomagnetism', *Science* **160**, 259–264.

Masters, G.: 1979, 'Observational Constraints on the Chemical and Thermal Structure of the Earth's Deep Interior', *Geophys. J. R. Astron. Soc.* **57**, 507–534.

Metchnik, V. I., Gladwin, M. T., and Stacey, F. D.: 1974, 'Core Convection as a Power Source for the Geomagnetic Dynamo: A Thermodynamic Argument', *J. Geomagn. and Geoelectr.* **26**, 405–415.

Parker, R. L.: 1972, 'Inverse Theory with Grossly Inadequate Data', *Geophys. J. R. Astron. Soc.* **29**, 123–138.

Rochester, M. G., Jacobs, J. A., Smylie, D. E., and Chong, K. F.: 1975, 'Can Precession Power the Geomagnetic Dynamo', *Geophys. J. R. Astron. Soc.* **43**, 661–678.

Sclater, J. G., Jaupart, C., and Galson, D.: 1980, 'The Heat Flow Through Oceanic and Continental Crust and the Heat Loss of the Earth', *Rev. Geophys. and Space Phys.* **18**, 269–311.

Urey, H. C.: 1951, 'The Planets, their Origin and Development', *Geochemica et Cosmochemica Acta* **1**, 209–277.

Verhoogen, J.: 1961, 'Heat Balance of the Earth's Core', *Geophys J. R. Astron. Soc.* **4**, 276–281.

Verhoogen, J.: 1980, *Energetics of the Earth*, National Academy Press, Washington.

THE CONTRIBUTION OF LABORATORY DYNAMO EXPERIMENTS TO OUR UNDERSTANDING OF THE MECHANISM OF GENERATION OF PLANETARY MAGNETIC FIELDS

I. WILKINSON

Dept of Geophysics, University of Newcastle upon Tyne

Abstract. The magnetic fields of the Earth, Jupiter, and Saturn are now accepted as originating in a dynamo mechanism in an electrically conducting fluid region of those planets.

Our extensive knowledge of the spatial and temporal variation of the geomagnetic field has been gained by observation in the recent past, and by inference from the remanent magnetisation of rocks for the distant past.

The theoretical problem of predicting what sort of magnetic field can be generated by motions in a homogeneous conducting fluid is extremely intractable. In order to obtain any solution at all the problem has to be idealised until it bears little resemblance to the situation existing in planetary interiors; consequently observation and theory have little common ground.

In the laboratory it is possible to construct homogeneous dynamos which, while they have a number of important differences from the mechanism which exists inside planets, nevertheless are considerably closer to reality than any theoretical model which can be shown to generate a magnetic field. Observations of the behaviour of such laboratory homogeneous self-exciting dynamos have, over the past twenty years in the Geophysics department at Newcastle University, together with theoretical predictions on one hand and palaeomagnetic observations on the other, helped towards the development of a consistent picture of both how the geomagnetic field is generated and of its morphology.

This review will attempt to show the part played by experimental homogeneous dynamos in the development of the subject.

1. Introduction

Historically, man's recognition that the Earth possessed a magnetic field, and of its nature, went hand in hand with the development of the magnetic compass for navigation. Recognition that it was primarily of internal origin followed Gauss's spherical harmonic analysis of 1838. Various mechanisms for the production of the field have been suggested since that time, including that it arose as a fundamental property of a rotating body, or that both thermo-electric and Hall currents could be responsible.

The first step towards the currently accepted position, that the field is generated by motions in the Earth's fluid interior, was the suggestion by Larmor in 1919 at a British Association meeting, that stellar magnetic fields might originate in such a way. The seismic evidence for a liquid iron core in the Earth provided, a little later, what some saw as a possible source region for the geomagnetic field generated by a similar mechanism.

2. The Palaeomagnetic Evidence

Having established the starting point for the present theory of origin of the geomagnetic field, let us now return to the historical development of the subject in some more detail.

Geophysical Surveys 7 (1984) 107-122. 0046-5763/84/0071-0107$02.40.
© 1984 *by D. Reidel Publishing Company.*

Palaeomagnetism – the study of the ancient magnetic field of the Earth through the remanent magnetism of rocks – had its origins in the realisation, again by mariners, that certain rocks had adverse effects on the directional properties of their compasses as a result of remanent magnetism. The subject developed over four centuries to the present position, where a rock sample can reveal both its own history and that of the geomagnetic field from its remanent magnetisation.

Briefly palaeomagnetism has shown that the oldest rocks known have recorded the presence of the geomagnetic field, and that presence has also been recorded by rocks of all ages. These are facts that a complete theory of the geomagnetic field must explain.

Of the two basic assumptions, that the geomagnetic field has been predominantly dipolar, and aligned with the axis of rotation of the Earth through most of geological time, the first is perhaps the more logical intuitively. Any dynamo mechanism in the core, producing a dipole component, will have a dipolar field at a sufficiently large distance (as it falls off least rapidly with distance of any multipolar field). The observational confirmation of this assumption relies on consistency of interpretation of global palaeomagnetic data. Less intuitively obvious is the axial character of the geomagnetic field. Coriolis forces might be expected to exert a marked influence on motions in the liquid core, but the extension of this to predictions about the geomagnetic field is a less sure process. Earlier this century palaeomagnetism began to evolve techniques which allow a quantitative description of the past field. The examination of recent lavas showed that for the past few million years the average field direction had been axial, leading to the suggestion, supported by palaeo-climatic evidence, that this was the case over all geological time.

The next refinement in our knowledge of the past geomagnetic field concerned its temporal variation. Early palaeomagnetic measurements showed that rocks of various ages were magnetised either in one direction, or an antiparallel one. Some initial confusion was caused by finding certain rocks which could, while cooling, spontaneously reverse their direction of magnetisation opposite to that of the applied field. But the observations on magnetised ocean sediments, from different oceans but covering the same time span, showed that the applied field had a consistent worldwide pattern of reversals. Further work led to a polarity timescale which has progressively extended our knowledge of the geomagnetic field back in time.

An extension of this work led to the recognition that for very short times indeed (compared to the normal polarity interval) the field direction could temporarily reverse, or even change direction without completely reversing and then return to its previous state. These are called 'events' and 'excursions' respectively.

Finding the magnitude, as opposed to the polarity, of the palaeomagnetic field is a much more complicated problem. Recent artifacts (pottery for example) proved more amenable to this sort of investigation initially, and then advances in techniques allowed field intensity determinations to be made on older rocks. This led to the conclusion that the geomagnetic field has always been of the same order of magnitude.

Returning to the time variation of the palaeomagnetic field, a range of values of the 'time spent undergoing reversal' have been estimated, giving a typical value of a few thousand years. This compares with an average time between recent reversals of 250 000 years. Concentrating on this short period during a reversal, several workers have tried, with mixed success, to define the spatial distribution of the field, mostly concluding that at least it is not simply dipolar.

That abbreviated history of the palaeomagnetic contribution has to be set in context both with the efforts of the theoreticians, to describe what sort of field can be generated in a homogeneous electrically conducting fluid, and the experimental work on homogeneous dynamos.

We will return to look at detailed palaeomagnetic evidence where relevant, but now it is appropriate to review the contribution of the theoreticians.

3. Theoretical Work on the 'Dynamo Problem'

Following Larmor's original suggestion, mentioned earlier, six decades of work have shown just how intractable is the theoretical analysis of homogeneous self-exciting dynamos.

It is a simple step from Maxwell's equations and Ohm's law to the conclusion that the time rate of change of magnetic field H, in fluid of conductivity σ, permeability μ and fluid velocity V is given by

$$\frac{\partial \mathbf{H}}{\partial t} = \frac{1}{\mu\sigma} V^2 \mathbf{H} + V \times (\mathbf{V} \times \mathbf{H}).$$

This is the basic dynamo equation, which cannot be solved analytically, and if we ignore the relevant Navier-Stokes equation and the problem of an energy source, it leaves what is called 'the kinematic dynamo problem'.

The rate of change of field depends on the relative sizes of the first (diffusion) term and the second (field generation) term. Ignoring the field generation term shows that a geomagnetic dipole field would have a decay time of 15 000 yr, illustrating the need for a continuous process of regeneration.

If the field is to have a positive growth rate, then it is also possible to deduce that $\mu\sigma VL$ (where L is the length scale of the system) must be greater than unity. Usually we are not sure precisely what are the correct values to use for V and L, so the dimensionless number $\mu\sigma VL$, which is called the 'magnetic Reynold's number' and given the symbol R_m, has to be considerably greater than unity to be sure that self-exciting dynamo action is possible. In the case of the Earth's core, we estimate that $R_m \sim 200$, thus allowing for a considerable degree of inefficiency in the dynamo mechanism for it yet to generate successfully the geomagnetic field.

There can of course be dynamo effects which are not self-exciting, in which case there is no critical value of R_m, but these are of no interest in a discussion of the mechanism of generation of the main geomagnetic field. We will now review the history of attempts to produce theoretical homogeneous self-exciting dynamos.

For the first two decades after Larmor's original suggestion a feeling lingered that homogeneous dynamos might not be capable of self-excitation, possibly due to the lack

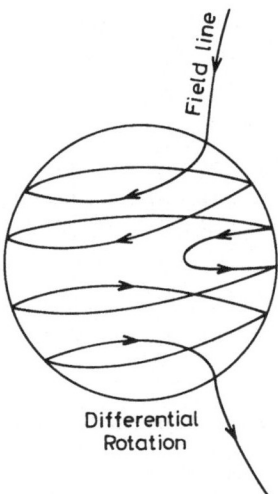

Fig. 1.   The generation of a toroidal field by differential rotation.

of insulation allowing the EMF's developed to 'short themselves out'. This was reinforced by the proof of Cowling (1934) that, in a particularly simple case, axisymmetric magnetic fields could not be maintained by fluid motion and would always decay.

Hampering the achievement of an 'existence proof' was the tendency to employ complicated models, which were thought to resemble possible core fluid motions. One useful approach, however, which allowed some insight into the relationship between fluid velocity distribution and field distribution was that of assuming infinite core conductivity. Field lines are then attached to the fluid and a 'hand waving' argument shows, for example, how a dipole field can be turned into a toroidal field by differential rotation in the core of the Earth (see Figure 1).

One of the most complicated models was that of Bullard and Gellman (1954), which involved a number of possible routes through fluid motion/field interactions leading back to the original field and hence a possible self-exciting dynamo mechanism, in fact it could not be shown to work in any of its forms. The key to a solution was to use fluid motions conceived so as to suppress a complicated series of interactions with higher order multipole and toroidal magnetic fields. Quite independently, Herzenberg (1958) and Backus (1958) used the device of attenuating these fields, in the former case by distance, and in the latter case by time, to leave only the least complicated of interactions. As the model of Herzenberg is most relevant to the experimental work, it will be discussed further. Figure 2 shows this dynamo. The way in which it works can be envisaged by imagining that it is superconducting and that field lines are stretched out by fluid motions. Any field line, passing through one of the small spherical rotating regions and parallel to its axis, will be stretched out into a toroidal field (i.e. into the form of concentric circles about the rotation axis). This toroidal field extends out into the rest of the model and can have a component along the axis of the second small

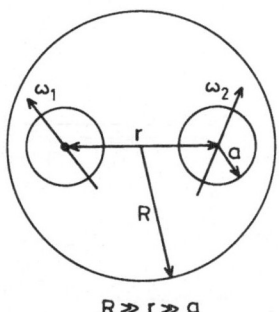

Fig. 2. The Herzenberg theoretical dynamo.

spherical rotating region, a second toroidal field is set up. Returning to the first rotating region (which will be abbreviated to 'rotor'): if the relative angle between the rotor axes and their senses of rotation are correct, then there will be a component of this second toroidal field back at the first rotor which will reinforce the original field. In a model with finite conductivity we have to make sure that R_m is big enough for the 'feedback' to make the field grow, that is, for the dynamo to self-excite.

In the Herzenberg dynamo, the rotor diameters are small, both compared to their separation and their distance from boundaries. Thus only the simplest fields will not have attenuated with distance, and be capable of interacting in an easily understandable way with the rotors. Whether or not this condition is also a necessary one for the dynamo to work at all is not clear, in fact experimental work showed that it was only a mathematical device and could be ignored in practice.

The next theoretical advance was that of Braginski (1964) who was able to show that much more complicated motions were able to generate fields. He and other workers have since shown that relatively large scale motions could act in such a way, but an interesting variation on the solution considered the field-generating effects of turbulent motion. Steenbeck et al. (1966) showed that turbulent motion, if ordered in a certain way to produce a 'preferred helicity' in motion of the fluid particles, could interact with a magnetic field to set up an EMF. This was called 'the α-effect'; later, in contrast, interactions involving large scale motions were referred to as 'the ω-effect'. So a dynamo could consist of a sequel of α and ω effects. The Herzenberg dynamo and experimental dynamos described later use only the ω-effect.

Nowhere, as has been said earlier, had the forces which exist in the Earth's core been considered. Inertial forces and electromagnetic forces are the least which must be included to describe the time variation of the magnetic field, other than by a simple 'rate of growth'. Introducing less realism in one sense, but allowing these neglected forces to be included, non-homogenous (theoretical) Faraday disk dynamos were investigated by a number of workers. The system of coupled disk-dynamos of Cook and Roberts (1970) is particularly interesting (Figure 3). This dynamo, whose working is obvious, could produce a magnetic field which varied irregularly with time as shown in the same figure. There are several grounds for criticising this model, for example the current flow

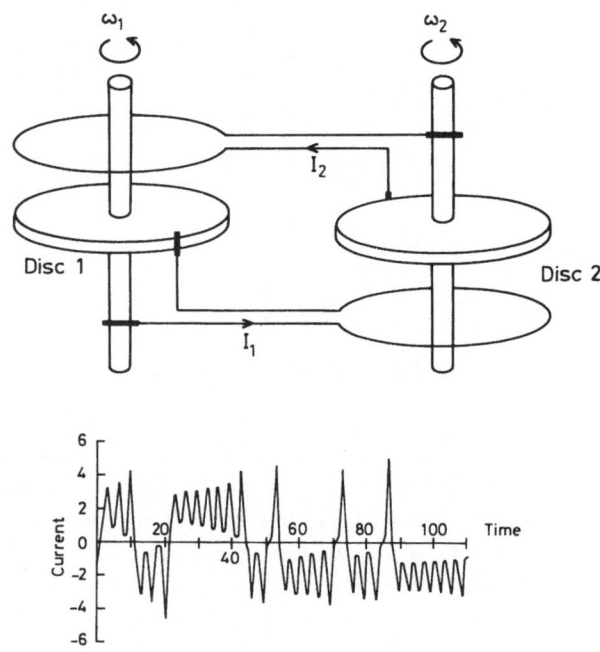

Fig. 3.   Coupled disk dynamos and the variation of their field with time.

is fixed in distribution, being constrained by insulators. Nevertheless the qualitative agreement with the palaeomagnetic record is interesting although it is not possible to solve the 'disk dynamo' equations if the correct scaling factor to the Earth's core is used.

Summarising the situation at this point, it has been shown that there is a serious gap between what it is possible to predict from theory, and the palaeomagnetic results.

4. The Contribution of Experiments on Laboratory Dynamos

The contribution of experimental work to filling that gap, albeit in a small way, will now be discussed.

Although the objectives of this work have been re-evaluated and extended as it progressed, listing them all at this stage will clarify the long term aims of the project.

In building laboratory dynamos it was hoped firstly, to verify the existence proofs of Herzenberg and of Backus. Secondly, observations of the time variation of the field produced could be related to that of disk-dynamos on one hand and the palaeomagnetic record on the other. As a development of the second objective was sought the answer to the question of how complicated a system of motions had to be, in order to show a time varying field as complicated as the palaeomagnetic data suggests.

So that the relationship between experimental observation and palaeomagnetic evidence can be seen at a glance, through a long sequence of experiments, Figure 4 has been drawn to illustrate the major points. This diagram will be referred to extensively in the text which follows. Note that driving motors have been omitted from each dynamo and in each case the dynamo drawn in a simplified, schematic form.

The experimental design, although differing in detail and sometimes scale over a succession of dynamo models, is basically the same and therefore will be described in detail for only the first, alterations will be mentioned where relevant in subsequent models.

The prime consideration is that R_m should be large enough to permit regeneration of a magnetic field. If the correct length scale L, to be used in the calculation of R_m was known, it was necessary only to achieve $R_m = 1$. However the best estimate of the necessary value of L is some typical dimension of the model. So, to allow for error, the minimum value of R_m, set as an objective by choice of the other parameters, was $R_m = 10$.

All liquid conductors have a sufficiently low value of $\mu\sigma$ to require impossibly large values of LV to achieve $R_m = 10$. Dimensions of several metres and power inputs of the order of 10^5 watts are needed to overcome viscous dissipation in mercury, for example.

The possibility of using solid conductors, with relative motion between various parts, and electrical contact being maintained by thin layers of mercury, was considered. It suffers from velocity variations being limited to that of magnitude only and not of distribution (in that respect resembling disk-dynamos), although different geometrical arrangements, to represent different velocity distributions, could be constructed. Even then it was not an easy task to design a dynamo for a value of $R_m = 10$. The use of a ferromagnetic material was attractive in this respect, it is easily possible to obtain value of $\mu\sigma$, for such materials, which are an order of magnitude greater than that for copper. Ferromagnetic materials would seem at first sight the least desirable for the construction of a model of the geodynamo, their ability to acquire a remanent magnetization being one factor which counts against them. However any dynamo needs a small field to 'trigger it off' and once generating a substantial magnetic field, it is its conducting property which is most important. In the event it was found to be saturation effects which were most troublesome.

The first dynamo was therefore constructed from an alloy known as 'perminvar', which has the property of a linear B/H characteristic passing through the origin for small fields; thus it was hoped to even eliminate remanence effects. Unfortunately, it was soon found in practice, the magnetic fields present in normal operation resulted in the usual hysteresis behaviour being displayed.

5. The First Dynamo

The first dynamo was essentially a distorted version of Herzenberg's dynamo. For ease of construction the spherical rotors became cylindrical, fitting into cylindrical cavities in the surrounding conductor (which was rather 'house-shaped' than spherical) and supported on externally mounted shafts. Electrical contact was established through a mercury layer less than one-tenth of a millimetre in thickness, with all surfaces lightly copper plated to allow the mercury to wet them. A section through this dynamo is

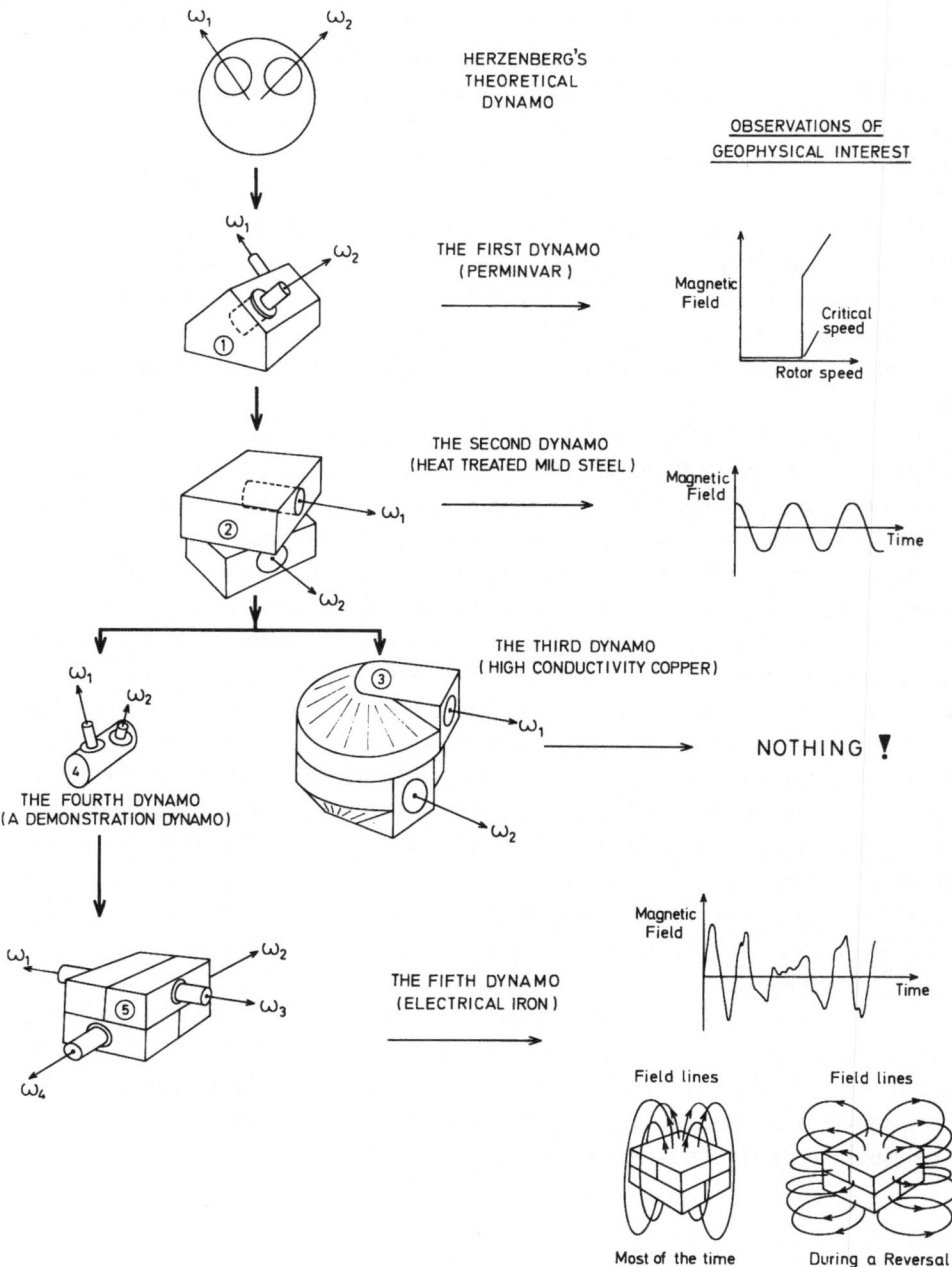

Fig. 4.   The evolution of laboratory homogeneous dynamos.

shown in Figure 5 and the general arrangement seen on Figure 4. It will be seen that the condition for validity of the existence proof, namely that the rotors should be small and far apart, has been violated. The inevitability, that retaining this condition would result

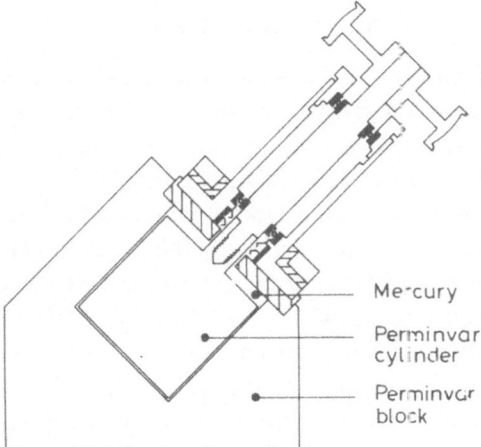

Fig. 5.   Section through one rotor of the first laboratory dynamo.

in a hopelessly inefficient experimental model, meant that the hoped-for improvement gained by placing the rotors close together had to be set against the risk that the dynamo would not function at all. The 'magnetic Reynold's number' for a two-rotor dynamo was taken to be

$$R_m = \sqrt{\omega_1 \omega_2}\, \mu \sigma L^2,$$

where ω_1 and ω_2 are the angular velocities of the rotors and L their diameter.

The power source for the geodynamo has been a subject of contention; in the models, electric motors of various sizes, drive the rotors through belts or shafts.

The $\frac{1}{4}$ H.P. motors used to drive the 7 cm diameter rotors of the first model were initially expected to reach the critical R_m at about 2000 rpm. Rotor speeds and power input, as well as magnetic field outside the dynamo, were measured and recorded.

It soon became apparent that this dynamo was not going to self-excite. Two possibilities existed, ignoring Herzenberg's criterion for his existence proof to be valid might mean the dynamo would not self-excite at any speed, or it might just be that the critical speed could not be reached with the available motors. Replacing the $\frac{1}{4}$ H.P. motors with 5 H.P. motors and installing a water cooling system to keep the dynamo's temperature to an acceptable level showed that the critical speed could not be reached even at 5000 rpm. By applying an external field, and measuring the small induced field, the critical speed was found to be finite, but an inaccessible 9200 rpm. The only approach was to insert a small amount of insulation to surpress a degenerative part of the interaction; this reduced the critical speed to a value easily attainable, in fact not too different from the design speed. So this first dynamo was known to be capable of excitation, in homogeneous form, at an inaccessible speed; it would excite with a little insulation above a certain critical speed, as shown in Figure 4. The field produced was always steady.

6. The Second Dynamo

The first objective had been reached in a rather round-about way, as just described. It would have been much more satisfactory to have verified the theoretical existence proofs with a completely homogeneous dynamo exciting at an attainable speed, the stability of the field generated was also slightly unexpected.

The second dynamo was designed to avoid the faults of the first. Quite obviously, the optimum geometry (i.e. the relative positions of the two rotors) had not been achieved from the use of a relatively qualitative argument, the only way to leave this as a variable was to design a new dynamo in two halves, with one rotor in each half. Electrical contact could be established through a thin mercury layers between the plane mating faces of these two halves, while allowing the relative angle between the rotor axes to take any value. A number of other modifications were incorporated, all to reduce the critical speed. First of all a 50% increase in scale should have halved the critical speed, but additionally, as an exotic alloy had no advantages, mild steel could be used as a constructional material. Mild steel has a conductivity about 50% greater than perminvar; it also has, with the correct heat treatment, a higher (but rather erratically so) initial permeability.

Finally, the thickness of the mercury layer around the rotors was reduced to the minimum value possible, as the relative reluctance of this layer becomes more important at the higher permeability values being achieved. The appearance of this dynamo is shown in Figure 4. The critical speed should have been reduced by all the modifications by a factor of four, with a further reduction by a completely unpredictable amount due to improvement in geometry.

In fact this dynamo was observed to excite at a range of speeds which depended on the relative angle between rotors, the lowest value being approximately 400 rpm (both rotors) with a total power input of about 200 W. The behaviour after excitation is rather complex; the permeability increases with increasing field before falling again. The result is that, once excited, the speed of the rotors fall, as electromagnetic drag on the rotors builds up. Under most conditions, the power dissipated in joule heating is much greater than the viscous power dissipation in the mercury layers; a comparable situation exists in the Earth, viscous energy dissipation is also negligible.

A range of geometries was observed where (given the correct senses of rotation for regeneration) either a steady field or a periodically reversing field could be produced, for constant power input levels. These were named 'non-reversing' and 'reversing' dynamos, respectively. A typical pattern of field variation with time is shown in Figure 4.

Taking stock at this point, what had been shown? The existence proofs had now been given a satisfactory verification. Field reversals were shown to be a feature of a very simple homogeneous dynamo, with a constant power input. What had not been observed was an irregular pattern of reversals, like the palaeomagnetic record. Also, rather disconcertingly, increasing power input lengthened the period of oscillation. Now disk-dynamo theory, whatever its deficiencies, predicts that greater input would increase reversal frequency. Examination of the detailed time variation of the magnetic field lead inescapably to the conclusion that saturation of the ferromagnetic material was modifying, and masking, the essential dynamo behaviour.

7. The Third Dynamo

Confident that a very efficient geometry could be attained, and as a consequence of some calculations which showed that a dynamo built from high-conductivity copper should be on the limit of achieving the critical magnetic Reynold's number, a larger dynamo of copper was planned. The linear scale was increased by a further factor of two, and with some alterations to the external appearance, dictated by constructional requirements, dynamo number three was built from copper. Viscous power dissipation in the mercury layers, around each rotor, was expected to increase dramatically (compared with previous models) and two 15 H.P. motors were therefore installed. The outcome of this approach was less than satisfactory, the critical magnetic Reynold's number could not be reached and no field was generated.

At this juncture it seemed that laboratory dynamo experiments had reached an impasse. While the construction of ferromagnetic models had been partly successful in achieving the objectives, using non-ferromagnetic material to avoid the introduction of irrelevant effects was not to be a practical solution. This line of research was discontinued for some time, although a small fourth 'demonstration' dynamo was built (of fixed geometry) to use as a lecture illustration.

8. The Fourth Dynamo

This fourth dynamo was built of non heat-treated mild steel, with rather thicker mercury annuli to ease construction by allowing wider machining tolerances. Using the extensive experience gained from dynamo number two, a geometrical arrangement was chosen that, despite reduction in efficiency permitted to make construction easier, satisfactorily operated, displaying the reversal behaviour as required.

After having been used as a demonstration in lectures and talks on geomagnetism many times, one of those chance observations, which in many fields of science have been responsible for new lines of investigation, was made. Normally the electric driving motors were fed from a constant voltage source, the frequency of reversals was therefore constant. On one occasion however, the voltage setting was increased and it was noticed that, surprisingly, the reversal frequency increased. Further investigation showed that the reversal frequency varied with power input as that of coupled disk-dynamos. Examination of the reversal waveform showed that the effects, earlier attributed to saturation in previous dynamos, had been eliminated.

9. The Fifth Dynamo

With the possibility of a resumption of experimental work on homogeneous dynamos, a fifth model was planned. This was to be in modular form, allowing the assembly of a number of units, each containing one rotor, into a homogeneous dynamo of variable complexity. The earlier two-rotor dynamos had illustrated the basic instability in the mechanism, but were of the simplest form possible, bearing more relationship to the theoretical model of Herzenberg than the Earth's core. It is not suggested that a laboratory dynamo could be built which would reproduce in detail the flow patterns in the Earth's core, indeed there is no concensus of opinion on the nature of these motions, beyond suggestions from idealised theory and model experiments as to their possible form. But the field generation process is most likely to involve a number of steps of field amplification, which each could be represented by one rotor in the laboratory dynamo. With up to four rotors, each interacting with one another, the intention was to study the variation of magnetic field with time (for constant power input to each rotor) as the geometrical relationship between various parts of the dynamo was altered. It was thought that various features, seen in the palaeomagnetic record, might appear as the number of rotors and complexity of the dynamo was increased.

A number of identical units were constructed of such a shape that they could be stacked together, with a mercury film everywhere between adjacent surfaces ensuring electrical contact, and hence homogeneity in conductivity. Four of the units contained cylindrical cavities with cylindrical rotors inserted. The thickness of the mercury filled annulus was similar to the fourth dynamo. Instead of using mild steel for their construction, a ferromagnetic material called 'electrical iron' was used. The reason for the change was partly that mild steel is not available to precise specification, and two samples can have noticeably different magnetic properties. Also 'electrical iron' is superior, in that the saturation magnetisation is higher, and the coercive force lower, than for mild steel. The general permeability characteristics are otherwise similar. With no heat-treatment to produce extreme values of permeability, it was hoped that the behaviour of the fourth dynamo would be reproduced. Once again $\frac{1}{4}$ H.P. motors provided the driving power, this time through flexible drive shafts originally intended for power tools. Instrumentation consisted of speed monitoring devices, and both fluxgate and Hall magnetometers for field measurement. In order to collect data in a suitable form for processing, it was also recorded digitally via an A to D converter and multiplexing system using an Apple II microcomputer.

To describe the results of these most recent experiments and their relationship to our knowledge of the geomagnetic field, the scheme will be adopted of starting with the simplest two rotor versions and progressing finally to its four rotor form. Figure 4 shows one possible way in which the units can be stacked together, the driving shafts and their mountings are omitted where fields are drawn. The two-rotor form of dynamo five was investigated first. As expected, the effects attributable to driving the ferromagnetic material into saturation did not appear and a simple periodic reversal of the magnetic field occurred above the critical speed, the frequency of which varied in the predicted way with power input.

Observation of the spatial distribution of the magnetic field produced by these laboratory dynamos was one of the original aims of the work. Refinements in palaeomagnetic and geochronological techniques were allowing the variation of the geomagnetic field, in the short time interval while it was in transition, to be deduced. Such data from several sites globally, and for a number of different (but identifiable) reversals of the field, have been used by various workers to try to infer the time variation of the spatial distribution of the geomagnetic field, while it is in transition from one polarity to the other. The 'palaeomagnetic' approach to the presentation of data is a rather confusing one (to someone who is not in that field) as it is based on the convention of referring all field measurements obtained from rock samples to Virtual Geomagnetic Poles (VGPs). During a field polarity transition, one has to determine if the movement of the VGP (dipolar) is consistent with the actual field being non-dipolar and of some specific form.

The literature on the subject is extensive, but one conclusion which emerges from all the efforts to reconcile the observations with possible models for the field, is that during a transition it departs substantially from the initial, and final, dipolar form. Whether or not the field is quadrupolar, or perhaps octupolar, cannot be decided from the data available at this time.

It should be stressed that the analysis of the geomagnetic field in terms of multipoles has no physical significance, the same field can be produced by an infinite variety of interactions; that is, there is no unique flow pattern defineable in the core from observations of the external field. Consequently the observations of similar behaviour in the laboratory dynamo's field and the geomagnetic field does not imply similarity of motions. The correct question to ask is, how does the complexity of spatial and temporal variation of the geomagnetic field depend on the complexity of the dynamo mechanism? If these variations can be successfully demonstrated with a laboratory dynamo, then it could be said, for example, that changes in the pattern (as distinct from changes in the amplitude) of core motions are not required. Such changes in pattern have been postulated to explain changes in the nature of the geomagnetic field.

With this justification for observation of the spatial variation, and its time variation, of the dynamo field, the experimental problems of data collection was assessed. Resolution of a quadrupole component would require a minimum of a 30° grid, 62 points in all. For experimental convenience they were on a spherical surface, proportionately as far from the dynamo as the Earth's surface is from the core. Simultaneous measurements at each of the 62 points was out of the question with the equipment available, however the periodic behaviour allowed separate measurements to be taken, correlation in time being achieved by the use of a fixed phase reference.

Only the radial component of field is needed to calculate the spherical harmonic coefficients, out the other two components were also recorded, so that a plot of the variation in time of the field vector at a point could be made without introducing any additional errors arising from the spherical harmonic analysis. It was observed that, while the field magnitude was large the dipole coefficient was large, but as the magnitude fell the distribution was seen to become dipolar. In general terms the field distribution changed in the following way – 'normal' dipole – transverse quadrupole – 'reverse' dipole – transverse quadrupole – 'normal' dipole, etc. The field vector, at any point, spent most of its time either in one direction, or an antiparallel one, rapidly tracing a path between them. In no position on the spherical surface did the field value fall to zero during any part of the cycle. In Figure 4, this behaviour is shown schematically as field lines representing a dipole for most of a cycle, while a quadrupolar set of field lines shows the situation as the dipole coefficient passes through zero in the middle of a reversal. The inference drawn from this particular observation was that, as a particularly simple two-rotor dynamo could model the geomagnetic field behaviour at the time of a reversal, no dramatic changes in core motion should be needed in the Earth to produce those effects.

The next question to answer concerned changes in an average rate of reversal. Recent reversals, as has been previously stated, occur every 250 000 yr, but 40 to 50 million years ago the rate dropped dramatically. At the same time a change occurred in the direction of motion of the Pacific plate. The obvious suggestion was that they were related effects. A two-rotor dynamo only changes its reversal rate slowly with input power, requiring considerable changes in physical conditions in the core, if it produced the geomagnetic field by a simple equivalent mechanism, to give that observed effect. What had to be ascertained was if a more complex dynamo, with more rotors, could be made much more sensitive in its behaviour to power input variations.

The three-rotor form of the fifth dynamo was investigated specifically to look for such sensitivity of reversal rate to power input levels. Two-rotor dynamos either excite or do not excite, depending on the senses of rotation. With three rotors there is a fundamental difference in conditions. Two of the three rotors can act together to generate a field, while the third interacts degeneratively; in another arrangement two rotors can interact to produce a steady field, while the third interacts with them to produce a reversing field. With these extra configurations available, it was possible to have a situation where small changes in driving power produced large changes in reversal frequency. Particularly interesting was the case where the field could be made steady by a small change in power input. It now becomes much more reasonable to suggest that changing conditions in the Earth's mantle could influence the reversal frequency of the core dynamo.

During a period when the geomagnetic field has been changing polarity at a certain rate, it nevertheless has an erratic character in detail. Three-rotor dynamos, when they are producing a reversing field, show regular reversals. Unless the power input to the core is varying on the same timescale as that between reversals, which seems unlikely, even more complexity in the core dynamo must be considered.

The four-rotor form of the fifth dynamo was assembled to give this extra complexity. The major observation was that for some geometrical arrangement of rotors, the time between reversals (here the use of the term 'period' is inappropriate) could be variable, for constant power input. An example is shown in Figure 4 of the field variation with time for this dynamo. The most extreme cases studied showed a variation, by a factor of about four, in time between reversals. Also, some features of magnetic field recordings resemble palaeomagnetic events and excursions (short and incomplete reversals), but they remain to be investigated further. As the number of adjustable parameters is enormous for the four rotor dynamo, many more observations remain to be made in the future.

10. Summary

From this series of experiments with homogeneous laboratory dynamos at Newcastle University Geophysics Department it can be said, with some confidence, that many major features observed (or inferred by palaeomagnetic methods) in the geomagnetic field record, are also characteristics of relatively simple dynamos.

The geomagnetic field, in recent geological times, had an average time between reversals of about twenty times the free decay rate of the dipole field; the same ratio can be seen in some laboratory dynamos.

Geomagnetic field reversals have an erratic character, so can the field of laboratory dynamos.

The geomagnetic field reversal rate seems to be sensitive to outside influences; laboratory dynamos can have a field reversal rate sensitive to power input. The geomagnetic field becomes non-dipolar during a reversal, so does the field of a laboratory dynamo.

A tentative model for the geodynamo would therefore be at least as complex as the last four-rotor dynamo. Such a dynamo would, under constant driving conditions, produce a reversing field of a slightly erratic character. Moderate variation of the power input could reduce the reversal rate dramatically. However the spatial characteristics of the geodynamo field would appear to be reproduced by the least complex of laboratory dynamos. (The four-rotor dynamo has not yet been investigated in that respect).

It is perhaps surprising that there is such good agreement between palaeomagnetic data and these experiments, as the coriolis force (which must have a dominant effect on core motions) has been ignored. A speculative explanation could be that the coriolis force acts to stabilise the pattern of core motions, restricting changes to those of magnitude rather than distribution, in which case the laboratory models may have more in common with the core dynamo than was originally envisaged.

References

Backus, G. E.: 1958, 'A Class of Self-Sustaining Spherical Dynamos', *Ann. Phys.* **4**, 372–447.

Braginskii, S. I.: 1964, 'Kinematic Models of the Earth's Dynamo', *Geomag. Aeron.* **4**, 572–583.

Bullard, E. C. and Gellman, H.: 1954, 'Homogeneous Dynamos and Terrestrial Magnetism', *Phil. Trans. Roy. Soc. Lond.* A **247**, 213–278

Cook, A. E. and Roberts, P. H.: 1970, 'The Rikitake Two-Disk Dynamo System', *Proc. Camb. Phil. Soc.* **68**, 547–569.

Cowling, T. G.: 1934, 'The Magnetic Field of Sun-Spots,' *Monthly Notices Roy. Astron. Soc.* **94**, 39–48.

An account of the experimental work described in this survey is given in the following literature:

Kerridge, D. J.: 1983, 'A Laboratory Study of Mechanisms Related to those Responsible for the Generation of the Geomagnetic Field', Ph.D. Thesis, University of Newcastle upon Tyne.

Johnson, R. B.: 1968, 'An Investigation into the Variation of Efficiency with Geometry of a Laboratory Homogeneous Dynamo', M.Sc. Thesis, University of Newcastle upon Tyne.

Lowes, F. J. and Wilkinson, I.: 1963, 'Geomagnetic Dynamos: A Laboratory Model', *Nature* **198**, 1158–1160.

Lowes, F. J. and Wilkinson, I.: 1968, 'Geomagnetic Dynamo: An Improved Laboratory Model', *Nature* **219**, 717–718.

Scouller, M.: 1972, 'An Investigation of a Homogeneous Copper Dynamo', M.Sc. Thesis, University of Newcastle upon Tyne.

Wilkinson, I.: 1963, 'Experimental Investigation of Homogeneous Dynamos', Ph.D. Thesis, King's College, University of Durham.

Bibliography

Moffat, H. K.: 1978, *Magnetic Field Generation in Electrically Conducting Fluids*, Cambridge University Press.

Tarling, D. H.: 1983, *Palaeomagnetism*, Chapman and Hall.

S. K. RUNCORN'S COMMENTARY

I went to Manchester University as an assistant lecturer in September, 1946 to work on cosmic rays and in the first term helped G. D. Rochester and C. C. Butler to get working a cloud chamber triggered by Geiger counters with which later they made notable discoveries in elementary particle physics. However, in November P. M. S. Blackett gave a colloquium on his theory of the geomagnetic field and I recall how surprised I, and others were, that such a commonplace natural phenomenon as the Earth's field had never been given an explanation which could stand up to the slightest critical evaluation. So many of us took Blackett's proposal that the field was a new fundamental property of rotating matter seriously – Einstein's efforts to unify gravitation and electromagnetism were much talked of – and we discussed possible experimental tests. Some did not: I think geophysicists were somewhat embarrassed that such an important problem had not been given the slightest attention. Stimulated by Bullard's suggestion that Blackett's theory might be tested because it would predict a different variation of the field with depth, I worked out a theoretical prediction (as did Chapman more exactly). We planned to test the theory by comparing the geomagnetic field at the surface and in deep mines. It was necessary to have quite a team and A. C. Benson and A. F. Moore and I carried through the experiment with the enthusiastic assistance of undergraduate students of whom F. J. Lowes and R. Hide gained their first interest in geophysics. Blackett's theory stimulated much interest (see my contribution in Notes of the Royal Society). Meanwhile W. M. Elsasser had written the fundamental papers of the dynamo theory. Frank Lowes soon got interested in the complexities of the dynamo and soon showed the relevance of laboratory experiments, first to the secular variation and then to the main field. Model experiments are difficult to make relevant in geophysics and I heard Sydney Chapman say that W. Gilbert with his terella had made the only successful model experiment in geophysics. Lowes and Wilkinson have been more successful than most, e.g. throwing light on reversals. Had they believed more in their model they could have predicted the short term geomagnetic events before they were discovered in the geological record! I am hardly in a position to criticize as I wrote papers in the early 1960's arguing for the existence of a lunar core but did not predict that the Apollo rocks would be magnetized: Velikovsky apparently has that achievement to his credit!

Geophysical Surveys 7 (1984) 123.

Spatial Statistics and Models

Edited by
GARY L. GAILE and CORT J. WILLMOTT

1984, x + 460 pp. + index
Cloth Dfl. 190,—/ US $ 69.00   ISBN 90-277-1618-8
THEORY AND DECISION LIBRARY 40

In the last decade, geographers have expanded their methodological bag of tools by developing quantitative techniques which are specifically directed towards analysis of explicitly spatial problems. The explicit incorporation of space into quantitative techniques has not, however, been the sole domain of geographers. Mathematicians, geologists, economists and regional scientists have shared the geographer's interest in the incorporation of explicitly spatial components into their analytical tools. This volume is a sampling of state-of-the-art papers on topics dealing directly or indirectly with spatial phenomena or processes. The interests of the contributors are highly diverse, e.g. geology, geomorphology, climatology, cartography, human geography, population and migration, urban geography, economic geography, and methodology. For purposes of presentation, various approaches have been segregated into predominantly process-based quantitative descriptions, i.e. 'models', and predominantly general or mathematically-based descriptions, i.e. 'statistics'. The vast majority of included papers fit equations to data.

Contents: I. **Spatial Statistics.** Contributors include: D. Griffith, P. Gould, M. Goodchild, R. Semple and M. Green, R. Balling, Jr., H. Moellering, D. Shepard, F. Agterberg, J. Burt, D. Mark, J. Miron, G. Gaile, R. Bennett, J. Schuenemeyer, R. Jarvis, R. Austin. II. **Spatial Models.** Contributors include: W. Krumbein, A. Pickles and R. Davies, J. Huff, H. Sheppard, R. White, J. Rayner, C. Wilmott.

 **D. Reidel Publishing Company**

P.O. Box 17, 3300 AA Dordrecht, the Netherlands
190 Old Derby St., Hingham, MA 02043, U.S.A.

Magnetic Resonance

Introduction, Advanced Topics and Applications to Fossil Energy
Proceedings of the NATO Advanced Study Institute on Magnetic Resonance
Introduction, Advanced Topics and Applications to Fossil Energy, Maleme,
Crete, Greece, July 3–15, 1983

Edited by
LEONIDAS PETRAKIS and JACQUES P. FRAISSARD

1984, xi + 807 pp.
Cloth Dfl. 255,– / US $ 98.00   ISBN 90-277-1752-4
NATO ADVANCED SCIENCE INSTITUTES SERIES
C. Mathematical and Physical Sciences 124

Containing lectures presented at a NATO ASI held on Crete in July, 1983, this volume deals with magnetic resonance in a unified and detailed manner, starting from the behaviour of spins in a magnetic field and proceeding to a density matrix consideration of the phenomena involved. The various important parameters of NMR and EPR are discussed in detail as well as their significance and the information that may be obtained from their proper consideration. Advanced topics are then considered, including line broadening interactions, coherent averaging and multiple quantum methods in NMR, ENDOR, electron spin echoes, dynamic nuclear polarization, problems of quantification, 2-Dimensional NMR, Mossbauer and instrumentation. This component of the volume can serve as an introduction to magnetic resonance and discussion of advanced topics regardless of applications. The volume also introduces the general fields of fossil energy, including resources, synthetic fuel processes and structural aspects of fossil energy materials. Finally, the issue of research problems and opportunities is discussed and specific recommendations are made.
Contents: Preface. Organizing Committee. I. Introduction and Advanced Topics in Magnetic Resonance. II. Topics in Fossil Energy. III. Paradigms of Magnetic Resonance in Fossil Energy Studies. IV. Research Prospects.

D. Reidel Publishing Company

P.O. Box 17, 3300 AA Dordrecht, the Netherlands
190 Old Derby St., Hingham, MA 02043, U.S.A.

Solar Terrestrial Physics - Principles and Theoretical Foundations

Based upon the Proceedings of the Theory Institute held at Boston College, August 9–26, 1982

Edited by
R. L. CAROVILLANO and J. M. FORBES
Dept. of Physics, Boston College, Chestnut Hill, MA, U.S.A.

1983, xvii + 859 pp.
Cloth Dfl. 265,– / US $ 115.00   ISBN 90-277-1632-3
ASTROPHYSICS AND SPACE SCIENCE LIBRARY 104

Several years ago there was a convergence of efforts to promote the role of theory in space plasma physics. Reports from the National Academy of Sciences and NASA advisory committees documented the disciplinary maturity of solar-terrestrial physics and recommended that theorists play a greater role in the continued development of the field. The so-called theory programme in solar-terrestrial physics was established by NASA in 1979 and implemented in accordance with the guidelines set forth by a panel of scientists, primarily theorists, in the field. The same panel motivated the Boston College programme. Published proceedings of the school will provide curricular materials for the training of graduate students in solar-terrestrial physics.

D. Reidel Publishing Company

P.O. Box 17, 3300 AA Dordrecht, the Netherlands
190 Old Derby St., Hingham, MA 02043, U.S.A.

Dynamics of the Middle Atmosphere

Proceedings of a U.S.-Japan Seminar, Honolulu, Hawaii,
8–12 November, 1982

Edited by
J. R. HOLTON and T. MATSUNO

1984, viii + 543 pp.
Cloth Dfl. 120,– / US $ 89.50   ISBN 90–277–1758–3
ADVANCES IN EARTH AND PLANETARY SCIENCES 18
*Available in Japan from TERRA Scientific Publishing Company,
Tokyo, Japan*

Containing both overview articles and original contributions treating
a wide variety of topics in dynamics of the middle atmosphere,
this book deals with, among other subjects, internal gravity waves,
planetary waves, tides, quasi-biennial oscillations, sudden warmings,
transport of water vapour, ozone and other tracers, and general
circulation in the middle atmosphere. Radiation and chemistry
are touched on only in the context of dynamical problems. A feature
is that this is the first book containing recent results of·both observational and theoretical studies of internal gravity waves in the
middle atmosphere. In the past couple of years there has been an
upsurge of research in breaking internal gravity waves and consequent generation of mean flows and turbulence, which now
forms the most active frontier of atmospheric sciences. The book
also contains comprehensive articles on the quasi-biennial oscillation, tracer transport mechanisms and general circulation models
in the middle atmosphere which will serve as references for research
scientists as well as graduate students.

Contents
Preface. Gravity Waves. Tides and Free Oscillations. Large-Scale
Waves and Wave, Mean-Flow Interaction. Radiation. Transport
of Tracers. Modeling. Author Index. Subject Index.

D. Reidel Publishing Company

**P.O. Box 17, 3300 AA Dordrecht, the Netherlands
190 Old Derby St., Hingham, MA 02043, U.S.A.**

NEW REIDEL TITLES

The Glaciers of Equatorial East Africa

STEFAN HASTENRATH

1984, 366 pp. + index
Cloth Dfl. 195,– / US $ 72.50   ISBN 90–277–1572–6
SOLID EARTH SCIENCES LIBRARY

The high mountains of East Africa are among the three regions of the world where glaciers still exist in the vicinity of the Equator. Tropical glaciers are extremely sensitive but complex indicators of climate. Accordingly, those involved in the fields of climate and environmental change have recognized the assessment of current ice extent and of long-term glacier fluctuations as tasks of high priority. This book reviews the environmental setting of the East African highlands, and summarizes the geomorphic evidence of Pleistocene to early Holocene glacier variations, which bears out the existence of a formerly much larger ice extent. Based on field observations, evaluation of aerial photographs and maps, and a variety of historical sources, an inventory of the current ice extent is presented, and glacier variations since the latter part of the 19th century are reconstructed. In the entire tropics, Lewis Glacier, Mount Kenya now possesses the most continuous and detailed historical documentation of long-term changes. It therefore offers the possibility of quantitatively inferring the unknown climatic forcing from the well-observed glacier response.

Contents
Preface. List of Figures. List of Photographs. List of Maps. List of Tables. 1. Introduction. 2. The Environmental Setting. 3. Pleistocene to Early Holocene Glaciations. 4. The Recent Glaciation. 5. Lewis Glacier, Mount Kenya. 6. East African Glaciers and the Global Tropics. 7. Abstract and Summary. Appendix 1: List of Topographic Maps, Air Photographs, and Satellite Imagery. Appendix 2: List of Historical Photographs and Drawings. Appendix 3: Data Supplied to World Glacier Inventory. Bibliography. Author Index. Subject Index. .

D. Reidel Publishing Company

P.O. Box 17, 3300 AA Dordrecht, the Netherlands
190 Old Derby St., Hingham, MA 02043, U.S.A.

GEOPHYSICAL SURVEYS / *Vol. 7 No. 2 March 1985*

REVIEW OF LAKE SEDIMENT PALAEOMAGNETIC DATA

(PART I)

K. M. CREER

Department of Geophysics, University of Edinburgh

Abstract. This first part of a two part review is essentially descriptive: palaeomagnetic data sets used to compile the published reference curves of geomagnetic secular variaticns (SV) through Holocene time are critically examined, particularly with respect to coring techniques used for sampling, reliability of the measured directions of remanent magnetization and transformation tc timescales. Also, progress made in extending SV records into Late Glacial time is reviewed. SV curves have been constructed for W. Europe and N. America going back respectively to 22000 and 31000 yr before the present (bp). The best quality SV records have invariably been obtained from homogeneous muds and clays which typically smooth the geomagnetic signal over intervals of a few hundred years. They show irregular oscillatory patterns in contrast with certain records obtained from some inhomogeneous sections which show sequences of strongly aberrant palaeomagnetic directions. The latter have been interpreted by some authors as evidence of abnormal geomagnetic field behaviour but such evidence is, on the whole, found to be unconvincing. If geomagnetic excursions or aborted attempts at polarity reversal have occurred, they must have been so rapid as to have been smoothed out in the quasi-continuous records obtained from homogeneous muds. The second part of the review, to be published shortly in this journal, will deal with analyses and interpretations of the data presented here.

1. Introduction

Clearly, a sound knowledge of the power distribution in the geomagnetic spectrum is a necessary prerequisite for a proper understanding of the processes by which the geomagnetic field is generated. On the positive side, a fairly detailed knowledge of the global behaviour of the field on the timescale of $\sim 10^2$ yr and less has been accumulated from observatory data and from measurements of the field direction recorded by mariners, especially during the era of wooden sailing ships. At the other end of the spectrum, a rather good picture of the pattern of polarity reversals of the main field which occurred in the timescales of 10^4 to 10^5 yr has been built up from sea floor magnetic anomaly studies and from the palaeomagnetic record carried by cores of ocean bottom sediments and land exposures of both volcanic and sedimentary rocks. However, until recently, our knowledge of that part of the geomagnetic spectrum with periods in the range of 10^3 to 10^4 yr has been noticibly deficient. Palaeomagnetic studies on lake sediments provide data to fill this gap. The methods and procedures are presented and discussed in 'Geomagnetism of Baked Clays and Recent Sediments', a multi-authoured book edited by K. M. Creer, P. Tucholka and C. E. Barton, published in 1983 and which will be referred to in this text as 'GBCRS'.

Although the first palaeomagnetic measurements on unconsolidated sediments were made more than 40 yr ago, starting with McNish and Johnson (1938), Griffiths (1955) and Ising (1943), it was not until the early seventies that Mackereth (1971) obtained the first convincing quasi-continuous SV records. Mackereth measured the magnetic remanence carried by the homogeneous Post Glacial muds deposited under relatively

Geophysical Surveys 7 (1985) 125-160. 0046-5763/85.15.

quiet conditions on the floor of Lake Windermere in N.W. England whereas the earlier workers had worked with varved clays. Although a floating timescale could be attached to the sequence of palaeomagnetic directions carried by many individual varve sequences, the high energy environment of deposition caused systematic errors in the alignment of the magnetic carrier grains along the geomagnetic field direction.

Following Mackereth's pioneer study, a steadily growing number of palaeomagnetic research groups have become involved in lake sediment studies particularly in W. Europe and N. America and, to a lesser extent, in Australia and Argentina. During the last decade, experimental methodology and techniques of data analysis have been developed to a satisfactory degree of sophistication and numerous snags and difficulties have been identified. It has become evident that not all lake sediments are reliable recorders of geomagnetic SV. One of the most important factors is the size of the magnetic carrier grains relative to the matrix grain size (Tucker, 1979, 1980): this affects the magnitude of the forces opposing rotation of the magnetic grains into the ambient field direction and hence influences the intensity of remanent magnetization acquired in a given geomagnetic field strength [see contribution by Tucker in GBCRS (1983)]. The stability of NRM is also affected by this parameter. Second, non-spherical magnetic carrier grains (which can be modelled by prolate or oblate ellipsoids) can settle with their magnetic moments misaligned from the ambient geomagnetic field direction. This has been well known since the time of some of the earliest investigations (e.g. King, 1955). Third, the NRM intensity also depends on the nature of the source rocks from which the particular lake sediment was derived: clearly a catchment area where predominantly igneous rocks are exposed provides a higher concentration of titanomagnetite grains than one in a sedimentary terrain. Fourth, pitfalls can arise because the virgin palaeomagnetic SV record may subsequently become distorted due to slumping of the sediment, induced for example by turbidity currents or by earthquakes. Fifth, erosion may have occurred at some horizons in sedimentary successions cored in lakes, particularly at times of low water level or high throughflow.

Cores previously collected for limnological or palaeoclimate studies have occasionally been made available to palaeomagnetic workers. However they have rarely been of satisfactory quality to provide good palaeomagnetic data. This is because palynologists invariably take narrow diameter ($\sim$2.5 cm) cores in short, unorientated, sections. To preserve the physical structure of the sediment satisfactorily, cores need to be more than about 5 cm in diameter and for palaeomagnetic work it is advisable to adopt a coring technique in which the rate of penetration is as slow and steady as possible, taking precautions to prevent the core twisting about its longitudinal axis. These requirements are very important and they have meant that much time has had to be devoted to devising time consuming (and therefore expensive) adaptions to standard methods of coring.

Another reason why the accumulation of palaeomagnetic SV data has been slower than was anticipated in the mid-seventies is that lake sediments are difficult to date by radiocarbon. This is the only technique at present capable of providing absolute ages to the required accuracy and it works best on highly organic sediments. But because these contain little detrital material, they carry only a weak palaeomagnetic signal. On the other hand, sediments rich in detritus usually are low in organic carbonaceous material so that while they may carry a strong and stable palaeomagnetic signal, they cannot be dated accurately. Furthermore, systematic errors can arise due to the incorporation of carbon from recycled fossil shells into the sediment or to the precipitation of calcium carbonate dissolved in the lake water and its subsequent incorporation into the sediment [see, for example, contribution by Hedges in GBCRS (1983)]. Thus one of the major difficulties encountered has been that many of the better quality palaeomagnetic records cannot be dated precisely, even with the help of indirect methods involving limnological and stratigraphic correlation. A further difficulty arises from the fact that the radiocarbon timescale can at present be corrected to calendar years only for the last 9000 yr (using tree ring data). This induces errors in any spectral analysis which may be carried out on palaeomagnetic SV records because even through Holocene time, the radiocarbon timescale has been shown to depart from the calendar year scale by up to 1000 yr (see Clark, 1975 for calibration tables). And we have to bear in mind that it is likely that discrepancies of at least this magnitude have occurred through Late Glacial time to which palaeomagnetic SV investigations are presently being extended.

2. European Data

2.1. HOLOCENE

Figure 1 shows the geographical localities of all core-sites discussed in this section.

2.1.1. *U.K. Type-Curves*

Type curves depicting geomagnetic SV in inclination and declination have been constructed for Holocene time (Figure 2) using data from U.K. These curves are reproduced from Figure 1 of Turner and Thompson (1983). The inclination curve traces out a series of irregular maxima and minima labelled, in Figure 2a with lower case letters from the Greek alphabet running from α starting in recent time, going back to ν. Declination traces out a sequence of oscillations, also irregular, the extreme values being labelled in Figure 2b with upper case letters of the Greek alphabet, from A to H from youngest to oldest.

Results from many European sites (geographic localities shown in Figure 1) including Greece, Poland, Switzerland and France, although less detailed than the U.K. results, are in good overall agreement with the patterns traced out by the U.K. type curves within the broad limits set by the less precise age control available for them. Many of these results are described in contributions to the book GBCRS (1983), where comprehensive lists of references to published papers in limnomagnetism and archaeomagnetism is given. In the sections which follow some explanation of how these type curves were developed is given and problems arising when attempts are made to extend them back into the Late Glacial are discussed.

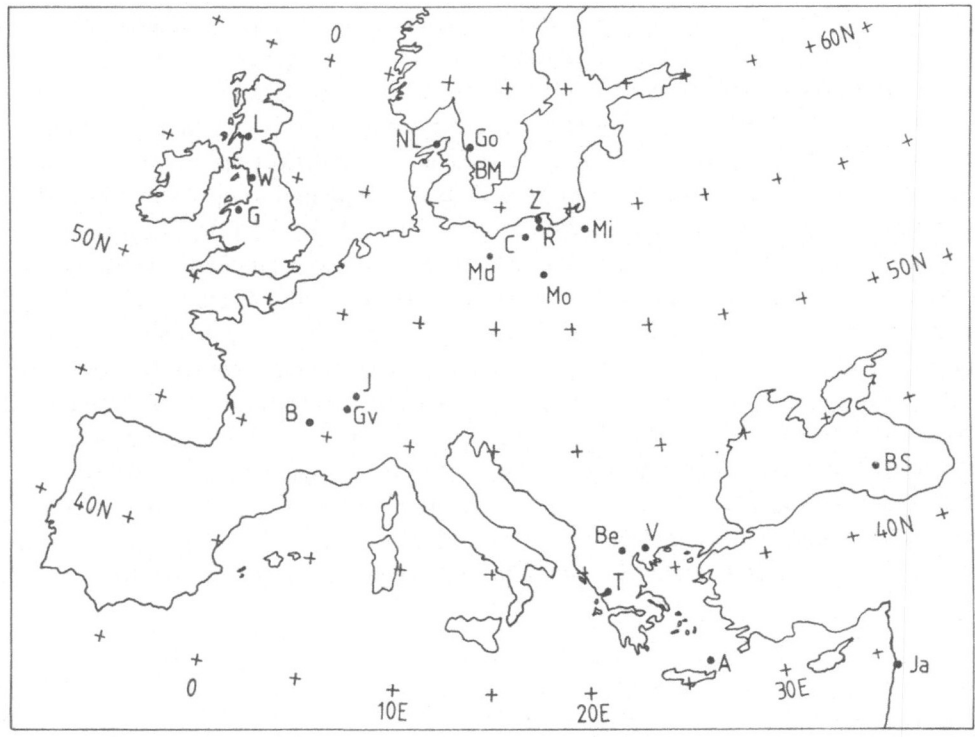

Fig. 1. Site location map for Europe. Key to labels as follows: - L = Loch Lomond, W = Lake
Windermere, G = Llyn Geirionydd, Gv = Lake Geneva, B = Lac du Bouchet, J = Lac de Joux, NL =
Norre Lyngby, Go = Gothenburg, BM = Bjorkerods Mosse, Md = Lake Miedwie, C = Lake
Charzykowskie, Z = Lake Zarnowieckie, R = Lake Radunskie, Mo = Lake Moszczonne, Mi = Lake
Mikolajskie, Be = Lake Bergoritis, T = Lake Trikhonis, V = Lake Volvi, BS = Black Sea, A = Aegean
Sea, Ja = Jeita Cave.

2.1.2. *Lake Windermere*

Mackereth (1971) constructed a composite curve of the variations in declination
recorded along two 6 m cores (Mackereth, 1958) and six 1 m cores (Mackereth, 1969)
taken from the bottom sediments of L. Windermere (2.8° W 54.4° N). The palaeomagnetic measurements were made using an astatic magnetometer on thin slices of sediment
cut horizontally, at right angles to the longitudinal axes of the cores. This produced a
high sampling density but the disk-shaped sub-samples were too thin to allow
measurement of inclination. The record shown here (WMERE71 in Figure 3b) has
been transformed to a timescale using the nine radiocarbon dates listed by Mackereth
(1971) in his Figure 2. These cores penetrated through the Post Glacial gel-like gyttjas
(10–15% organic carbon content – base dated at about 10000 bp, rate of deposition
~0.5 mm yr⁻¹) through about 0.5 m of finely laminated clay (Younger Dryas) and
then into some 0.3 m of highly compacted, partly organic deposits laid down during the
period of Late Glacial climatic amelioration (Allerød and Bølling). Seven and a half
declination swings were recorded in the Post Glacial gyttjas and the period of these

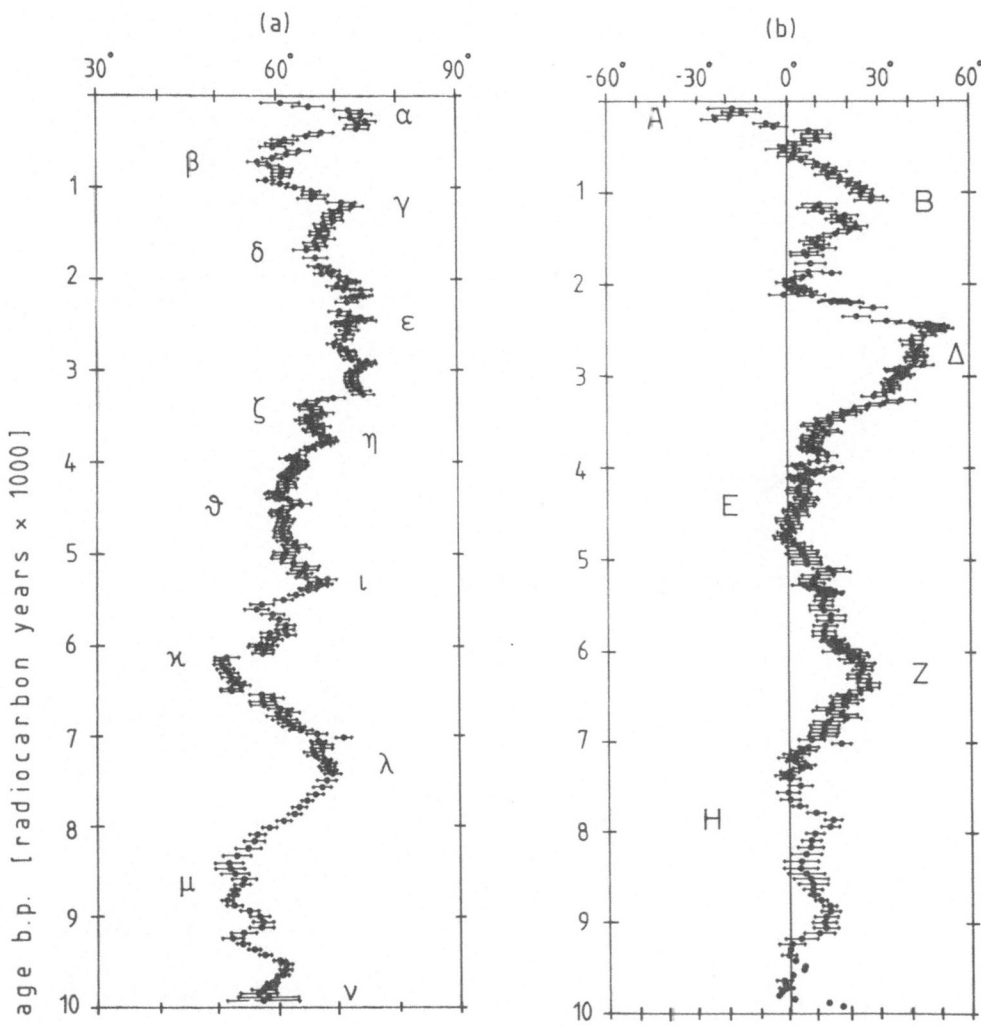

Fig. 2. UK type curves (a) inclination (b) declination. Scales calibrated in ten degree intervals. Time scale in radiocarbon years bp (before present).

seemed to be almost regular at 2700 calendar years. [Note that all the palaeomagnetic records illustrated in this review are plotted on timescales calibrated in radiocarbon not calendar years.]

The following year, Creer *et al.* (1972) obtained inclination as well as declination records from another 6 m core. They took approximately cubic shaped sub-samples of side ~20 mm (which has become standard practice) and these were measured with a Digico fluxgate magnetometer. Although the declination record they obtained from this single core was not so good as Mackereth's composite record it served to date the core because the extreme easterly and westerly values of the two curves could easily be correlated. The corresponding inclination record (WMERE72) is shown in Figure 3a.

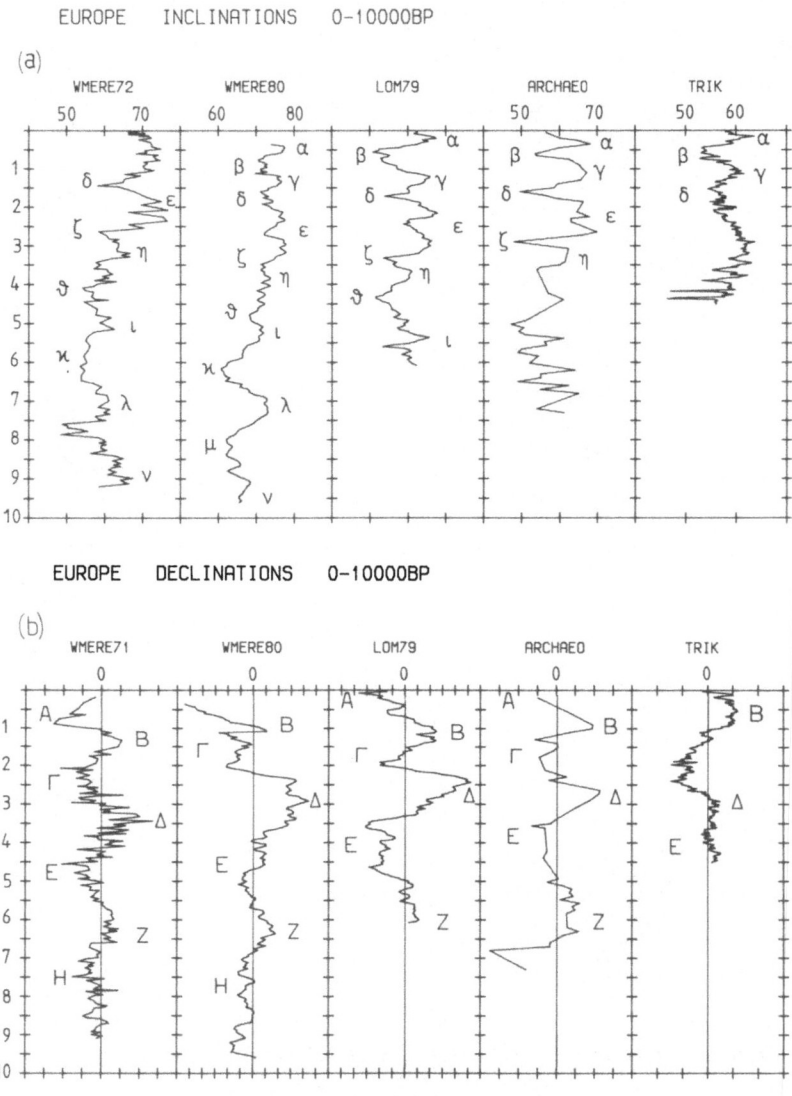

EUROPE   INCLINATIONS   0-10000BP

EUROPE   DECLINATIONS   0-10000BP

Fig. 3.   Holocene inclination (a) and declination (b) curves for several localities in Europe. WMERE 71 is from Mackereth (1971); WMERE 72 is from Creer *et al.* (1972); WMERE 80 and LOM 79 are detransformed stacked records from Thompson and Turner (1979); Turner and Thompson (1979, 1981); ARCHAEO is from Kovacheva (1980) and TRIK is from Creer *et al.* (1981). Features characterizing both curves are labelled with letters from the Greek alphabet using lower case for inclination and upper case for declination. Timescale in thousands of radiocarbon years before present.

This core did not penetrate through the Younger Dryas (Dryas III) finely laminated clays into the underlying organic material deposited during the period of climatic amelioration penetrated by Mackereth's (1971) cores.

2.1.3. *U.K. Lakes*

The next step, carried out during the late seventies, comprised a comprehensive study of three U.K. lakes, Windermere, Lomond and Gerionydd. Turner and Thomson (1979) reported results for four cores from Loch Lomond which were correlated using 16 characteristic features identified along the susceptibility profiles. The measured parameters were transformed to a timescale using ten radiocarbon age determinations made in two different laboratories. The resulting stacked curves which are shown in Figure 3 (LOM79), provide the most precise and detailed record presently available for the last 7000 yr. A preliminary account of these second generation results was presented by Thompson and Turner (1979) and subsequently a full account by Turner and Thompson (1981). The stacked curves for the three Windermere cores which were dated with three new radiocarbon ages and by correlation with the Lomond cores are shown in Figure 3 (WMERE80).

2.1.4. *S.E. Europe*

The most comprehensive archaeomagnetic data available are those compiled for S.E. Europe (area centred on 41.5°–42.5° N, 15.8°–27.8° E): data from Table I of Kovacheva (1980) are presented here in Figure 3 (ARCHAEO), in which the ages have been converted from calendar to radiocarbon years. Stacked inclination and declination SV curves from three cores from Lake Trikhonis, Greece (Creer *et al.*, 1981) are also presented in Figure 3 (TRIK).

2.1.5. *Comments*

Let us discuss first the declination curves which are illustrated in Figure 3b. In Mackereth's original Windermere record, the swings appeared to be of regular duration with an estimated period of ~ 2700 calendar years. The newer records for both Lake Windermere (WMERE80) and Loch Lomond (LOM79) exhibit the same major characteristic fearures as WMERE71. However extra detail, repeatable between cores, is clearly visible. But, most important, the new timescale destroyed the apparent regularity of duration of the declination cycles. The topmost westerly swing A has become much younger (~ 150 bp instead of ~ 750 bp); westerly swing Γ became younger by ~ 500 yr and easterly swing Δ became younger by ~ 1500 yr. The new ages of easterly swing Z and of westerly swing H have not been appreciably altered on the new timescale. The overall patterns of the newer declination records (especially LOM79) agree well with the archaeomagnetic record for S.E. Europe. Some important differences in detail between the stacked records will be noticed: westerly swing Γ is much more pointed as recorded along LOM79 than along WMERE80 or WMERE71 or along both S.E. Europe records, ARCHAEO and TRIK. Westerly swing Δ is 'V'-shaped in the LOM79 and ARCHAEO records while it is 'U'-shaped along WMERE80 and WMERE71.

Turning now to the corresponding inclination records (Figure 3a), we note that, with hindsight, all of the more important features, labelled α to ν, exhibited by the stacked

records WMERE80 and LOM79 can be identified along the single core record
WMERE72. The duration of the successive swings is clearly irregular and their
individual shapes differ in detail from one record to another. Let us first consider the
three U.K. curves on the left of Figure 3a. The Loch Lomond record shows the best
developed swings, with α, β, γ, and δ all of larger amplitude than in the Lake
Windermere record. Next, ε appears as a double maximum in WMERE80 and LOM79
rather than the single maximum shown by WMERE72 or by ARCHAEO. Variations
in patterns registered by different cores from the same lake are on a similar scale to the
variations under discussion here (refer to Figure 1 of Turner and Thompson, 1979). The
amplitudes of the Lomond records are typically larger than those of the Windermere
records. The agreement between the LOM79 and ARCHAEO records is again
strikingly good, considering the distance between the sites, though the former carries
more detail because of the higher sampling density. Another point that will be noted is
that the amplitudes of the oscillations registered by the lake sediment records are in
some cases noticibly smaller than those recorded by archaeological material: this is due
to the finite time ($\sim 200\,\mathrm{yr}$) taken for the magnetic signal to be fixed in lake sediments.
The bottom part of the ARCHAEO record below about 4500 bp carries no coherent
signal probably due to lack of dating precision, but in principle this curve could be
much improved by additional measurements. Finally, it would seem that the timescale
of the Trikhonis record needs to be rechecked in that the bottom of feature ε starts at
about 4500 bp which now appears to be too old.

Attempts to estimate rates of geomagnetic drift by comparing the ages of similarly
labelled inclination or declination swings, as recorded in U.K. and S.E. Europe, yield
unreliable results because the errors inherent in radiocarbon age determinations on
sediments are of the same order of magnitude as the age differences to be expected
between sites. If, for example, we were to suppose the observed swings were caused by
sources in the earth's core drifting westwards at the historically observed rate of about
$0.15°\,\mathrm{yr}^{-1}$ we would expect the U.K. records to lag in phase behind those for S.E.
Europe by only ~ 150 to $\sim 200\,\mathrm{yr}$ which is equal or less than experimental or systematic
errors associated with each radiocarbon age determination. For example, the directly
determined age of the top part of LOM79 is younger by $\sim 800\,\mathrm{yr}$ than the age of the top
part of WMERE80, determined directly on that core. This has been attributed to the
inwash of old soils into the post-elm decline Windermere sediments producing
anomalously old ages (Turner and Thompson, 1979).

2.1.6 *Other European Lakes*

Results from Poland (Creer *et al.*, 1979), Switzerland (Creer *et al.*, 1980; Thompson
and Kelts, 1974), Finland (Stober and Thompson, 1977), Lebanon (Creer and Kopper,
1976), France (Hogg, 1978; Bonifay *et al.*, 1984) and Iceland and Israel (Thompson,
unpublished) agree well with the U.K. results discussed above. However, in general,
neither the detail of pattern carried by these palaeomagnetic records, nor the dating
controls over them are of a sufficiently high quality to warrant incorporating them into

any broader based declination and inclination type-curves designed to supercede those illustrated in Figure 3 which are fully discussed in GBCRS.

2.2. LATE GLACIAL

2.2.1. *Scandinavian Excursions?*

Data for the Late Glacial of northern Europe are sparse and somewhat confusing. Several authors have reported large amplitude departures of measured palaeomagnetic directions from the present axial dipole field direction (ADF) and have claimed that these reflect either large amplitude geomagnetic SV or even aborted attempts at polarity reversal of the main field. These claims have been discounted by other workers who argue that they have been caused directly by the changing environment during the complex transition from the last glacial to the present interglacial period.

Short sequences of aberrant palaeomagnetic directions may be termed palaeomagnetic excursions and the crucial question is whether any of these can be shown to have a real geomagnetic origin. Some authors have set down minimal criteria to be satisfied before records exhibiting palaeomagnetic excursions (as defined here) should be accepted as records of sequences of real aberrant ancient geomagnetic field directions. These are: first, the records should be repeatable at at least two (preferably more) adjacent sites: second, the results should have been derived from homogeneous lithologies with fine grain size (less than ~0.06 mm) since these indicate a low energy depositional environment: third, the NRM directions should be stable to AF demagnetization in at least 100 oe peak field. Furthermore, to be acceptable as a record of a real geomagnetic excursion, any given palaeomagnetic excursion should be observed in at least two different depositional basins because water movements capable of producing large deflections of non-spherical magnetic carrier grains can be coherent at separate localities in the same basin.

One of the most widely discussed palaeomagnetic excursions is the Gothenburg excursion which was first reported by Mörner *et al.* (1971) and interpreted by them to represent aberrant behaviour of the geomagnetic field direction. Upward palaeomagnetic inclinations (Figure 4a – GOBURG) were measured in 5 samples of varved marine silty clay, assigned by Mörner (1977) to the Agard Interstadial (12 400–12 700 bp), sampled from the bottom of core B873, taken from the Botanical Gardens in Gothenburg. The corresponding declinations were essentially reversed (Figure 4b). Core B873 also revealed a narrow band of upward and shallow inclinations accompanied by almost reversed declinations (note that the lines connecting points with declinations near ±180° run right across the field of the plot in Figure 4b) in sediments of Older Dryas (Dryas II) age. Mörner refers to this particular feature as the 'Gothenburg Flip'.

Thompson (1976) and Thompson and Berglund (1976), have argued that Mörner's aberrant palaeomagnetic directions were caused by mechanical sedimentation processes such as slumping or weathering and they therefore maintained that these directions are not reliable indicators of the ancient geomagnetic field direction. To

Northern Europe Late Glacial INCS 9000–15000 bp

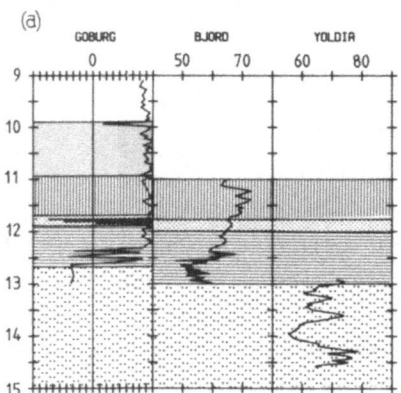

Northern Europe Late Glacial DECS 9000–15000 bp

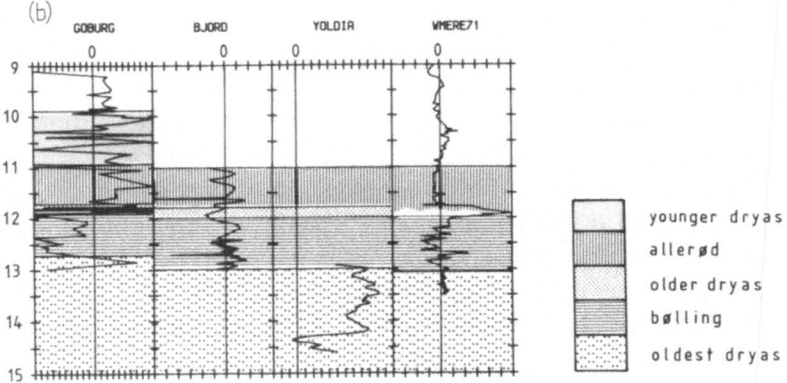

Fig. 4. Late glacial inclination (a) and declination (b) records for Scandinavia and UK, 9000–15000 bp. GOBURG = Core B893 from Botanic Gardens, Gothenburg (Morner *et al.* (1971), Morner (1977); BJORD = Bjorkerods Moss (Thompson and Berglund, 1976); YOLDIA = Younger Yoldia Clay (Abrahamsen and Readman, 1980), WMERE 71 = Lake Windermere (Mackereth, 1971). Biozones shown by hatching see key – ages are those given by the respective authors. Timescale in thousands of years before present.

substantiate their claim, these workers presented the results of a detailed study (408 samples) of two cores collected at Bjorkerods Mosse in southern Sweden (56.2° N, 52.6° E). They chose this site because its chronology is well known from both pollen and radiocarbon dating. Smooth curves drawn through the many individual data points from their Figure 1 are shown here in Figure 4 (BJORD) and it is clear that Morner's palaeomagnetic excursion is not reproduced. A few individual aberrant directions (not revealed by the smoothed curves of BJORD) were identified by Thompson and Berglund with lithological discontinuities or thin coarse grained layers.

However, Morner's is not the only report of aberrant Late Glacial palaeomagnetic directions. Abrahamsen and Readman (1980) measured a sequence of declinations directed along azimuths between $\sim 60°$ and $\sim 110°$ east of north in samples from a 12.5 m cliff section of Younger Yoldia Clay exposed in northern Jutland. The corresponding inclinations were perfectly normal, lying between $\sim 55°$ and $\sim 80°$, so they referred to their anomalous result as the 'Norre Lyngby Declination Excursion'. Age control was provided by a radiocarbon date of 14270 ± 180 bp for in-situ shells from a horizon 5–10 cm thick at 8 m elevation above sea level. The sampled section was estimated to span some 1000–1500 yr but it was inferred that the rate of deposition was probably not uniform, being about twice as rapid $(13 \, \text{mm yr}^{-1})$ below the shelly horizon as above it $(5 \, \text{mm yr}^{-1})$. The timescale attached to the palaeomagnetic logs shown in Figure 4 (YOLDIA) is based on the rather tenuous controls described by Abrahamsen and Readman (1980). On AF demagnetization, MDF's were found to be in the range 250–600 oe. Also there was a viscous component attributed to a small proportion of multidomain grains: its presence is apparent when the uncleaned and cleaned logs are compared (see Figure 9 of Abrahamsen and Readman's paper). The strongly oblique declinations however, are not associated with this viscous component, but rather with the hard stable component (after removal of the viscous component directional changes were no greater than $\sim 5°$). The path traced out in space by the succession of cleaned palaeomagnetic vectors is essentially linear. This suggests that if the excursion were of geomagnetic origin, it must have been caused by a standing, fluctuating source in the earth's core rather than by a drifting source (Creer, 1983). However it remains unproven whether the anomalous declinations were produced by the streaming effect of fast flowing water. Magnetic anisotropy measurements would be helpful in this respect.

2.2.2. *Lake Windermere*

The Late Glacial part of Mackereth's (1971) declination curve (WMERE71) is also shown in Figure 4a. This core penetrated through the Post Glacial gyttjas, through finely laminated clays of the Younger Dryas (Dryas III), then through highly compacted organic sediment which Mackereth assigned to the Allerod and Bolling chronozones and finally into coarsely laminated varved clay of Dryas I (Middle Weichselian) age. Mackereth (1971) estimated the age of the base of the Bolling sediments to be about 15 800 bp, but chronological tables published since then (e.g. Figure 2 of Lowe and Gray, 1980) date this horizon at 13 000 bp. The older Dryas (Dryas II) zone, situated between the Allerod (11 000–11 750 bp) and the Bolling (11 750–13 000 bp) zones in chronological tables, would appear to be absent. Mackereth assigned his highly compacted organic layer to the Allerod and Bolling zones combined, making no mention of Dryas II. Thus the following dates have been used here to define the timescale used for the Late Glacial part of Mackereth's (1971) cores: 10 000 bp = early Flandrian (Pre-boreal)/Younger Dryas (Dryas III) boundary, (ii) 11 000 bp = Younger Dryas/Allerod boundary and 13 000 bp = Bolling/Dryas I boundary.

2.2.3. *Lacs de Joux and du Bouchet*

Creer *er al.* (1980) described records for two cores from Lac de Joux (Switzerland – 46.6° N, 6.3° E) which penetrate into the Late Glacial. These cores were dated by equating the local biozones defined on the basis of pollen assemblages with the accepted biostratigraphy of the Jura Mountains illustrated in Figure 3 of the above paper, where it was noted (p. 142) that both pine and birch zones, ascribed classically to the Allerød/Older Dryas complex, are missing. This explains why the inclination and declination records for the interval 8000–15 000 bp shown in Figure 5 (JOUX3 and JOUX4) have a gap between 10 800 and 12 300 bp.

Figure 5 also illustrates (BOUCH5) a declination and inclination record for a single core (No. 5) from Lac du Bouchet (45.2° N, 3.8° E, France), where a comprehensive study of more than fifty cores (in press) is currently being undertaken (Bonifay *et al.*, 1984). The timescale attached to these records was established by defining chronozones on the basis of palynology and diatom analysis as follows: 10 300 bp = Pre-boreal/Recent Dryas boundary, 10 700 bp = Recent Dryas/Allerød-Bølling complex boundary, 13 000 bp = Bølling/Dryas boundary, and ~15 250 ± 290 bp = growth of Artemesia curve (also radiocarbon date of ~15 800 ± 900 bp marking the Dryas I/Würm boundary. Note that the ages used here for the boundaries between these bio-chronozones differ slightly from those used in the Lac de Joux study as will be evident from an inspection of Figure 5.

2.2.4. *The Black Sea*

Records for core 1474 from the Black Sea (Creer 1974) are shown in Figure 5 (BLSEA). The timescale was constructed using the seven radiocarbon dates given in the original paper (p. 35). Since orientation was not retained between the successive sections into which this piston core was cut immediately after collection, on board ship, the declination record shown here had to be constructed by subjectively matching the trends of the measured values across adjacent section ends (these are located at 12 200, 14 800, 16 700, and 20 400 bp) and therefore has to be viewed with some caution.

2.2.5. *Aegean Sea*

Opdyke *et al.* (1972) described measurements of inclination along core V10–58 collected from a depth of 2283 m in the Aegean Sea (35.7° N, 26.3° E). Three dating horizons were used. First, a layer of sapropelic mud, occurring in this core between 1.82 and 2.26 m, was correlated with the upper of three such layers found in other eastern Mediterranean cores. Three radiocarbon dates had been obtained for this upper layer sampled in another core (V10–64) viz. 7400 ± 200 bp, 8400 ± 200bp and 8700 ± 1000 bp but the sapropelic mud from core V10–58 was not dated directly. Second, the upper of three volcanic ash layers found in core V10–58 was correlated with a tephra overlying Late Minoan I sediments on Thera on the basis of physical and chemical properties of glass shards. This tephra had already been dated at 3370 ± 100 bp and 3527 ± 44 bp, but again, no direct age measurements were made on

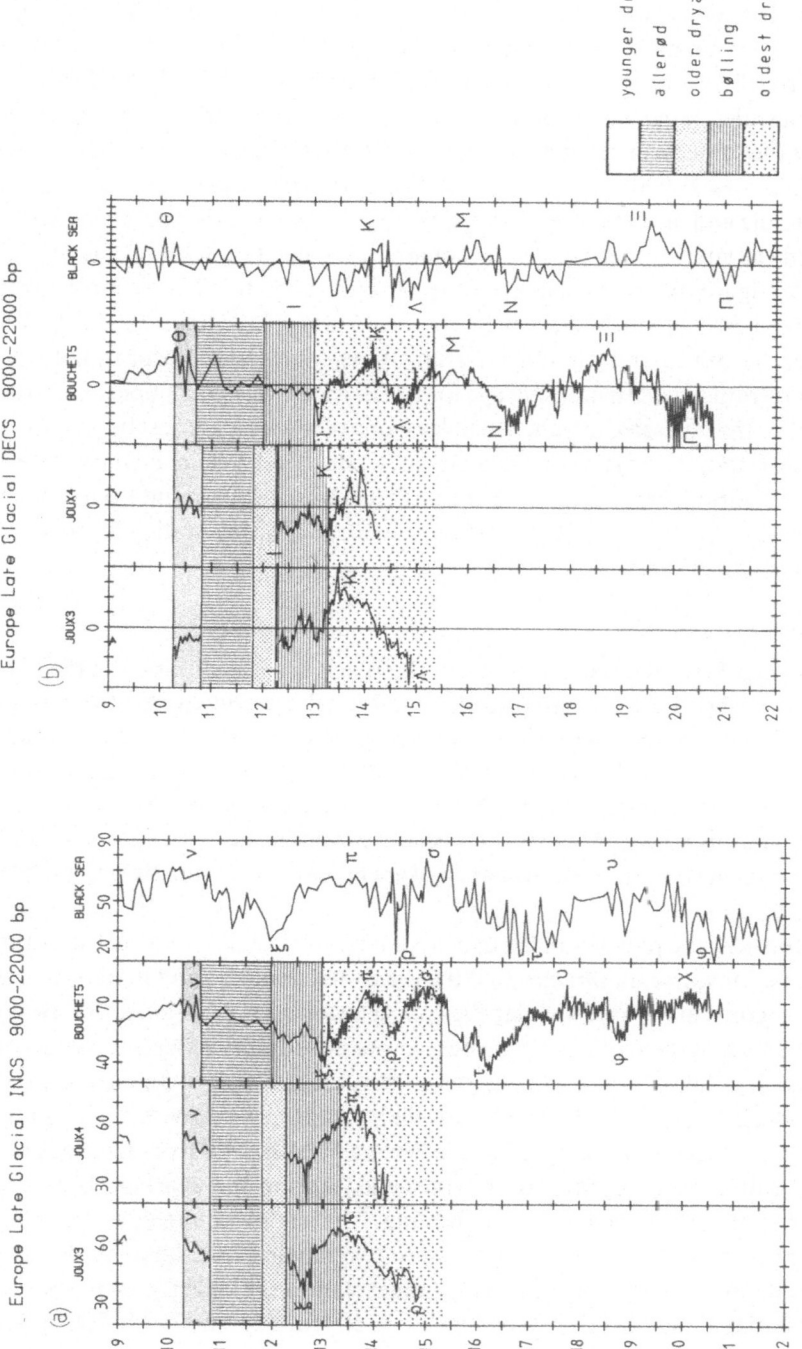

Fig. 5. Late glacial (9000–22 000 bp) inclination (a) and declination (b) curves for four sites in Europe. JOUX 3 and JOUX 4 are individual cores from Lac de Joux (Creer *et al.*, 1980), BOUCHET 5 is an individual core (patterns substantiated by many other cores) from Lac du Bouchet (Bonifay *et al.*, 1984) and BLACK SEA is from core 1474 (Creer, 1974). Biozones indicated by hatching – the ages shown are those given by the respective authors. Features characterizing both curves are labelled with letters from the Greek alphabet. The timescale is in thousands of years before present.

this ash layer from core V10–58. Third, the lower ash layer was correlated with an ash layer dated at 23 000 ± 800 bp in yet another core (V10–67) and also with an ash layer found on Lipari island which had been dated with a piece of wood at 24 000 ± 2000 bp. Again no direct measurements were made on core V10–58.

Core site V10–58 was located only some 1300 km from the site of Black Sea core 1474. The timescales attached to each of these cores suggest that they represent approximately the same span of time viz. ~0 to ~26 000 bp in the case of the Aegean core and ~3000 – ~24 000 bp for the Black Sea core. Both exhibit a sequence of inclination maxima and minima: the former are rather rounded in shape and convex to high values and the latter rather pointed, in some cases, almost cusp like in shape (e.g. see Creer 1974, Figure 8), yet the average time per cycle is quite different for the two cores, being about 6000 yr for the Aegean Sea core and about 2800 yr for the Black Sea core. Moreover, the overall trend is different along the two inclination records, showing a decrease in magnitude with increasing age for the Black Sea and the opposite for the Aegean Sea. Thus the timescale attached to at least one of these cores must be grossly in error and in this respect two facts should be noted: (i) the Black Sea inclination record agrees well with the Bouchet records over the period 9000–22 000 bp and (ii) no direct age determinations were made on Aegean Sea core V10–58. The inclination log for core V10–58 is not shown in this review.

2.2.6. *Comments*

In all logs shown in Figures 4 and 5, except for the Black Sea, the duration of the bio-chronozones from the Würm through to the Pre-boreal are shown. All the results obtained from sites outside northern Europe reveal oscillations of moderate amplitude which can reasonably be attributed to ordinary secular variations of the geomagnetic field. No excursions are recorded – refer to Figure 5. Evidence of excursions is restricted to results from sites in northern Europe, especially Scandinavia, where the effects of glaciation in producing environmental changes were much more severe than in France and Switzerland.

Let us first consider the patterns of SV of inclination and declination revealed by the results for Lac de Joux, Lac du Bouchet and the Black Sea. Maxima and minima along the inclination records of Figure 5a are labelled using lower case Greek letters following on from v at the base of the Holocene type-curve. The maximum v extends down right through Dryas III, Allerød and Bølling zones in the Bouchet record and leads on to minimum ξ which occurs at the Dryas I/Bølling boundary. However, the same minimum ξ occurs in sediments assigned to the Bølling in both cores from Lac de Joux. The transition from v to ξ appears to be represented also in the Bjorkerods Mosse record where ξ occurs near the base of the Bølling. The age of feature ξ, as recorded along Black Sea core 1474, is interpolated between two radiocarbon dated horizons at 8690 ± 150 and 14 010 ± 210 bp and is younger (age ~12 200 bp) than the independently estimated ages for the other sites, ~12 500 for Joux and ~13 200 bp for Bouchet. Below minimum ξ, a maximum π and then another minimum ρ occur along all three records (from de Joux, Bouchet and the Black Sea). The Bouchet and Black

Sea records exhibit a further maximum σ, and then a minimum τ, though there is some discrepancy in the ages estimated for these (15 000 and 15 500 bp; 16 200 and 17 000 bp respectively, Bouchet ages given first). Below this, any correlation between the two records becomes unconvincing.

Extreme westerly and easterly declination values are labelled in upper case letters of the Greek alphabet from H (following on from the curves in Figure 3b) to Π at the bottom of the Bouchet and Black Sea records. The easterly double maximum labelled K runs across the Bolling/Dryas I boundary in the Joux records, but lies entirely within the Dryas I in the Bouchet record, confirming the discrepancy in the two adopted timescales suggested by the non alignment of inclination features ξ and π. Extrema Λ, M, N, Ξ, and Π appear both in the Bouchet and Black Sea records, their ages being, in all cases, slightly older along the latter record. The same age discrepancy is to be inferred from correlation of the corresponding inclination features.

Turning now to the results obtained from northern European sites, we note that no consistent pattern emerges. Mörner's 'flip' which occurs within Dryas II consists of a broad range of declination values, the most divergent of which are directed easterly, in common with the diverging easterly declinations which occur within the combined Allerød/Bølling zone of Mackereth's Lake Windermere core (Figure 4a – WMERE71). Unfortunately, no inclination measurements exist for this zone in L. Windermere so there is no check on the negative inclinations representing the 'flip'. However we should stress that no abnormal behaviour is recorded at Bjorkerods Mosse which covers the same span of time. The anomalous easterly declinations recorded by the Yoldia Clay are not reproduced in any of the records illustrated in Figure 5a either at the expected age or at any other age.

3. North American Data

The geographical locations of all the sites discussed in this section are illustrated in the sketch map which comprises Figure 6. The type curves constructed by Creer and Tucholka (1982a) are shown in Figure 7 and will be further discussed in Section 3.1.3.

3.1. 0–14 000 BP

3.1.1. *Great Lakes*

In this section, the successive stages of the development of limnomagnetism to the point where type-curves depicting geomagnetic SV of inclination and declination (Figure 8) for the east-central part of North America could be constructed (Creer and Tucholka, 1982a). Fuller information is given in GBCRS.

The first investigations were carried out on sediments from the Great Lakes (Creer *et al.*, 1976a, b). Several other studies soon followed and while a fairly consistent pattern of directional variations, particularly of inclination, was obtained by the different research groups involved, problems arose in attempts to establish correlations between the different data sets. This was largely because radiocarbon did not provide an

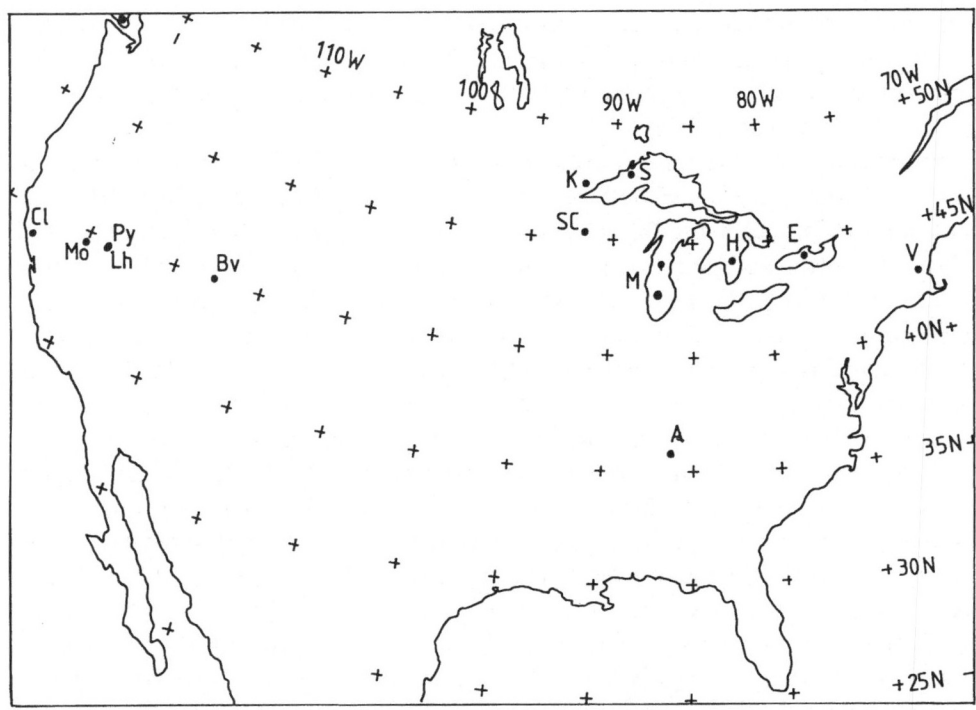

Fig. 6.   Site location map for North America. Key to labels as follows: - V = varve sites (McNish and Johnson, 1936); E = Lake Erie (Creer *et al*., 1976a); H = Lake Huron (Mothersill, 1981); M = Lake Michigan (Creer *et al*., 1976b; Vitorello and Van der Voo, 1977; Dodson *et al*., 1977); S = Lake Superior (Mothersill, 1979); K = Kylen Lake and SC = St. Croix Lake (Lund and Banerjee, 1979, 1983; Banerjee *et al*., 1979; Lund, 1981); A = Anderson's Pond (Lund, 1981; Lund and Banerjee, 1983); Bv = Lake Bonneville, Lh = Lake Lahontan (Liddicoat, 1976); Py = Pyramid Lake (Verosub *et al*., 1980); Mo = Mono Lake (Denham and Cox, 1971; Denham, 1974; Liddicoat, 1976; Liddicoat and Coe, 1979); Cl = Clear Lake (Verosub, 1977); Sp = speleothems (Latham *et al*., 1982); B = Bessette Creek (Turner *et al*., 1983).

effective or reliable method of dating these sediments which consist of silty clays with low organic carbon contents, typically less than 2%, often as little as 0.5%. Note that the gyttjas deposited through Post-Glacial time in most of the lakes studied in Europe are much richer in organic carbon ($\sim 5\%$).

Creer, Anderson and Lewis (1976a) reported results from cores of the bottom sediments of Lake Erie. These were collected using corers and ships belonging to the Canada Centre of Inland Waters (CCIW). Three types of corer were used, a Benthos corer (cores of ~ 1.5 m length) to sample the topmost, very watery sediments, an Alpine piston corer which penetrated through the Post Glacial sediments (cores of ~ 9 m length) and a Christensen corer to penetrate into the underlying glaciolacustrine clays and tills (down to 19 m). A rather loose dating control was provided by the palynologists: for example, the horizon of decline in pine and increase in oak near the bottom of the Post Glacial muds (accepted age ~ 8000 bp) could be identified, as could the horizon of the Nipissing Transgression (accepted age $\sim 5750 \pm 180$ BP, i.e. 4750 bp),

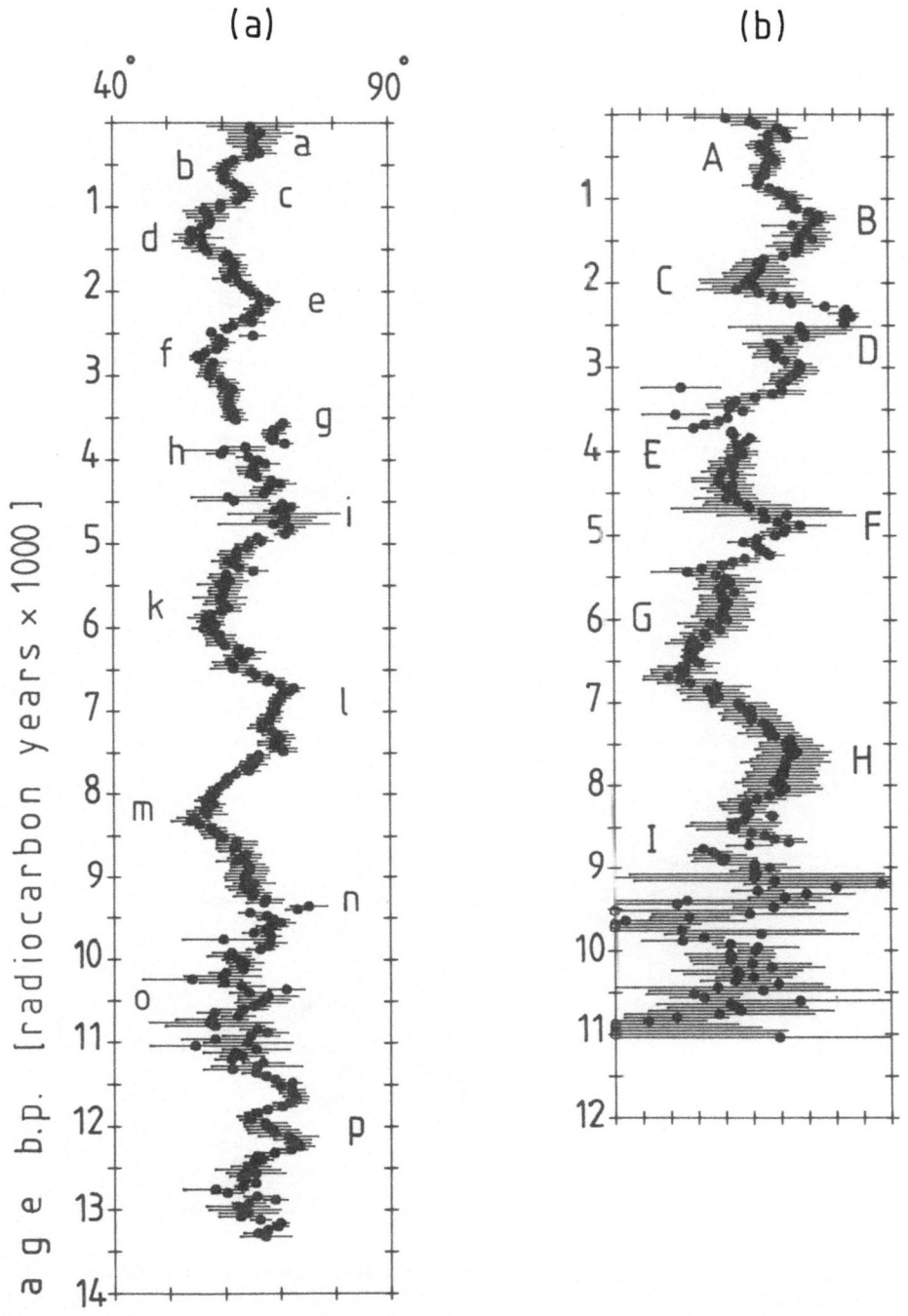

Fig. 7. Type curves for N. America (after Creer and Tucholka, 1982a). Characteristic features of inclination (a) and declination (b) curves labelled with letters from the Roman alphabet.

North America INCS 0-14000 BP

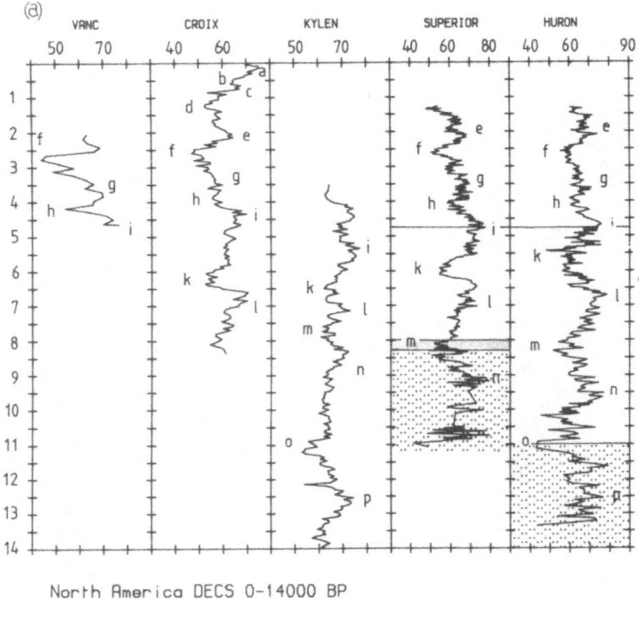

North America DECS 0-14000 BP

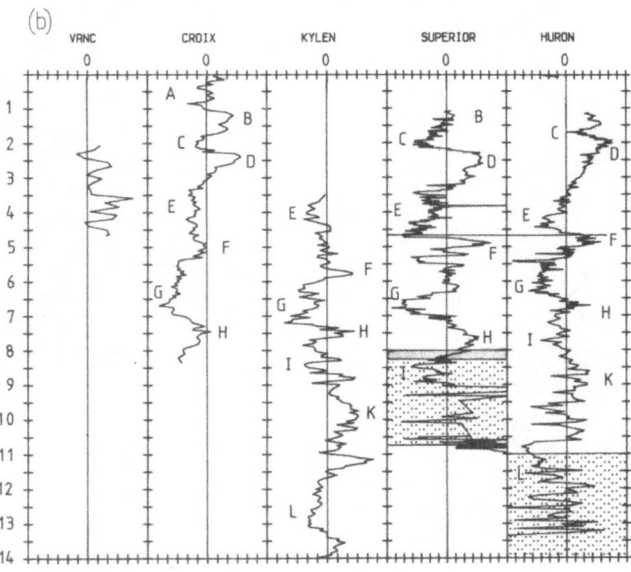

Fig. 8. Inclination (a) and declination (b) curves for several N. American sites, 0–14 000 b.p. VANC = speleothems from Vancouver Island (Latham *et al.*, 1982); CROIX = Lake St. Croix and KYLEN = Kylen Lake, (Lund and Banerjee, 1979; Banerjee *et al.*, 1979); SUPERIOR = Lake Superior (Mothersill, 1979) and HURON = Lake Huron (Mothersill, 1981). Characteristic features labelled with letters from the Roman alphabet, lower case for inclination and upper case for declination. The following additional information is given on the logs for Lakes Superior and Huron (see text): – (i) the horizon of the Nipissing Transgression (~ 4500 bp); (ii) the base of the Post-Glacial Sediments (~ 8000 bp for L. Superior and ~ 11 000 bp for L Huron); (iii) the horizons of late glacial varved clays occurring in L. Superior (just older than 8000 bp) and (iv) the unvarved glacial clays ('dotted' shading).

which occurred when the rate of sedimentation increased due to the transfer of the Nipissing drainage from the North Bay to the Sarnia outlets (note BP and bp indicate respectively calandar and radiocarbon years before present). In this study, as with all the other Great Lakes studies which followed, the quality of the declination records was adversely affected due to the corers twisting about their longitudinal axis on penetrating the sediment, sometimes through a complete revolution or more. The tangential stresses thus exerted on the outside skin of sediment caused horizontal cracks to develop across the whole core diameter. Thereafter, torsional relaxation produced differential rotation between the ends of adjacent sections of core on either side of each crack. Since it is usually difficult, and sometimes impossible, to estimate the magnitude of each increment of differential rotation, errors are inevitably introduced when declination curves for individual core sections are aligned to a common azimuth. These unwanted effects were alleviated on board ship by always allowing the cable time to untwist each time before the corer was dropped to take a core. This same technique was later used by Mothersill (1979, 1981) in coring Lakes Superior and Huron, also using a CCIW ship and corers.

As a result of this first study on Lake Erie, it was found that declination oscillated about true north through amplitudes of some 30° east and west of north, taking on average about 2000 yr to complete a cycle. Independent age control (using radiocarbon, palynology or stratigraphy etc.) was insufficient to prove whether or not the cycles were regular. Nevertheless the 'magnetic' age (estimated on the assumption of regular cyclicity) for the horizon of the Nipissing transgression was in satisfactory agreement with its accepted age determined directly at type localities by radiocarbon dating.

There followed three independent studies on the bottom sediments of Lake Michigan. Creer et al. (1976b) observed oscillations in declination and deduced an average duration of slightly more than 2000 yr per cycle in agreement with the Lake Erie result. Radiocarbon dating again proved ineffective. Subsequently, Creer and Tucholka (1982a) correlated these Lake Michigan data with results for Lakes Superior and Huron (Mothersill, 1979; 81) so defining an improved timescale for them. Vitorello and van der Voo (1977) reported results from two cores and their records were relatively compressed in the younger parts (above the Wilmette Bed) than Creer, Gross and Lineback's and more spread out from the Wilmette Bed down to the Wadsworth Till (age > 13 000 bp). So these two results complemented one another well. Dodson, Fuller and Kean's (1977) results are discussed in Section 3.3.

Next, Mothersill (1979) studied three cores taken with Alpine gravity piston corers from Thunder Bay in Lake Superior. He was able only to attach a very tenuous timescale since radiocarbon dates were demonstrably affected by 'old' carbon: for example modern sediment from near the top yielded an age of 1700 yr. The horizon recording the Nipissing transgression, which occurred when the North Bay controlled the level of Lake Superior – accepted age ~4750 bp, could be identified. Also the boundary between the Post Glacial olive, silty, non-calcareous clays and the underlying (~30 cm) glacial varved and then unvarved calcareous clays could be placed at about 8000 bp, possibly slightly older. There is some fairly loose stratigraphic control over the bottom parts of the Lake Superior cores in that the varves are known to have been deposited between ~11 000 bp when ice began to retreat from the lake and ~9000 bp when ice no longer existed in the vicinity of the lake.

Results are shown in Figure 8 for one of Mothersill's cores (LU–77–3). The timescale used is that described by Creer and Tucholka (1982a). In the Post Glacial sediments, inclination maxima and minima from 'e' to 'm' (Figure 8a) are identified (note that the top-most sediments appear to be missing) and a corresponding sequence of declination swings labelled 'B' to 'I' (Figure 8b) can also be identified. Inclination maximum 'n' is recorded in the underlying varves. The corresponding declinations, though rather scattered, possibly reveal easterly maximum 'K'.

Later, Mothersill (1981) studied two cores from Lake Huron. These two cores were correlated using seven datum horizons observed along the NRM intensity (J) logs, four susceptibility (k) horizons and seven Q-ratio (J/k) horizons. Again, radiocarbon dates on the sediments themselves were of little practical use and the cores were tenuously dated by recognizing a diastem and seven 'dry' zones within a depth range of about 80 cm at around 11 m depth. These were interpreted as having been produced when the North Bay outlet opened to the Champlain Sea, an event dated in the literature (Prest 1970) at about 9700 bp. This event resulted in a maximum northern regression of the southern shoreline of Lake Stanley. Ice began to retreat from the Huron Basin at about 12 900 bp (earlier than from the Superior Basin) and remained in the vicinity until ~11 800 bp. Identification of the glacial outwash deposits thus provides a rather tenuous dating control over the lower parts of the Huron cores.

The results for Mothersill's core LU–79–2 are shown in Figure 8, the timescale being the improved one of Creer and Tucholka (1982a). Inclination features 'e' to 'o' and declination features 'D' to 'K' are recorded in the Post Glacial sediments. A further inclination maximum 'p' and a further declination swing 'L' are recorded in the glacial outwash deposits.

Mothersill's Lake Superior and Lake Huron cores had been twisted as evidenced by his declination logs (Figures 2 and 3, Mothersill 1979 and 1981 respectively), even though precautions had been taken to reduce this effect during the coring operation. Creer and Tucholka (1982a) attempted to untwist the declination logs by fitting one knot cubic spline curves to each data set, and then subtracting individual measured values so allowing the recorded oscillations to be plotted on the same as the other declination curves of Figure 8b.

3.1.2. *Minnesota Lakes*

Sediment cores from the Minnesota Lakes St. Croix and Kylen were studied by Lund and Banerjee (1979); Banerjee, Lund and Levi, (1979) and Lund (1981). The sediments in both these lakes are rich in organic carbonaceous material $(10\% \pm 3\%)$ and consequently could be reliably dated by radiocarbon. Two 19 m cores were taken from Lake St. Croix and three 4 m cores from Lake Kylen in 1 m sections, using a modified Livingston piston corer (Livingston, 1955). Each core provided an independent inclination record but the declination record for each lake had to be built up by cross correlation of the two or three individual core records because of loss of orientation between adjacent 1 m sections. The timescales were based respectively on nine radiocarbon age determinations made on the St. Croix cores and four of the Kylen cores. In both lakes, deduced rates of deposition are inferred to have increased slowly with passage of time; from ~ 1 mm yr^{-1} at 9000 bp to ~ 4 mm yr^{-1} at ~ 500 bp at St. Croix and from ~ 0.3 mm yr^{-1} at $\sim 16\,000$ bp to ~ 0.8 mm yr^{-1} at ~ 8000 bp at Kylen.

The within-lake agreement between the individual core records was good and, as can be seen in Figure 8, the records for the two lakes agree reasonably well for the time interval of overlap (~ 4000–8000 bp). The St. Croix record exhibits inclination features 'a' to 'l' and declination features 'A' to 'H'. Features 'i' to 'l' and 'F' to 'H' can be identified also along the Kylen records where they appear to be slightly older due to the independently established timescale. The Kylen Lake records extend to 14 000 bp as far as inclination minimum 'p' and easterly declination swing 'L'.

3.1.3. Construction of Type-Curves

The results from the Great Lakes cores complement those from the Minnesota lakes in that while the former yield a larger amplitude and better formed pattern than the latter, the associated timescale is relatively poor. Nevertheless, age control over the Great Lakes cores is sufficiently good to allow the more pronounced features characterizing the recorded cyclic pattern of inclination SV to be correlated uniquely with corresponding features (labelled 'a' to 'n' through the Holocene) as recorded along the Minnesota cores. Creer and Tucholka (1982a) adjusted the timescales for the Great Lakes cores to the Minnesota lakes' timescale using these criteria (see Figure 7).

The agreement between the declination patterns was noticibly less good than that between the inclination patterns though it may be considered satisfactory bearing in mind the scale of the corrections that had to be made to the Great Lakes declination records to straighten them and the additional subjective corrections made to correct for discontinuities caused by differential rotation at horizons where the cores had been sheared. Similar, though smaller adjustments were also made at the joins between the Minnesota core sections by the workers concerned. In both studies the availability of several cores from each lake provided, at most joins, an invaluable check on each individual correction because the discontinuities in the declination records occurred at different levels in different cores.

3.1.4. Speleothems

Recently, methods and techniques have been developed to obtain dated SV records from speleothems (stalagtites and stalagmites). These objects are derived from precipitation of calcium carbonate from groundwater which has seeped into limestone or dolomite caves. Their rate of growth depends on the ambient temperature because this controls the readiness of bicarbonate groundwaters to dissolve and reprecipitate calcium carbonate. Speleothems have been observed to show signs of rythmic growth and increased deposition during warmer interglacial times. And, specifically, they have been found to carry a weak remanent magnetization due to the incorporation of detrital impurities including magnetite. Each layer can be dated, using the 230 Th–234U method, if it contains sufficient uranium and was formed free of thorium provided the age is less than about 350 000 yr. Thus, this method is potentially capable of providing 'windows' of geomagnetic SV through late Pleistocene time well beyond the range of radiocarbon dating. But, as yet, it is still in its infancy and the only data which can be compared with classical limnomagnetic data are those of Latham *et al.* (1982) obtained from a stalagmite from Vancouver Island for the interval 5400 to 2100 bp (Figure 8–VANC). The agreement in inclination is remarkably good, features '*f*', '*g*', and '*h*' being detectable at the expected ages but the declination results do not correlate with the corresponding lake sediment curves.

3.2. 14 000–31 000 BP

3.2.1. *Secular Variation Records*

In Section 3.1 we discussed palaeomagnetic records carried by Great Lakes sediments (Figure 8) which extend to $\sim$11 000 bp in Lake Superior (as far as the top of inclination minimum '*o*') and to $\sim$13 000 bp in Lake Huron (as far as the base of inclination maximum '*p*'). Also, we have seen that the Kylen Lake record extends back to almost 15 000 bp (to top of inclination minimum '*t*').

Beyond this only two reliable data sets have been obtained. The first of these runs from $\sim$13 000 bp to $\sim$19 000 bp and was derived from two 7 m cores taken from Anderson's Pond, Tennessee (Lund, 1981; Lund and Banerjee, 1983). Inclination decreases (Figure 9a) from about 65° at $\sim$13 500 bp (feature '*p*') down to about 30° at 16 000 bp (feature '*r*'). This part of the Anderson's Pond record correlates well with the bottom part of the Kylen Lake record where the inclination decreases from about 75° at 12 500 bp (maximum '*p*'), through minimum '*r*' and maximum '*s*' towards a further more pronounced minimum '*t*' at $\sim$14 700 bp. The timescale for the Kylen Lake record was constructed by linear regression through the four measured radiocarbon ages. In fact, the lowest of these ages was 16 500 $\pm$ 1500 bp which occurs at 14 000 bp on the timescale of Lund and Banerjee (1983) at the level of the small maximum '*s*'. A very similar maximum '*s*' occurs at $\sim$15 000 bp in the Anderson's Pond record rather than at $\sim$14 000 bp as along the Kylen record, serving to illustrate the importance of being able to compare timescales attached to different data sets: systematic errors can only be checked by such comparisons. In the Anderson's Pond record below the pronounced minimum '*t*' there is a sharp increase in inclination to values of around 65° (labelled '*u*') followed by a decrease to about 50° (at '*v*'). Declination exhibits about 5 minor maxima and minima of magnitude $\sim$25° peak to peak and with duration $\sim$1000 yr. These are superimposed on a steady trend from $\sim$15° W at $\sim$13 000 bp to $\sim$25° E at $\sim$19 000 bp from westerly extreme '*L*' also indentified near the base of the Kylen Lake record to easterly extreme '*O*' at 19 500 bp. Smaller amplitude extrema '*M*' and '*N*' between '*L*' and '*O*' are also labelled.

The oldest reliable data set yet obtained in N. America runs from $\sim$19 500 bp to $\sim$31 000 bp. It was obtained from a 20 m surface exposure of fine-grained organic muds

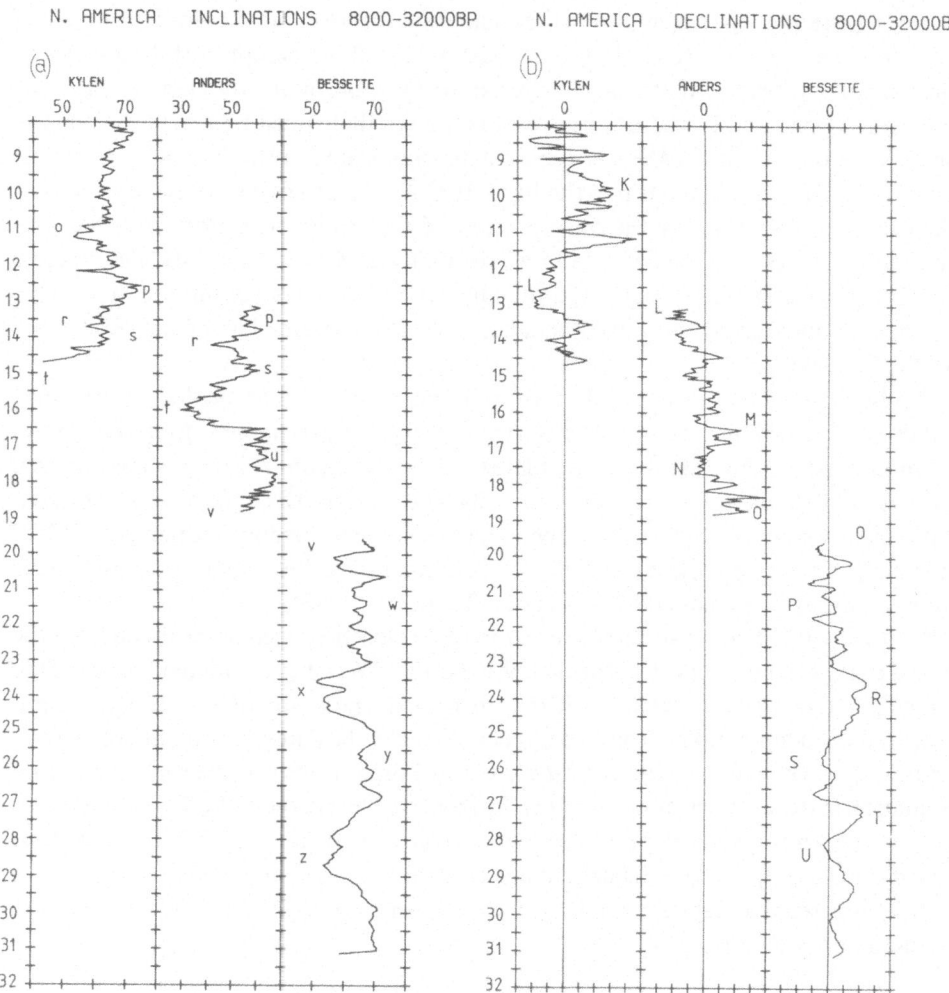

Fig. 9.   Inclination (a) and declination (b) curves for three sites from North America, 8000–31 000 bp. KYLEN = Kylen Lake and ANDERS = Anderson's Pond (Banerjee *et al.*, 1979; Lund and Banerjee, 1979, 1983; Lund 1981); BESSETTE = Bessette Creek, British Columbia (Turner *et al.*, 1983). Timescale in thousands of years before present.

and silts from Bessette Creek, a stream-cut escarpment located in British Columbia at (50° N, 118.8° W). Transformation to a timescale was achieved by constructing the best fitting quadratic using five horizons at which radiocarbon age determinations had been made (Turner *et al.*, 1983). 212 subsamples were taken, representing average time increments of about 50 yrs. On this basis, it was concluded that the sedimentation rate increased from ~ 0.85 mm yr^{-1} at $\sim 31\,000$ bp to ~ 2.8 mm yr^{-1} at $\sim 19\,500$ bp.

Unfortunately the Bessette Creek and Anderson's Pond records do not overlap, but it is noted that the slow easterly trend in declination (going back in time) recorded at Anderson's Pond appears to be continued back to $\sim 23\,000$ bp in the Bessette Creek record (Figure 9b). The upper part of the inclination record exhibits some high values of ~65° followed by a short minimum at ~20 000 bp. If we accept both timescales as approximately correct, there would appear to be little or no overlap between the Anderson's Pond and Bessette Creek records though there must be some possibility of this because the two timescales were constructed independently and each one is subject to error. The mimum at the top of the Bessette Creek record could be the same as that labelled 'v' just starting at the very bottom of the Anderson's Pond record. This minimum is preceded in the Bessette Creek record by three broad maxima. The first two are labelled 'w' and 'y'. They are separated by two rather narrower minima 'x' and 'z'. Declination undergoes about two and a half cycles the extrema being labelled 'O', 'P', 'R', 'S', 'T', and 'U'.

The Bessette Creek logs record an overall bias of some 3° to shallow inclinations (relative to the ADF inclination of 67.3°), caused largely by the large amplitudes (25°) of the minima which separate the broad plateau like maxima on which are superimposed undulations of some 10° amplitude with ~500 yr duration. Declination is also biassed to the east by some 6°. Both these phenomena were mentioned by Turner et al. (1983) and must, at least in part, be due to the use of unit vectors rather than total vectors in the statistical analyses (Creer, 1983; Creer and Tucholka, 1983).

Palmer et al. (1979) reported results from three closely spaced cores taken from the central part of Lake Tahoe, California (39.1° N, 120° W). They estimated the age of the lowest parts of these cores at ~30 000 bp by extrapolation of radiocarbon dates measured on samples taken from the upper sections of the cores. A comparison of their Figure 7 (p. 132) with the records illustrated in Figure 9 of this review suggests that inclination features 'o' through to 'z' can be identified along the Lake Tahoe cores and also, at the appropriate depths, declination features 'L' through to 'U'. It is thus important that improved independent age control be obtained for these cores.

Other less reliable data are available for N. America going back to 40 000 bp: these are discussed in Section 3.4.

3.3. EXCURSIONS

There have been a number of reports in the published literature of palaeomagnetic excursions recorded in Late Glacial and even in Post Glacial sedimentary successions. Some authors have argued that these represent aberrant geomagnetic behaviour and it is therefore necessary that some discussion be given of these claims.

(a) Dodson, Fuller and Kean (1977) described records from Lake Michigan which extend back to about 13 500 bp. They reported two 'palaeomagnetic' excursions which they maintained could represent excursions of the geomagnetic field. The first of these was recorded in about 25 cm of cored sediment from the Sheboygan Member of the Lake Michigan Formation. It comprises a band of westerly directed declinations extending to reversed values (~180°). These are associated with negative inclinations of up to −60°. However, Creer et al. (1976b) observed no large amplitude oscillations of either declination nor inclination which could possibly be described as anything other than normal SV in any of their cores which passed through the Sheboygan

Member. Suspiciously, this 'palaeomagnetic' excursion occurs near the bottom of a core and it is not substantiated by measurements in parallel cores. A second palaeomagnetic excursion, recorded in glacial till underlying the Carmi Member of the Equality Formation (age ~ 13 500 bp), was also described by these authors. It comprises reversed declinations but none of the associated inclinations are shallower than about +40°. Again any interpretation of these anomalous directions in terms of aberrant geomagnetic behaviour must be regarded as suspect because they occur near the bottom of the core and only in that particular core.

It is perhaps pertinent to note that Dodson.et al. (1977) carried out their palaeomagnetic measurements on unopened cores. Therefore they would not have been able to identify discontinuities in the cored sediment inside the tubes containing them. For example, Mothersill (1981) noticed that a short segment of his core LU79–1 from L. Huron was physically separated by short columns of water above and below. He noted breaks in the declination curve which less cautious authors would have described as an excursion (refer to Figure 4 of Mothersill 1981).

(b) Creer *et al.* (1976a) described a thick sequence of shallow and inverted inclinations (~ −80°) recorded in over 8 m of glacial tills and glacio-lacustrine clays sampled by two cores, Nos. 13 193 and 13 194 (lengths 19 and 27 m respectively), taken in Lake Erie at sites separated by ~ 19 km (their Figure 6 illustrates). There is good agreement between the patterns recorded by the two cores but unfortunately no declination records could be recovered because the Christensen corer used sampled individual short unorientated sections. No sediment was recovered from between 19 and 26 m depth in the longer core, but between 26 and 27 m depth the inclinations were clearly seen to be normal (~ +60°). The rate of deposition of these glacial tills and clays is difficult to estimate: certainly deposition was not slow and continuous between the accepted age of the till/clay boundary (~ 14 000 bp) and the top of the glaciolacustrine clays (~ 10 500 bp). So the duration of the palaeomagnetic excursion cannot be estimated satisfactorily. Since this excursion, which Creer *et al.* (1976a) called the Erieau excursion, is recorded at two sites located some 19 km apart, large scale slumping would have to be postulated to explain it away. Furthermore, since the underlying tills at the base of core 13 194 carry normal inclinations, it becomes necessary to devise some process capable of overturning a thick slab of material with linear dimensions some 100 times greater than its thickness without physically disturbing the structure of the sediment (because it carries a coherent palaeomagnetic signal) and without disturbing the underlying tills. For these reasons, Creer *et al.* (1976a) interpreted this palaeomagnetic excursion as evidence of abnormal geomagnetic field behaviour. However, if a geomagnetic field excursion really occurred at the time when the glaciolacustrine clays and glacial tills were being deposited in Lake Erie, we gave to explain why this excursion was not recorded by Late Glacial sediments deposited in Lakes Huron and Superior. If it really occurred, it must have been at a time when no deposition was taking place in either of these two lakes. The older of the two Lake Michigan excursions of Dodson *et al.* (1977) discussed above in (a) has a quite different signature and since it's origin is unlikely to be geomagnetic it should not be quoted to reinforce any argument in favour of a geomagnetic origin for the Erieau excursion.

(c) Finally, let us consider briefly the excursions which Clark and Kennett (1973) claim are recorded in the upper parts of 8 out of 15 cores taken from the Gulf of Mexico. Sedimentation rates were estimated to have been between 0.1 and 0.2 mm yr^{-1}, i.e. an order of magnitude slower than most lake sediments. They estimated the age of this excursion to be between $\sim$12 500 and $\sim$17 000 bp on palaeontological grounds. However inspection of their plots (see their Figure 2) reveals that excursion occurs in only two or three cores, not 8 as claimed. In core K67 the excursion comprises an 0.7 m band of inclinations which decrease through zero, attaining upward values of $-40°$ together with a broader ($\sim$2 m) band of westerly directed ($\sim$90°) declinations with values becoming directed southerly at the levels of the inverted inclinations. Weaker excursions are also registered along cores K8 and K136 but the pattern is not consistent. There is no evidence that the anomalous palaeomagnetic directions reported were not been produced by the action of natural (e.g. slumping) or technical (e.g.action of corer) forces. Thus the evidence they present is not convincing.

3.4. DRY LAKES OF THE BASIN AND RANGE

3.4.1. *Mono Lake Excursion*

About 300 m of sediment have been deposited in the pluvial lakes of the Mono Basin during the last 170 000 yr. Deposits that were formed during the last major high stand between 11 000 and 30 000 bp occur in stream cuts on the margin of the present lake. Denham and Cox (1971) sampled the Wilson Creek Formation from its type locality on the north shore of Mono Lake (38° N, 119° W). It is a glacially derived silt with flat-lying varve-like laminae interbedded with flat-lying ash layers. Two ostracod horizons were radiocarbon dated (13 300 ± 500 bp and 23 300 ± 300 bp) and the above authors gave an extrapolated age for the bottom of the Wilson Creek section of 30 400 bp, having assumed a steady rate of deposition. Although the authors maintained that their results (MDFs $\sim$ 250 oe, stable end points reached) were of acceptable stability, the directions obtained were rather scattered as compared with those which comprise currently acceptable palaeomagnetic records as obtained from wet lake sediments (see Figure 10 – MONODC). This being so, with hindsight, it is clear that the sampling interval was too coarse for the sequence of measured directions to have revealed any hint of cyclicity, even if it had existed and furthermore the sediments themselves appear to be rather poor quality palaeomagnetic recorders. Nevertheless, the Mono Lake study became well known because of a large amplitude departure of declination from true north to about 120° E accompanied by abnormally high inclinations reaching nearly 90° (down) as compared with the overall average value of about 55° and the ADF value of 57°. This feature became known as the 'Mono Lake Excursion': its age was estimated to have run from 24 600 to 24 000 bp, during which time the NRM vector rotated in a counter-clockwise sense (with advancing time). Denham (1974) later observed the same phenomenon at another site which he correlated with the first site

NORTH AMERICA   DRY LAKES INCLINATIONS   13000–40000BP

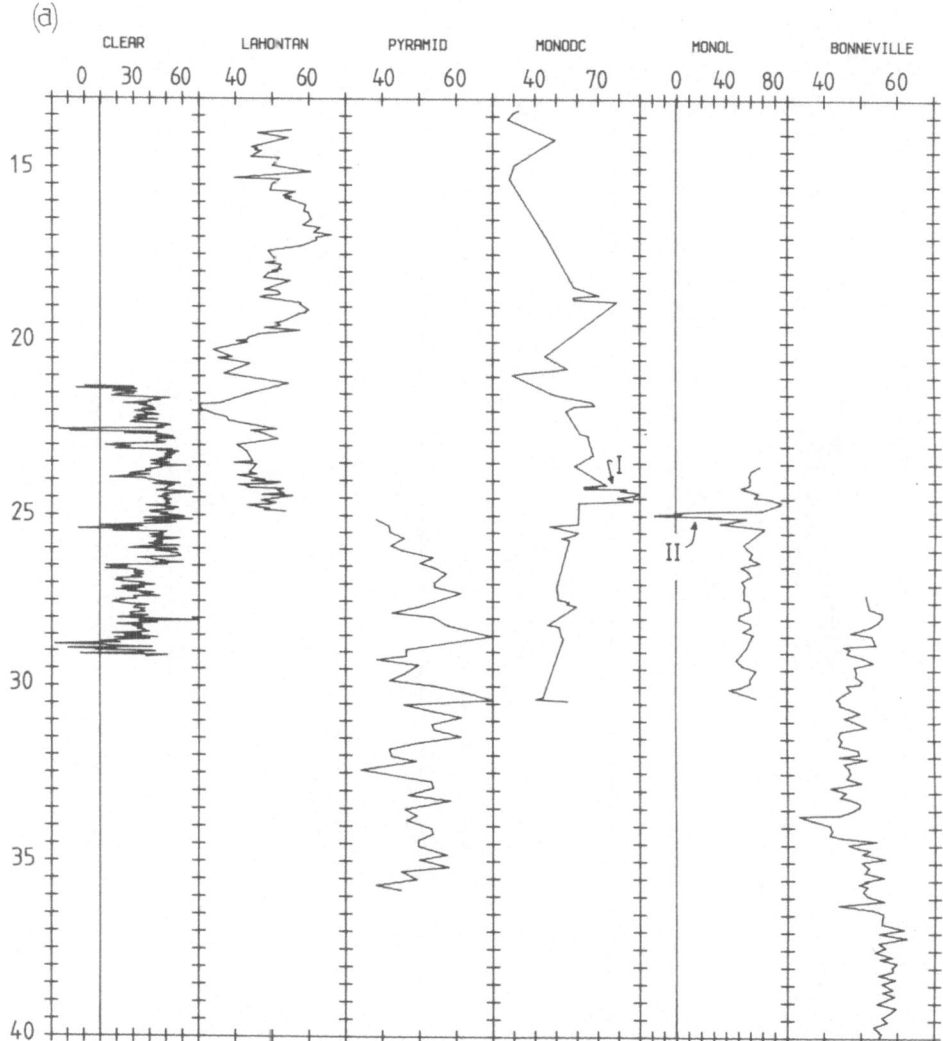

Fig. 10.   Inclination (a) and declination (b) curves for dry sediments from the plurial lakes of the Great Basin and Range Province, western U.S.A. CLEAR = Clear Lake (Verosub, 1977); LAHONTAN = Lake Lahontan (Liddicoat, 1976); PYRAMID = Lake Pyramid (Verosub *et al.*, 1980); MONODC and MONOL = Mono Lake (Denham and Cox 1971 and Liddicoat, 1976 respectively); BONNEVILLE = Lake Bonneville (Liddicoat, 1976). Timescale in thousands of years before present.

over a distance of 17 km using rhyolite and ash layers as marker horizons.

Later, Liddicoat (1976), carried out a further study (also reported in Liddicoat and Coe 1979), employing a higher sampling density at a third site, ~ 20 m north of the first site. Their results are also shown in Figure 10 (MONOL) and reveal not only the

NORTH AMERICA DRY LAKES DECLINATIONS 13000–40000BP

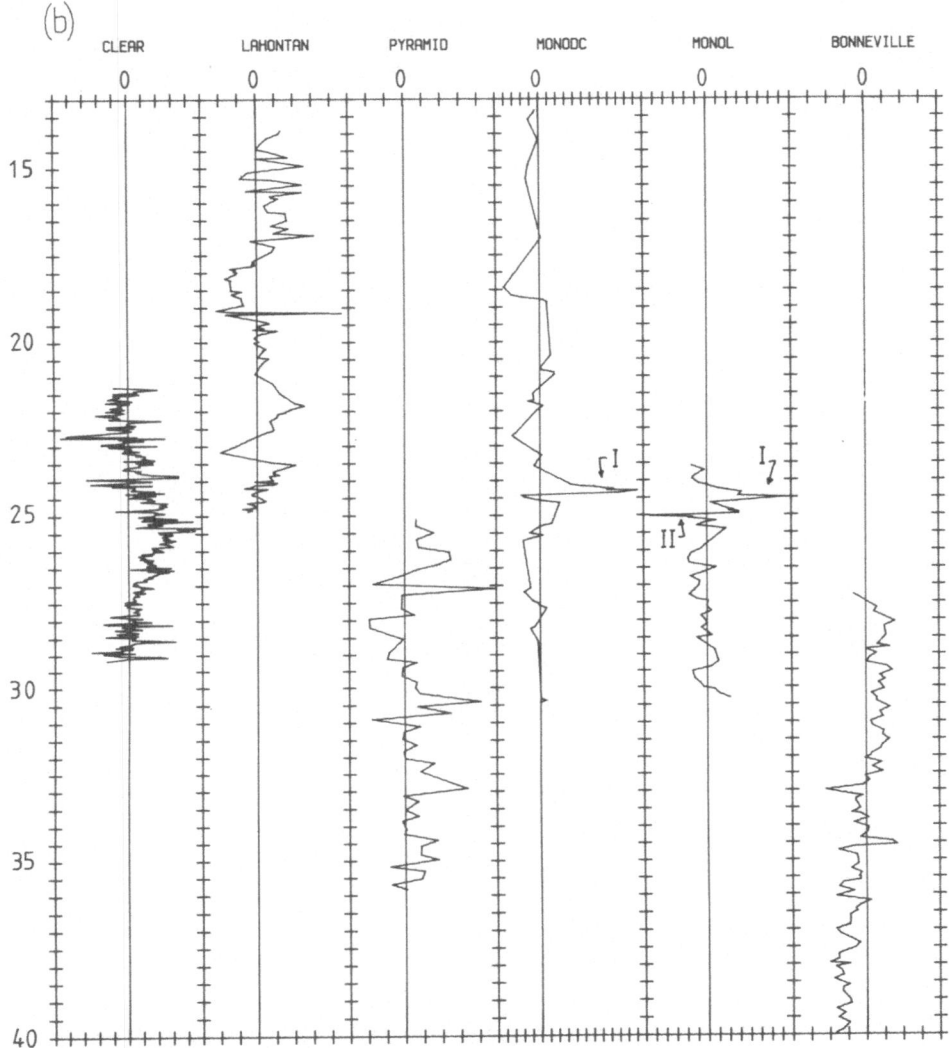

'excursion' (I) described in the last paragraph, but also an additional 'excursion' (II) comprising westerly declinations extending as far as ∼ 55° W recorded by two samples, and by a long tail of shallow inclinations leading to an upward value of − 20° recorded by half a dozen or more samples. It was deduced that these represented a span of time running from about 25 000 to 24 600 bp, again supposing a steady rate of deposition throughout the section. During this slightly older 'excursion' (II), the NRM vector rotated clockwise (witg advancing time). Prior to about 25 000 bp, the NRM vector recorded by the Wilson Creek Formation underwent small amplitude movements from the overall mean about which it rotated, forming alternately small clockwise and counter-clockwise loops.

Should these 'excursions' of NRM directions be accepted as records of actual geomagnetic field excursions as claimed by these authors? The absence of any abnormal behaviour along the record from Bessette Creek which, as we have seen in Section 3.2.1, exhibits oscillations in direction with amplitudes typical of those recorded by Holocene and Late Glacial sediments (Figures 3, 5, 8, and 9); would suggest not. Furthermore, since the Mono Lake records are all rather scattered and lack any coherent pattern, it would seem most probable that the 'excursions' in NRM directions, like those occurring along the Gothenburg cores, were produced by some sudden change in environment at the time when the sediments were magnetized rather than to abnormal geomagnetic behaviour.

In fact, the Mono 'excursion' is not observed at any other site in the Basin and Range Province where palaeomagnetic studies have been carried out. Since individual oscillations (as labelled in Figures 3 and 8 respectively) observed in declination and inclination through Holocene time remain recognizable over distances of a few thousand kilometres in both Europe and North America, it would seem unreasonable to argue that the Mono Lake 'excursion' could have been restricted to a particular geographical area much smaller than these in order to be able to explain away its absence at Bessette Creek and also at Fargher Lake ($\sim 46°$ N, 122.5° W) in Washington State (Doh and Steele, 1983).

3.4.2. Clear Lake

At Clear Lake, which is located some 320 km from Mono Lake, Verosub (1977) sampled a 5 m section of core at closely spaced intervals, each representing about 25 yr on the adopted timescale which was defined by a linear regression through 6 radiocarbon dates. The inclination curve (Figure 10a – CLEAR) will be seen to consist of several portions which are convex to high inclination values $\sim 60°$, with narrower minima extending to about $-20°$. The mean inclination of 37° is markedly lower than the ADF value of 58° and this low value is unlikely to be due to inclination error of the type described by King (1955) for two reasons. First, the NRM was most likely acquired by a bioturbation process and second, negative inclination values cannot be produced simply by settling of 'flat' grains. It is most likely that the pattern of magnetic directions has been strongly influenced by physical disturbance during coring: the low inclinations appear to occur at the ends of core sections. Declination (Figure 10b – CLEAR) undergoes one cycle only, changing from $\sim 30°$ W to 40° E and back again during the time represented by the core, estimated to run from 30 000 bp to 21 000 bp. Neither inclination nor declination patterns show any of the well defined features exhibited by the Bessette Creek records which cover the same span of time.

3.4.3 Lake Lahontan

The second record (from the left) illustrated in Figure 10 was obtained from lacustrine clays and silts of the Sehoo Formation. These were sampled in a 7.5 m section exposed in the Truckee Canyon about 50 km ENE of Reno, Nevada (118.5° W, 39.7° N). The timescale was constructed from a linear interpolation between ages estimated for the top and bottom of the section. The top was dated by two radiocarbon age determinations (12 910 ± 110 bp and 13 110 ± 140 bp) on a dendritic tufa overlying the sampled section. A stratigraphically equivalent tufa had previously been dated at 13 700 ± 300 bp. Another date (16 800 ± 500 bp) for a silty ostracod bed, possibly equivalent to another sandy bed rich in ostracods occurring in the sampled section at 3.2 m elevation was not used as a time control because of uncertainty about the correlation. There are no radiometric dates fot the base of the Sehoo formation at the sampling site, though stratigraphic work combined with radiometric dates available elsewhere, suggest a maximum age of about 25 000 bp. However, Liddicoat (1976) mentions that there is some possibility that the time represented by the Sehoo formation could be quite different, perhaps running from ~20 000 to ~100 000 bp.

On the basis of the timescale adopted by Liddicoat (1976), the average sedimentation rate was estimated to have been ~0.625 mm yr^{-1}. This gives an average sampling interval of 150–200 yr. The overall appearance of the patterns of the inclination and declination logs are similar to those observed for the equivalent Holocene and Late Glacial logs. Inclination varies between about 30° W and 30° E and the mean value is slightly biassed towards the east (2.8° E). The patterns of variations recorded at this Lahontan section do not correlate with those obtained from Kylen Lake, Anderson's Pond and Bessette Creek discussed in Section 3.2.1. The most likely explanation of this lack of agreement is that the age control over the Lahontan section is inadequate.

3.4.4. Pyramid Lake

Along the shores of Pyramid Lake, some twenty km to the north of Liddicoat's Lahontan site, Verosub et al. (1980) sampled a 3.3 m section of deep lake clays also from pluvial Lake Lahontan. The section was dated by three radiocarbon ages made on two wood fragments (24 480 ± 430 bp and 33 650 ± 1720 bp) and on one sample of disseminated carbon (29 000 ± 980 bp) to which a simple linear regression was fitted to define a timescale. The sampling interval is thus estimated to be about 180 yr and the average rate of sedimentation ~0.27 mm yr^{-1}. The inclination and declination records obtained (Figure 10 – PYRAMID) are therefore taken to span the interval 25 000 bp to 36 000 bp and thus would appear to follow on from Liddicoat's Lahontan section with no overlap. The results illustrated in Figure 10 – PYRAMID are after AF demagnetization in 150 oe. The mean inclination is 50.1°, i.e. 9.1° shallower than the ADF inclination. The mean declination is 5.2° E and values range from ~30° W to ~40° E. Two maxima (~60°) and two minima (~35°) and another maximum (~50°) in inclination (extreme range = 73°) can be recognized and the pattern recorded is typical of what we would expect of normal SV at mid-latitudes, for example as depicted in Figures 6 and 7 for N. America. Nevertheless, these curves do not correlate with the Bessette Creek curves (Section 3.2) over the 6000 yr of overlap (25 000 bp − 31 000 bp). Furthermore there is no indication of the Mono Lake 'excursion'.

3.4.5. *Lake Bonneville*

Liddicoat (1976) also reported results (see Figure 10 – BVILLE) from the upper member of the Alpine formation, 10 m thick siltstone which belongs to the Lake Bonneville Group. A sand containing gasteropods from the top of the middle member of the Alpine Formation was dated at 40 000 bp by the 230Th method. The upper member of the Alpine formation is overlain by the Bonneville Formation for which a caliche in the Promontory Soil near its base was assigned a radiocarbon age of 23 400 ± 1000 yr. Liddicoat (1976) also quoted 27 700 bp as being the oldest age assigned to the Bonneville Formation. Thus the age control of this sampled section from the Alpine Formation is rather vague, and tentatively a linear relationship between age and depth was assumed in assigning the timescale which runs from 40 000 bp to 27 000 bp. This gives an average rate of deposition of ~ 0.77 mm yr^{-1} for the 10 m section and an average sampling interval of ~ 130 yr. Inclination varies by about $\sim 10°$ around an average of $50°$ in the upper part of the section: beneath this there are some low values of between $30°$ and $40°$ at $\sim 33\,500$ bp and below this the values are mainly contained within the range $50°$ to $60°$. Declination is east of north in the top part of the section down to about 33 000 bp and westerly below this, with variations of minor amplitude ($\sim \pm 10°$) along the whole record. MDF values for samples from this section are typically ~ 150 oe and changes in direction of remanence are insignificant below peak fields of ~ 400 oe. Thus, although the stability of magnetization would appear satisfactory, the record does not provide a convincing magnetic 'signature'. Bearing in mind the poor age control, it is possible that the section represents a rather longer span of time than tentatively supposed.

3.4.6. *Comments*

Clearly these four sites from the Basin and Range Province do not yield consistent results. One reason for this could be that the adopted timescales are substantially in error, but more importantly it could be that the sediments themselves do not constitute good recorders of the geomagnetic SV signal. In particular, it is possible that any signal which may have been initially carried when the samples were wet, was weakened or destroyed by grain rotation when the sediments eventually dried out. Laboratory experiments on wet lake sediments reveal that drying out causes the intensity of remanent magnetization do decay appreciably (Stober, 1978). Since these dry lake sediments do not carry a recognizable palaeomagnetic SV signal, it is difficult to accept that the anomalous directions in a thin layer of Mono Lake sediment represent a true record of anomalous geomagnetic behaviour.

4. Conclusions

'How can we select material which we can be sure will yield good results?' This question has frequently been posed by those thinking of entering this field of research. The answer is that one can never be absolutely sure, but that several guidelines can be applied which will substantially enhance the chances of success.

(a) In order to obtain a quasi-continuous SV record a first requirement is that the sediments should be homogeneous with rates of deposition as uniform and steady as possible. It is thus better to avoid big lakes through which large volumes of water have flowed, such as for example, Lake Geneva where large variations in influx and eflux occur following the annual climatic cycle and where layers of the bottom sediments have been scoured out at times when the throughflow of water was abnormally high. The best results to date have been obtained from smaller lakes such as Lake Windermere, Loch Lomond, Lake St Croix, Lac du Bouchet etc. (see Sections 2 and 3). Where acceptable results have been obtained from large lakes, coring sites located in sheltered coves have invariably been selected – for example Thunder Bay in Lake Superior (Section 3.1.1) or Brazo de Campanario in Lago Nahuel Huapi in Argentina (Creer *et al.*, 1983).

(b) Rates of deposition are also very important. Rates of about 0.5 to 1.0 mm yr^{-1} have been found optimal to resolve SV with periods of the order of a few thousands of years. Lakes Windermere and St Croix provide good examples (Sections 2.1.2 and 3.1.2).

(c) For acceptably accurate radiocarbon dating, the organic carbon content should be around 10%. For example, the gyttjas of Lake Windermere contain 10% to 15% and those of Lake St Croix, 10% $\pm$ 3% organic carbon. However, this is just about the upper limit for the signal/noise ratio to be high enough to produce a good quality SV record as this resides in the detrital component of the sediment. Note how the quality of the SV record carried by the Late Glacial sediments of Lac du Bouchet for which the organic carbon content is only a few per cent, is very noticibly superior to that of the Post Glacial record carried by gyttjas which contain more than 10% organic carbon. The price that has to be paid is inaccuracy in dating sediments with little organic carbon (Section 2.2.3).

(d) A substantial number of measurements on at least one core from each lake is desirable to construct an adequate age/depth transform function. For example, 9 dates were measured for the interval 0–8000 bp for Lake St Croix (Section 3.1.2) and 10 dates for Loch Lomond for the interval 0 to 7000 bp (Section 2.1.3). In the Kylen study, 4 dates were measured between 8000 and 16 000 bp and this was possibly not enough (Section 3.1.2). Barton and McElhinny (1981) obtained 18 radiocarbon dates along a 4.3 m long core (0–10 000 bp) from Lake Keilambete in Australia. Most studies made to date have had to rely on fewer radiocarbon age determinations than those quoted above for transformation to a timescale. This has been due, at least in part, to a certain reluctance on the part of some radiocarbon laboratories to include sediments in their work schedules.

(e) Even then, systematic errors may still be present in the age/depth transform adopted for any single lake (e.g. see contribution by Hedges in GBCRS 1983): the top sediments of Lake Superior have a finite age of 1700 yr (Section 3.1.2); also the radiocarbon ages by Mackereth (1971) in his original Lake Windermere study had to be modified later (Section 2.1.3). Systematic errors are difficult to detect if the age determinations are made on sediments from only one lake. For this reason it is essential that investigations be carried out in more than one lake in any given geographical region. Also, it is crucial that palynological and sedimentological studies be carried out in parallel with the palaeomagnetic work.

(f) Lake sediment records have to be adjusted for amplitude (see 'Epilogue' in GBCRS 1883). This can only be done by comparison with archaeomagnetic data as illustrated for Lake Windermere by Turner and Thompson (1983). The recorded signal is also filtered with a low period cut-off governed by the rate of deposition, the signal being frozen in when the sediment overburden reaches ~ 20 cm (see contribution by Tucker in GBCRS 1983). This means that any changes in geomagnetic direction which took place in less than a few centuries would not be recorded in those lakes which have yielded the best quality SV records.

(g) This leads to the subject of geomagnetic excursions. If such a phenomenon as Morner's 'Flip' really occurred it could only have been recorded in very rapidly deposited sediments which would inevitably be coarse-grained. These are notoriously poor recorders of palaeomagnetic SV. Much of the argument and discussion in the published literature concerning the reality of geomagnetic excursions stems from this dilemma.

(h) Excursions have been defined as departures of the VGP beyond some critical latitude limit: $\sim 45°$ has been suggested by consensus. However, low or even slightly negative inclinations can be caused by factors causing normal SV – for example, by drifting or standing core sources with intensities only about twice as great as the strongest NDF foci of the contemporary geomagnetic field. A problem for the future is to demonstrate clearly whether or not any of the excursions reported in the literature in fact have a geomagnetic origin (e.g. see Verosub and Banerjee, 1977), and if so, whether they should be regarded as aborted attempts at polarity reversal or merely larger than normal amplitude SV. Ideally the same excursion should be observed recorded by both igneous and by sedimentary rocks. The search for the Laschamp Event (Bonhommet and Zahringer 1969; Guerin et al., 1984) in the Lac du Bouchet sediments (see Sections 2.1.6 and 2.2.3) should therefore be considered a prime objective.

(i) The best quality SV records have been obtained from wet sediments: results from the dried-out sediments of the Great Basin and Range Province (see Section 3.4) have not provided satisfactory SV data. This is understandable when one considers the internal stains imposed on the sediment fabric during the process of drying out. Sediments dried out artificially in the laboratory can be demonstrated to lose a substantial proportion of any remanence they carried initially when wet.

(j) Part 2 of this review, now under preparation, will be concerned with analyses and interpretation of the results presented here and with comparisons between different geographical areas as previously attempted by Creer and Tucholka (1982b).

References

Abrahamsen, N. and Readman, P. W.: 1980, 'Geomagnetic Variations Recorded in Older ($\geq$ 23 000 bp) and Younger Yoldia Clay ($\sim$ 14 000 bp) at Nørre Lyngby, Denmark', *Geophys. J. Roy. Astr. Soc.* **62**, 329–344.

Banerjee, S. K., Lund, S. P., and Levi, S.: 1979, 'Geomagnetic Record in Minnesota Lake Sediments – Absence of the Gothenburg andErieau Excursion', *Geology* **7**, 588–591.

Bonhommet, N. and Zahringer: 1969, 'Palaeomagnetism and Potassium Argon Age Determinations of the Laschamp Geomagnetic Polarity Event', *Earth Planet. Sci. Lett.* **6**, 43–46.

Bonifay, E., Creer, K. M., de Beaulieu, J. L., Casta, L., Delibrias, G., Permet, G., Pons, A., Reille, M., Servant, S., Smith, G., Thouveny, N., Truze, E. and Tucholka, P.: 'A Study of the Recent Sediments of Lac du Bouchet (Haute-Loire), France: First Results', 1984, (in press).·

Clark, H. C. and Kennett, J. P.: 1973, 'Palaeomagnetic Excursion Recorded in Latest Pleistocene Deep-Sea Sediments, Gulf of Mexico', *Earth Planet. Sci. Lett.* **19**, 267–274.

Clark, R. M.: 1975, 'A Calibration Curve for Radiocarbon Dates', *Antiquity* **50**, 61–63.

Creer, K. M.: 1974, 'Geomagnetic Variations for the Interval 7000–25 000 bp as Recorded in a Core of Sediment from Station 1474 of the Black Sea Cruise of "Atlantis II"', *Earth Planet. Sci. Lett.* **23**, 34–42.

Creer, K. M., Anderson, T. W., and Lewis, C. M. F.: 1976a, 'Late Quaternary Geomagnetic Stratigraphy Recorded in Lake Erie Sediments', *Earth Planet. Sci. Lett.* **31**, 37–47.

Creer, K. M., Gross, D. L., and Lineback, J. A.: 1976b, 'Origin of Regional Geomagnetic Variations Recorded by Wisconsin and Holocene Sediments from Lake Michigan, U.S.A. and Lake Windermere, England', *Geological Society of America Bulletin*, **87**, 531–540.

Creer, K. M., Hogg, T. E., Malkowski, Z., Mojski, J. E., Niedziolka-Krol, E., Readman, P. W., and Tucholka, P.: 1979, 'Palaeomagnetism of Holocene Lake Sediments from North Poland', *Geophys. J. Roy. Astr. Soc.* **59**, 287–314.

Creer, K. M., Hogg, T. E., Readman, P. W. and Reynaud,·C.: 1980, 'Palaeomagnetic Secular Variation Curves Extending Back to 13 400 yr bp recorded by Sediments Deposited in Lac de Joux, Switzerland', *J. Geophys.* **48**, 139–147.

Creer, K. M. and Kopper, J. S.: 1976, 'Secular Oscillations of the Geomagnetic Field Recorded by Sediments deposited in Caves in the Mediterranean Region', *Geophys. J. Roy. Astr. Soc.* **45**, 35–58.

Creer, K. M., Readman, P. W., and Papamarinopoulos, S.: 1981, 'Geomagnetic Secular Variations in Greece Through the last 6000 Years Obtained from Lake Sediment Studies', *Geophys. J. Roy. Astr. Soc.* **66**, 193–219.

Creer, K. M., Thompson, R., Molyneux, L., and Mackereth, F. J. H.: 1972, 'Geomagnetic Secular Variation Recorded in the Stable Magnetic Remanence of Recent Sediments', *Earth Planet. Sci. Lett.* **14**, 115–127.

Creer, K. M. and Tucholka, P.: 1982b, 'Secular Variation as Recorded in Lake Sediments: A Discussion of North American and European Results', *Phil. Trans. Roy. Soc.* **A306**, 87–102.

Creer, K. M. and Tucholka, P.: 1982a, 'Construction of Type Curves of Geomagnetic Secular Variation for Dating Lake Sediments from East Central North America', *Can. J. Earth Sci.* **19**, 1106–1115.

Creer, K. M.: 1983, 'Computer Synthesis of Geomagnetic Palaeosecular Variations', *Nature* **304**, 695–699.

Creer, K. M. and Tucholka, P.: 1983, Epilogue, Chapter 5 of *Geomagnetism of Baked Clays and Recent Sediments*, Creer, K. M., Tucholka, P., and Barton, C. E. (eds.), Elsevier, 324 pp.

Denham, C. R. and Cox, A.: 1971, 'Evidence that the Laschamp Polarity Event did not Occur 13 300–30 400 Years ago', *Earth Planet Sci. Lett.* **13**, 181–190.

Denham, C. R.: 1974, 'Counterclockwise Motion of Palaeomagnetic Directions 24 000 Years Ago at Mono Lake, California', *J. Geomag. Geoelectr.* **26**, 487–498.

Dodson, R. E., Fuller, M. D., and Kean, W. F.: 1977, 'Palaeomagnetic Secular Variations from Lake Michigan Sediment Cores', *Earth Planet. Sci. Lett.* **34**, 387–395.

Doh, Seong-Jae and Steele, William K.: 1983, 'The Late Pleistocene Geomagnetic Field as Recorded by Sediments from Fargher Lake, Washington, U.S.A., *Earth Planet Sci. Lett.* **63**, 385–398.

GBCRS: 1983, 'Geomagnetism of Baked Clays and Recent Sediments', K. M. Creer, P. Tucholka, and C. E. Barton (eds.), Elsevier, 324 pp.

Griffiths, D. H.: 1955, 'The Remanent Magnetism of Varved Clays from Sweden', *Monthly Notices Roy. Astr. Soc. Geophys. Suppl.* **7**, 103–114.

Guerin, G., Gillot, P. Y.,˙Reyss, J-L., and Valladas, G.: 1984, 'Datation Par Thermoluminescence et Polassium-Argon de Coulées Volcaniques Recentes. Application a la Chaine de Phys.', *Earth Planet. Sci. Lett. (in press.)*

Hedges, R. E. M.: 1983, 'Radiocarbon Dating of Sediments', Section 2.3 (pp. 37–44), of *Geomagnetism of Baked Clays and Recent Sediments*, Creer, K. M., Tucholka, P., and Barton, C. E. (eds.), Elsevier, 324 pp.

Hogg, T. E.: 1978, 'The Holocene Geomagnetic Field in Europe', Ph.D. Thesis, University of Edinburgh, 98 pp.

Ising, G.: 1943, 'On the Magnetic Properties of Varved Clay', *Ark. Mat. Astron. Fys.* **29A**, 1–37.

King, R. F.: 1955, 'Remanent Magnetism of Artificially-Deposited Sediments', *Monthly Notices Roy. Astron. Soc., Geophys. Suppl.* **7**, 115–134.

Kovacheva, M.: 1980, 'Summarised Results of the Archaeomagnetic Investigation of the Geomagnetic Field Variation for the Last 8000 Years in South–Eastern Europe', *Geophys. J. Roy. Astr. Soc.* **61**, 57–64.

Latham, A., Schwarcz, H. P., Ford, D. C., and Pearce, G. W.: 1982, 'The Palaeomagnetism and U-Th Dating of Three Canadian Spelethems: Evidence for the Westward Drift, 5.4–2.1 Ka bp', *Can. J. Earth Sci.* **19**, 1985–1995.

Liddicoat, J. C.: 1976, 'A Palaeomagnetic Study of Quaternary Dry Lake Deposits from the Western U.S.A. and Basin of Mexico', Ph.D. Thesis, Univ. of California – Santa Cruz, Parts I and II.

Liddicoat, J. C. and Coe, R. S.: 1979, 'Mono Lake Geomagnetic Excursion', *J. Geophys. Res.* **84**, 261–271.

Livingstone, D. A.: 1955, 'A Lightweight Piston Sampler for Lake Deposits', *Ecology* **36**, 137–139.

Lowe, J. J. and Gray, J. M.: 1980, 'The Stratigraphical Sub-Division of the Late Glacial of N.W. Europe – A Discussion', in Lowe, J. J., Gray, J. M., and Robinson, J. E. (eds.), *Studies of the Late Glacial of N.W. Europe*, Pergamon Press, Oxford, pp. 157–175.

Lund, S. P.: 1981, 'Late Quaternary Secular Variation of the Earth's Magnetic Field as Recorded in the Wet Sediments of Three North American Lakes', Doctoral dissertation, University of Minnesota, Minneapolis, 300 pp.

Lund, S. P. and Banerjee, S. K.: 1979, 'Palaeosecular Variations from Lake Sediments', *Rev. Geophys. and Space Phys.* **17**, 244–249.

Lund, S. P. and Banerjee, S. K.: 1983, 'Late Quaternary Secular Variations Recorded in Central North American Wet Lake Sediments', Section 4.3 (pp. 211–222) *Geomagnetism of Baked Clays and Recent Sediments*, Creer, K. M., Tucholka, P., and Barton, C. E. (eds.), Elsevier, pp. 324.

Mackereth, F. J. H.: 1958, 'A Portable Core Sampler for Lake Deposits', *Limnol. Oceanogr.* **3**, 181–191.

Mackereth, F. J. H.: 1969, 'A Short Core Sampler for Subaqueous Deposits', *Limnol. Oceanogr.* **14**, 145–151.

Mackereth, F. J. H.: 1971, 'On the Variation in the Direction of the Horizontal Component of the Magnetization in Lake Sediments', *Earth Planet. Sci. Lett.* **12**, 332–338.

McNish, A. G. and Johnson, E. A.: 1938, 'Magnetization of Unmetamorphosed Varves and Marine Sediments', *Terr. Magn. Atmos. Elec.* **53**, 349–360.

Morner, N.-A.: 1977, 'The Gothenburg Magnetic Excursion', *Quaternary Res.* **7**, 413–427.

Morner, N.-A., Lanser, J. P., and Hospers, J.: 1971, 'Late Weichselian Palaeomagnetic Reversal', *Nature, Phys. Sci.* **234**, 173–174.

Mothersill, J. S.: 1979, 'The Palaeomagnetic Record of the late Quaternary Sediments of Thunder Bay', *Can. J. Earth. Sci.* **16**, 1016–1023.

Mothersill, J. S.: 1981, 'Late Quaternary Palaeomagnetic Record of the Goderich Basin, Lake Huron', *Can. J. Earth Sci.* **18**, 448–456.

Opdyke, N. D., Ninkovich, D., Lowrie, W., and Hays, J. D.: 1972, 'The Palaeomagnetism of two Aegean Deep-Sea Cores', *Earth Planet. Sci. Lett.* **14**, 145–159.

Palmer, D. F., Heyney, T. L., and Dodson, R. E.: 1979, 'Palaeomagnetic and Sedimentological Studies at Lake Tahoe, California – Nevada', *Earth Planet. Sci. Lett.* **46**, 125–137.

Prest, V. K.: 1970, 'Quaternary Geology of Canada', in *Geology and Economic Minerals of Canada*, Douglas, R. J. W. (ed.), Geol. Surv. of Canada, Economic Geology Report 1, pp. 675–764.

Stober, J.: 1978, 'Palaeomagnetic Secular Variation Studies on Holocene Lake Sediments', Ph.D. Thesis, University of Edinburgh.

Stober, J. C. and Thompson, R.: 1977, 'Palaeomagnetic Secular Variation Studies of Finnish Lake Sediments and the Carriers of Remanence', *Earth Planet. Sci. Lett.* **37**, 139–149.

Thompson, R.: 1976, 'The Palaeomagnetism of Varved Clays from Blekinge, Southern Sweden: A Comment', *Geol. Foren. Stockh. Fohr.* **98**, 283–284.

Thompson, R. and Berglund, B.: 1976, 'Late Weichselian Geomagnetic "Reversal" as a Possible Example of the Reinforcement Syndrome', *Nature* **263**, 490–491.

Thompson, R. and Kelts, K.: 1974, 'Holocene Sediments and Magnetic Stratigraphy from Lakes Zug and Zurich, Switzerland', *Sedimentology* **21**, 577–596.

Thompson, R. and Turner, G. M.: 1979, 'British Geomagnetic Master Curve 10 000–0 yr bp for Dating European Sediments', *Geophys. Res. Lett.* **6**, 249–252.

Tucker, P.: 1979, 'Selective Post-Depositional Alignment in a Synthetic Sediment', *Phys. Earth Planet. Inter.* **20**, 11–14.

Tucker, P.: 1980, 'A Grain Mobility Model of Post-Depositional Realignment', *Geophys. J. Roy. Astr. Soc.* **63**, 149–163.

Turner, G. M. and Thompson, R.: 1979, 'The Behaviour of the Earth's Magnetic Field as Recorded in the Sediments of Loch Lomond', *Earth Planet. Sci. Lett.* **42**, 412–426.

Turner, G. M. and Thompson, R.: 1981, 'Lake Sediment Record of the Geomagnetic Secular Variation in Britain during Holocene Times', *Geophys. J. Roy. Astr. Soc.* **65**, 703–725.

Turner, G. M. and Thompson, R.: 1982, 'Detransformation of the British Geomagnetic Secular Variation Record for Holocene Times', *Geophys. J. Roy. Astr. Soc.* **70**, 789–792.

Turner, G. M., Evans, M. E., and Hussin, I. B.: 1982, 'A Geomagnetic Secular Variation Study (31 000–19 500 bp) in Western Canada', *Geophys. J. Roy. Astr. Soc.* **71**, 159–171.

Verosub, K. L.: 1977, 'The absence of the Mono Lake Geomagnetic Excursion from the Palaeomagnetic Record of Clear Lake, California', *Earth Planet. Sci. Lett.* **36**, 219–230.

Verosub, K. L. and Banerjee, S. K.: 1977, 'Geomagnetic Excursions and Their Palaeomagnetic Record', *Rev. Geophys. and Space Phys.* **15**, 145–155.

Verosub, K. L., Davis, J. D., and Valastro, S. Jr.: 'A Pallaeomagnetic Record for Pyramid Lake, Nevada and Its Implications for Proposed Geomagnetic Excursions', *Earth Planet. Sci. Lett.* **49**, 141–148.

Vitorello, I. and Van der Voo, R.: 1977, 'Magnetic Stratigraphy of Lake Michigan Sediments Obtained from Cores of Lacustrine Clay', *Quaternary Research* **7**, 398–412.

S. K. RUNCORN'S COMMENTARY

At every stage in the development of palaeomagnetism the geophysical and geological community has been sceptical as to whether the rocks in question were truthfully recording the direction of the geomagnetic field at the time of formation or magnetization. More than scientific caution was involved: it seemed too good to be true that archaeological specimens from hearths and bricks, then lavas, then red sandstones, then sediments generally, then granites, ocean bottom cores and even coal, yielded stable remanent magnetization. At some stage surely palaeomagnetists' luck must run out! So when Mackereth from the Freshwater biological station at Windermere contacted our group in Newcastle wondering whether the corer he had designed for sampling lake sediments could be used for palaeomagnetism, few would have thought that such studies would be done in the Great Lakes and the Rift Valley lakes within a decade. This is the achievement of Ken Creer and his group at Edinburgh who, after work in European lakes, carried it much further a field e.g. in Argentina. Now these studies proceed in many laboratories.

The development of these lake sediment studies holds out the hope that the properties of the secular variation will be determined. Creer himself was a pioneer of studies of the dependence of amplitude, as measured by Fisher's parameter k, of the secular variation with palaeolatitude. But it seems as if this does not give as much insight, into core processes as would the discovery that there are certain definite periods present in the secular variation. For this accurate dating is needed and perhaps the new and powerful accelerator techniques for C14 dating will supply it.

ON THE EXCITATION OF SHORT-TERM VARIATIONS IN THE LENGTH OF THE DAY AND POLAR MOTION

RAYMOND HIDE

Geophysical Fluid Dynamics Laboratory, Meteorological Office (Met O 21), Bracknell, Berkshire RG12 2SZ, England, U.K.

Abstract. Variations in the distribution of mass within the atmosphere and changes in the pattern of winds produce fluctuations in all three components of the angular momentum of the atmosphere on time-scales upwards of a few days. It has been shown that variations in the *axial* component of atmospheric angular momentum during the Special Observing Periods in the recent 'First GARP Global Experiment' (FGGE, where GARP is the Global Atmospheric Research Programme) are well correlated with short-term changes in the length of the day. They are consistent with the total angular momentum of the atmosphere and 'solid' Earth being conserved on short timescales (allowing for lunar and solar effects), without requiring significant angular momentum transfer between the Earth's liquid core and solid mantle on timescales of weeks or months. It has also been shown that fluctuations in the equatorial components of atmospheric angular momentum make a major contribution to the observed wobble of the instantaneous pole of the Earth's rotation with respect to the Earth's crust. A necessary step in the investigation was a re-examination of the underlying theory of non-rigid body rotational dynamics and angular momentum exchange between the atmosphere and solid Earth. Since only viscous or topographic coupling between the atmosphere and solid Earth can transfer angular momentum, no atmospheric flow that everywhere satisfied inviscid equations (including, but not solely, geostrophic flow) could affect the rotation of a spherical solid Earth. New 'effective angular momentum functions' were introduced in order to exploit the available data and allow for rotational and surface loading deformation of the Earth. A theoretical basis has now been established for future routine determinations of atmospheric angular momentum fluctuations for the purpose of meteorological and geophysical research, including the assessment of the extent to which movements in the solid Earth associated with very large earthquakes contribute to the excitation of the Chandlerian wobble.

1. Introduction

In the absence of internal energy sources or mechanical, gravitational, or radiative interactions with other astronomical bodies, the whole Earth would move as a rigid body, with its solid parts (crust, mantle, and inner core) and fluid parts (atmosphere, hydrosphere, and outer core) all rotating together at a constant rate about a fixed axis of maximum moment of inertia through the Earth's centre of mass. Positional astronomers equipped with perfect telescopes and clocks would find no variation in the astronomical latitude of any observatory, nor in the angular rate at which all fixed stars appeared to circle the Celestial Pole. The successful use of the rotation of observatories fixed on the Earth's crust as the basis of early attempts to provide a practical scale of time attests to the validity of this picture as a good first approximation to the truth (see, for example, Gaposchkin and Kolaczek, 1981). However, over the years, as clocks based on other periodic physical phenomena were invented and the methods of positional astronomy were improved, there came to light tiny fluctuations in the length of the day, as measured by the time interval between successive transits of a particular star across the meridian, and slight movements of the Earth's pole of rotation (usually called polar motion or 'wobble'), manifested at a given station by variations in its

Geophysical Surveys 7 (1985) 163–167. 0046-5763/85.15.

astronomical latitude. Typically the length-of-day shows variations of up to a few milliseconds on timescales from days to years, while the rotation pole executes rough ovals, a few metres in size and somewhat longer than a year in period, in the vicinity of the geographical reference pole (the Conventional International Origin).

Interpreting variations in the magnitude the direction of the Earth's rotation vector in terms of energetic processes and angular momentum transfer within the Earth-Moon system is a fascinating scientific problem, which brings together many diverse areas of study, notably solid Earth geophysics, geodesy, meteorology, oceanography, hydrology, glaciology, geomagnetism, palaeoclimatology, astrometry, and even aspects of historical scholarship. Such studies of the Earth's rotation go back to the last century. The first thorough review of the subject was presented by Munk and MacDonald (1960), and Lambeck (1980) has given an up-to-date discussion taking into account recent advances in geodynamics, instrumentation, and international co-operation.

2. Atmospheric Effects

During the past few years, the Geophysical Fluid Dynamics Laboratory of the U.K. Meteorological Office has undertaken a systematic investigation of the extent to which short-term changes in the magnitude and direction of the rotation vector of the solid Earth can be accounted for in terms of angular-momentum exchange with the atmosphere (Hide, 1977; Hide et al., 1980; Barnes et al., 1983, cited as 'BHWW'), and the present summary of the main findings of that work is based on the introductory section to BHWW.

Denote by H the total angular momentum of the atmosphere about the Earth's centre of mass. The magnitude of H is about 10^{-6} that of the whole Earth. All three of its components exhibit variations on timescales from a few days to years, reflecting changes in the distribution of mass in the atmosphere and in the pattern of winds, particularly the strength and location of the major mid-latitude jet-streams. Fluctuations in H are of interest to meteorologists concerned with the general circulation of the atmosphere, since they are gross indicators of changes in the strength of the zonal circulation and in the surface pressure distribution. The consideration of these fluctuations can provide useful constraints on numerical models of the general circulation, since the time rate of change of angular momentum, $\dot{H}$, must equal the torque exerted at the surface, and this requires the satisfactory representation both of motions in surface boundary layers and of topographic effects on air flow. Atmospheric angular momentum fluctuations are of interest also to geophysicists and astronomers concerned with the structure and dynamics of the Earth, who must make allowances for the meteorological contribution to the variable rotation of the solid Earth when dealing with effects due to 'non-meteorological' processes (see, for example, Mansinha et al., 1970; Hide, 1977; Lambeck, 1980; Runcorn et al., 1982; Brosche and Sündermann, 1982). (Unless otherwise stated, we shall use the term 'Earth' to mean the core, mantle, crust and hydrosphere, i.e. the whole Earth minus the atmosphere, and the term 'solid Earth' to mean the crust and mantle.)

Variations in the rotation of the solid Earth can be caused by (i) changes in the total angular momentum due to the application of external torques (lunar and solar gravitational attraction on the equatorial bulge, bodily tides, the solar wind), (ii) changes in the inertia tensor (earthquakes), and (iii) exchange of angular momentum with the overlying oceans and atmosphere and underlying fluid core. External effects (i) are now largely calculable and so may adequately be subtracted from astronomical observations, leaving the determination of the relative importance of the internal effects (ii) and (iii) as a geophysical problem.

Comparison of astronomical observations of changes in the rotation of the solid Earth with variations of **H** as determined from meteorological data elucidates the significance of atmospheric contributions to changes in length-of-day and to the excitation of polar motion. Several recent studies (Lambeck and Cazenave, 1977; Lambeck and Hopgood, 1981) have related length-of-day variations to changes in the axial component of on timescales ranging from months to a few years. The atmospheric contribution to the forced motion of the pole has been investigated by several workers (Munk and Hassan, 1961; Sidorenkov, 1973; Wilson and Haubrich, 1976), who concentrated on the effects of the redistribution of mass on the atmosphere's inertia tensor, and thus found evidence of meteorological excitation of the annual component of polar motion. All these studies used mean monthly or longer period atmospheric data.

There are also rapid and irregular variations in length-of-day and polar motion on timescales of days and weeks. Hide *et al.* (1980) compared length-of-day data from the Bureau International de l'Heure (BIH) with determinations of daily values of the axial component of **H** from wind data collected during the Special Observing Periods of the First GARP Global Experiment (FGGE, where GARP is the Global Atmospheric Research Programme; see Fleming *et al.* (1979)). The correlations found on these short timescales could (within the small errors involved) be fully accounted for on the basis of angular momentum exchange between the atmosphere and solid Earth, implying that angular momentum transfer between the Earth's liquid core and solid mantle, which is considered to be substantial and even dominant on timescales upwards of a few years (see, for example Munk and MacDonald, 1960; Hide, 1977; Lambeck, 1980; Runcorn *et al.*, 1982), is probably not significant on much shorter timescales. After removing known tidal effects from BIH length-of-day values for 1979, Feissel and Gambis (1980) noticed a persistent fluctuation on a timescale of about 55 days, with an amplitude near 0.4 ms. The work of Hide *et al.* (1980) implied that the axial component of **H** should exhibit a similar fluctuation, and this expectation was confirmed by Langley *et al.* (1981) and Rosen and Salstein (1983), from their studies of several years of axial atmospheric angular momentum values, as evaluated from the daily global wind data of the U.S. National Meteorological Center.

BHWW made the first quantitative attempt to relate short term variations in the equatorial (non-axial) components of **H** to irregularities in polar motion. While changes in the distribution and strength of zonal winds provide the main contribution to fluctuations in the axial component of **H** and hence to changes in length-of-day, it is the redistribution of air mass that is largely responsible for altering the direction of **H** and thus exciting a wobble in the orientation of the Earth with respect to its rotation axis. BHWW calculated the axial and equatorial components of **H** for the period 1 January 1981 – 30 April 1982, using the surface pressure and eastward and northward wind fields of the 'initialised analysis global database' archived by the European Centre for Medium-Range Weather Forecasts (ECMWF). The results were used in a comparison with the values of length-of-day and polar motion published by the Bureau International de l'Heure. Identical atmospheric quantities are now being evaluated by Dr R. D. Rosen and Dr D. A. Salstein from the wind and surface pressure fields of the U.S. National Meteorological Center's (NMC's) initialized analysis global database, and it is hoped that comparison of the two sets of atmospheric angular momentum values will furnish further information on the accuracy and reliability of our results. (This approach to the problem of error estimation proved to be the most practicable in the work of Hide *et al.* (1980) on the axial component.)

BHWW found it necessary to re-examine the theory of wobble excitation, by considering carefully the dynamical coupling between the atmosphere and the underlying solid Earth. The excitation functions **Ψ** used by the previous workers are proportional to the equatorial components of the total frictional and pressure torque on the solid Earth. Thus **Ψ** cannot be evaluated reliably from meteorological data by applying inviscid forms of the equations of atmospheric motion or by neglecting topography. Instead of attempting to evaluate **Ψ** directly, values of a new atmospheric equatorial effective angular momentum (E.A.M.) function χ were calculated. This function includes Love number corrections for rotational and surface loading deformation of the Earth and can be evaluated reliably from available meteorological data. Full allowance for the response of the oceans to atmospheric surface pressure changes was not made in the main part of the study, but it was shown that the use of an 'inverted barometer' correction would not substantially change the results. The results confirm (see Hide *et al.*, 1980) that length-of-day changes observed during the period 1 January, 1981 – 30 April, 1982 could be accounted for on the basis of transfer of axial angular momentum between the atmosphere and solid Earth (i.e. crust and mantle). This transfer exhibits persistent fluctuations on the time-scale of about 7 weeks with an amplitude of about 15 % of the total relative angular momentum of the atmosphere, the corresponding changes in length-of-day being near 0.5×10^{-3} s. Respective contributions to this 7-week fluctuation from the northern and southern hemispheres are comparable in magnitude and show little systematic difference in phase. These findings strongly imply that the fluctuation is of intrinsic origin, and driven by dynamical atmospheric processes occurring in the Tropics.

3. Concluding Remarks

During the period 1975–1980 the amplitude of the polar motion (wobble) diminished gradually from about 9 m to a minimum of about 3 m towards the end of 1980, and then began to increase slowly. Our evaluation of the atmospheric equatorial effective angular momentum functions from meteorological data for the period 1 January 1981 – 30 April 1982 indicates that atmospheric excitation alone was sufficient to account for the observed polar motion over that period. There is apparently no need to invoke substantial excitation either by the fluid core of the Earth, or movements in the mantle associated with earthquakes, of which, admittedly, there were no major instances during the interval covered by our study. There were 7 earthquakes exceeding 7.2 in magnitude during the period covered by our study, and none exceeded 7.9 in magnitude. If we are successful in our efforts to establish a programme for applying the results of the present paper to the routine monitoring of the atmospheric contribution to changes in the angular momentum of the solid Earth, then it will be possible to determine by subtraction any non-atmospheric contributions that might be present from time to time, and in particular to assess whether or not processes associated with major earthquakes are important in this connection. It will be possible in this way to find further evidence bearing on our preliminary finding that angular momentum transfer between core and mantle does not appear to occur to any significant extent on timescales much less than a few years.

As one of many scientists who were lucky enough to become infected at an early stage of their careers by Professor Keith Runcorn's enthusiasm for geophysics, I am grateful to the organizers of this meeting on magnetism, rotation and convection in the solar system for inviting me to take part.

References

Barnes, R. T. H., Hide, R., White, A. A., and Wilson, C. A.: 1983, *Proc., R. Soc. Lond.* **A387** 31–73 (cited as BHWW).

Brosche, P. and Sündermann, J. (eds.): 1982, *Tidal friction and the Earth's Rotation* II. Berlin, Springer-Verlag, 345 pp.

Bureau International de l'Heure (BIH): *Annual Reports for 1977–1980 and Circulars D* 172–187, Paris.

Feissel, M. and Gambis, D.: 1980, *C.r. hebd. Séanc. Acad. Sci. Paris* **B291**, 271–273.

Fleming, R. J., Kaneshige, T. M., and McGovern, W. E.: 1979, *Bull. Am. Met. Soc.* **60**, 649–659.

Gaposchkin, E. M. and Kolaczek, B. (eds.): 1981, *Reference Coordinate System for Earth Dynamics'*, D. Reidel Publ. Co., Dordrecht, Holland, 396 pp.

Hide, R.: 1977, *Phil. Trans. R. Soc. Lond.* **A 284** 547–554.

Hide, R., Birch, N. T., Morrison, L, V., Shea, D. J., and White, A. A.: 1980, *Nature, Lond.* **286**, 114–117.

Lambeck, K.: 1980, *The Earth's Variable Rotation: Geophysical Causes and Consequences*, Cambridge University Press, 449 pp.

Lambeck, K. and Cazenave, A.: 1977, *Phil. Trans. R. Soc. Lond.* **A 284**, 495–506.

Lambeck, K. and Hopgood, P.: 1981, *Geophys. J. Roy. Astr. Soc.* **64**, 67–89.

Langley, R. B., King, R. W., Shapiro, I. I., Rosen, R. D., and Salstein, D. A.: 1981, *Nature, Lond.* **294**, 730–732.

Madden, R. A. and Julian, P. R.: 1971, *J. Atmos. Sci.* **28**, 702–708.

Mansinha, L., Smylie, D. E., and Beck, A. E. (eds.): 1970, *Earthquake Displacement Fields and the Rotation of the Earth*, D. Reidel Publ. Co., Dordrecht, Holland, 308 pp.

Munk, W. H. and Hassan, E. M.: 1961, *Geophys. J. Roy. Astr. Soc.* **4**, 339–351.

Munk, W. H. and MacDonald, G. J. F.: 1960, *The Rotation of the Earth*, Cambridge University Press, 323 pp.

Rosen, R. D. and Salstein, D. A.: 1983, *J. Geophys. Res.* **88**, 5451–5470.

Runcorn, S. K., Creer, K. M., and Jacobs, J. A. (eds.): 1982, *Phil. Trans. R. Soc. Lond.* **A 305** 1–289.

Sidorenkov, N. S.: 1973, *Izv. Acad. Sci. USSR, Atmos. Ocean. Phys.* **9**, 339–351.

Wilson, C. R. and Haubrich, R. A.: 1976, *Geophys. J. Roy. Astr. Soc.* **46**, 745–760.

S. K. RUNCORN'S COMMENTARY

Raymond Hide cut his teeth in geomagnetism by working as a student assistant, as also did Frank Lowes and Ken Creer, with Alan Moore, Tony Benson and me, on the measurements in deep coal mines to determine the radial variation of the geomagnetic field. Such useful experience for students and such welcome assistance with research is now virtually impossible as a result of University's bureaucratic attitudes to temporary employment and to unionisation.

When, as a research project, I suggested to Hide to do convection experiments in a rotating annulus where the Coriolis force could be easily made dominant and where the Proudman theorem (of which I learnt from Sir Harold Jeffreys) would hold, I was thinking of modelling the core. In 1950 we were only just starting to adsorb Elsasser and Bullard's papers on the dynamo theory, and thought how nice it would be to begin to understand what sort of motions it would be reasonable to think about. What a great success he made of this experiment – the wavy jet stream was the source of much interest: Sir Ronald Fisher used always to refer to it as 'Hide's worm'! He set off a whole series of experimental studies of this kind in many parts of the world, which are continuing to bear fruit.

Professor Hide is therefore a most distinguished exponent of geophysical fluid dynamics, a discipline which has fitted him for the Presidencies of both the Royal Meteorological Society and the Royal Astronomical Society – and who knows, in view of geologists' acceptance of mantle convection, for the Presidency of the Geological Society of London!

PLANETARY ROTATION AND INVERTEBRATE SKELETAL PATTERNS: PROSPECTS FOR EXTANT TAXA

W. WILLIAM HUGHES

Biology Department, Andrews University, Berrien Springs, MI 49104, U.S.A.

Abstract. It has been two decades since Wells proposed that the duration of absolute geologic time could be estimated using growth patterns found in fossil corals. Since then, the temporal and environmental records encoded in the accretionary skeletons of other invertebrates also have been studied. Although extensive research on the significance of skeletal patterns has been done on bivalved molluscs, other taxa such as the brachiopods, bryozoans, cephalopods and echinoderms are in need of further study. For two taxa, the nautiloids and brachiopods, additional growth pattern data are presented here. These data indicate that erroneous geophysical conclusions about the length of the synodic lunar month were previously reached using what now appear to be unfounded assumptions about the temporal significance of their growth patterns. Assumptions regarding the temporal and environmental information contained in the skeletons of these poorly studied taxa need to be replaced by more extensive analyses using standardized techniques. Only then will we arrive at correct conclusions about the dynamical history of the Earth-Moon system.

1. Introduction

"Can paleontology give any support to the shaky chronometric creation of the geophysicists and astronomers?" Although this symposium is in celebration of Professor Runcorn's 60th birthday, 1983 also represents the 20th anniversary of J. W. Wells' (1963) paper in which he asked and attempted to answer the above question. Prior to 1963, astronomical data and radioactive isotopes were the primary geochronometers (i.e., measurements of geologic time), with fossils being useful only as sources of relative, not absolute, geologic time. Wells (1963) challenged this limited application of fossils and wrote: "How nice it would be if instead of paying a large sum for an isotopic analysis we could examine a fossil and estimate directly, with luck, not only its relative, but also its absolute age – every paleontologist a geochronometrist; every fossil a geochronometer."

In agreement with current astronomical theory, Wells postulated that the Earth's periods of rotation was decreasing (2 msec/century) as a result of tidal friction. This meant that the length of the solar day had been increasing throughout geologic time. Wells knew what was needed were fossils that faithfully preserved the number of days per year in the distant past in their skeletons. Then, because the length of the year (in modern hours) has remained constant throughout geologic history (for geophysical reasons), the length of the day in the distant past could be simply calculated.

He assumed that the major annulations and fine ridges observed on the Paleozoic rugose and the Recent scleractinian corals represented annual and daily periodicities respectively. "I have tested it indirectly on one or two Recent corals the annual linear growth-rate of which is fairly well known, and to my gratification found that the number of ridges on the epitheca of the living West Indian scleractinian *Manicina areolata* hovers around 360 in the space of a year's growth." Observations on 'limited'

Geophysical Surveys 7 (1985) 169–183. 0046-5763/85.15.

fossils from the middle Devonian of New York and Ontario showed in every instance more than 365 fine ridges between major annulations, averaging 400 and ranging from 385–410. Wells concluded that fossils may well supply "a third stabilizing and much cheaper clue to the problem of geochronometry".

Wells' (1963) paper is generally recognized as the starting point in the use of invertebrate skeletal patterns for revealing the Earth's rotational history. This does not mean that the work done by researchers predating Wells is not important. Studies by Isley (1913), Orton (1923), Weymouth *et al.* (1931), Ma (1934), Tang (1941), Shuster (1951), and Craig and Hallam (1963) provided the foundation for further study. Wells' contribution was his integration of geophysics (e.g., tidal friction), paleontology, and biology in attempt to answer questions about the rotational history of the Earth (and Moon). Additionally he hoped "that further search for diurnal or circadian records in groups other than corals may result in strengthening this weak anthozoan prop". And, indeed, it is now known that the skeletons of many other organisms, such as bivalved molluscs, barnacles, brachiopods and stromatolites, contain a record which reflects temporal and environmental events.

The early years following Wells' (1963) paper were filled with optimism as organisms other than corals were studied – optimism shared also by Professor Runcorn (1966): "If these (fossil algae) and other marine organisms have recorded time in the same way as corals, we shall indeed have factual information on the early history of the earth."

During the last 20 years the temporal and environmental significance of skeletal growth patterns has been studied in numerous invertebrate taxa. Excellent reviews are available (see Neville, 1967; Pannella and MacClintock, 1968; Rhoads and Pannella, 1970; Pannella, 1972; Scrutton and Hipkin, 1973; Clark, 1974; Scrutton, 1978; Lutz and Rhoads, 1980). The first review by Neville (1967) totals 20 pages and includes data on vertebrates and plants, whereas the last review which deals only with growth patterns in molluscan shells is 51 pages (Lutz and Rhoads, 1980). The increase in total pages suggests and parallels accumulating interest and data in this area of study.

Two recent books devoted to skeletal growth patterns also reflect the growing interest. Rosenberg and Runcorn (1975) co-edited the proceedings of the 'Interdisciplinary Winter Conference on Biological Clocks and Changes in the Earth's Rotation: Geophysical and Astronomical Consequences', (*Growth Rhythms and the History of the Earth's Rotation*). The second work, *Skeletal Growth of Aquatic Organisms* co-edited by Rhoads and Lutz (1980), which covers the biological records of environmental change, is enhanced by appendices containing information about preparation techniques and methods of analysis.

A major concern among researchers has been the difficulty in detecting growth lines so as to yield consistent, repeatable, and objective data (Crabtree *et al.*, 1980; Hughes and Clausen, 1980). Although Wells (1963) relied on measurements of external growth increments, there are now a variety of recently developed or newly applied techniques to examine invertebrate skeletons (see Table I). The use of these techniques (e.g., densitometer, chemical analysis) should improve the quality and consistency of the data among researchers.

TABLE I

Selection of techniques used in the study of invertebrate skeletons

Technique	Organism	Author(s)
External measurements	Corals Brachiopods	Wells, 1963 Mazzullo, 1971
Thin sections	Molluscs	Clark, 1980a
Acetate peels	Bivalved Molluscs	Kennish *et al.*, 1980
Scanning electron microscopy (critical point drying)	Molluscs	Clark, 1980b
Fourier analysis	Molluscs	Dolman, 1975: Rosenberg *et al.*, 1980
Statistical test for contemporaneity	Bivalved Molluscs	Dillon and Clark, 1980; Hughes and Clausen, 1980
Chemical analysis	Bivalved Molluscs	Rosenberg and Jones, 1975; Rosenberg, 1980
Densitometry	Bivalved Molluscs	Dolman, 1975
Radioautography	Corals Bivalved Molluscs	Knutson *et al.*, 1972 Bonham, 1965
X-radiography	Corals	Dodge, 1980
Physiology (in vivo)	Bivalved Molluscs	Gordon and Carriker, 1978
Controlled growth experiments	Barnacles Bivalved Molluscs	Bourget, 1980 Ursula, 1981
Radioactive and stable isotope analysis	Cephalopods Corals	Cochran *et al.*, 1981 Emiliani *et al.*, 1978

It has been 20 years. Has the 'weak anthozoan prop' been corroborated as envisioned by Wells (and others)? As is typical in science, definitive results are often elusive. There are now at least 20 temporal and environmental parameters known to be recorded in invertebrate skeletons (see Lutz and Rhoads, 1980; Scrutton, 1978; Clark, 1974; Rhoads and Pannella, 1970), with the possibility of more to be added. Wells' (1963) assumption that fine ridges counted between major annulations on corals *equal* the number of days per year is now recognized as an oversimplification, if not for the corals that Wells studied, at least for other taxa with accretionary skeletons. Growth patterns are now seen as a complex combination of any or perhaps all parameters (e.g., solar daily, lunar tidal, spawning events, predation, intertidal position, latitude) in many of the invertebrate taxa studied.

While the complexity of growth patterns has become increasingly evident during the past two decades, it was only recently that a mechanism was theorized to explain their formation in bivalved molluscs. Lutz and Rhoads (1977) proposed that growth lines result from the 'dissolution' of shell material during periods of anaerobiosis. In their model, growth patterns reflect respiratory changes, such as occur when bivalved

TABLE II

Selection of accomplishments relating invertebrate skeletal patterns to the Earth's
rotational history during the past 20 years

Author(s)	Contribution
Wells, 1963	Solar daily growth increments in corals used to determine days/year in distant past.
Scrutton, 1965	Solar daily growth increments used to determine days/month in past.
Runcorn, 1964	Days/year and days/month used to distinguish effects of lunar tidal friction from moment of inertia effects.
Barker, 1964	Systematic description of growth patterns and varying periodicities in bivalves.
Evans, 1972	Demonstrated lunar daily (tidally controlled) increments in the cockle.
Dolman, 1975	Use of automated techniques for measuring and analyzing growth patterns.
Rosenberg and Jones, 1975	Demonstrated chemical, as well as structural, skeletal patterns in bivalves.
Berry and Barker, 1975	Systematic examination of bivalves throughout the Phanerozoic.
Lutz and Rhoads, 1977	Proposed shell dissolution model to explain the formation of growth increments in bivalves.

molluscs switch from aerobic to anaerobic respiratory pathways during periods of shell closure. Succinic acid produced during anaerobic metabolism is hypothesized to be neutralized by $CaCO_3$ from the shell, leaving behind an organic-rich layer and thus forming a growth line. Lutz and Rhoads' model is in opposition to Barker's (1964) suggestion that growth lines in bivalved molluscs resulted from periodic episodes of $CaCO_3$ 'deposition' alternating with secretion of conchiolin. It is important to note that 'dissolution' does not explain the formation of *all* growth lines (see Gordon and Carriker, 1978), and that continued research is needed on mechanisms of growth line formation.

Bill O'Reilly, the organizer of this symposium, has reminded us to place as much emphasis on 'prospect' as 'retrospect'. So, with the realization that the future use of invertebrate skeletal patterns is dependent on a contemplation of past accomplishments (both successes and failures), I will conclude this brief and biased retrospective survey and move on to the prospects for extant taxa. I list in Table II some of the major accomplishments in the field since Wells' (1963) classic paper.

2. Prospects for Extant Taxa

Bivalved molluscs have received the most study during the past 20 years, and it is in this taxon that the temporal and environmental complexity of growth patterns can best be seen (see overview by Lutz and Rhoads, 1980). But there are other taxa, such as barnacles (Bourget, 1980), echinoderms (Weber, 1969), gastropods (Ekaratne and Crisp, 1982), polychaetes (Olive, 1980) and bryozoans which may merit the same effort given to bivalved molluscs. For example, Bartley and Anstey (personal communi-

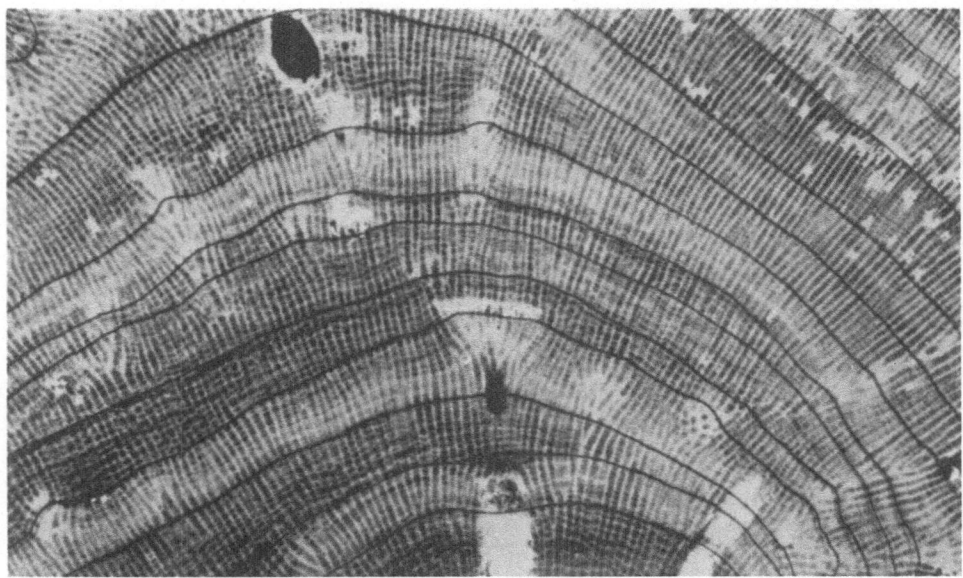

Fig. 1. Photograph from thin-section of the Ordovician trepostome bryozoan *Amplexopora filiasa* showing endozone-exozone couplets. Boundaries between the thick-walled exozone layer and thin-walled endozone layer are drawn in. Typically the diaphrams in the exozone are more closely spaced, suggesting a reduction in growth rate during their formation. Horizontal width = 50 mm. Photomicrograph provided by John Bartley and Robert Anstey.

cation) are studying the growth patterns in extant cyclostome bryozoans and also in *Amplexopora*, a Paleozoic trepostome (Figure 1). Conclusions as to the temporal and environmental information contained in these growth patterns is in need of further study. On the other hand, in the case of the brachiopods (Mazzullo, 1971) and nautiloids (Kahn and Pompea, 1978), the use of skeletal patterns to support geophysical theories may have been premature. Given our understanding of the possible complexity of these skeletal patterns, I am suspicious of the simple 'fine ridges' *equal* 'days' application in these taxa. The following examples document the basis for my concern.

2.1. BRACHIOPODS

Mazzullo (1971) proposed new values for the length of the year during the Silurian and Devonian periods. In his study, 'daily' growth increments and 'monthly' markings were directly counted on each fossil brachiopod using a 15X binocular microscope. The assumption that the fine growth increments in brachiopods are formed daily came from Wells' (1963) coral study. However, subsequent to Mazzullo's (1971) study of external growth increments on brachiopods, MacKinnon and Williams (1974) casually suggested that the growth increments preserved within the primary shell layer of terebratulid brachiopods may represent a daily periodicity. Pope (1976) made a similar suggestion for strophomenid brachiopods.

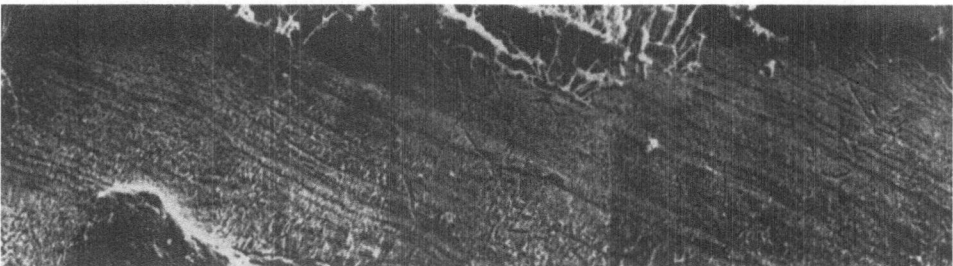

Fig. 2.   S.E.M. photomicrograph of growth increments in the primary shell layer of the extant brachiopod *Terebratalia transversa*. Note that the growth increments narrow and widen cyclically and may represent responses to tidal fluctuations in this intertidal specimen. The cyclic variation in increment widths is similar to that observed in intertidal bivalved molluscs known to have grown under tidal influence. Direction of growth is to left. Horizontal width = 280 μm.

To date only limited study has been given to brachiopods to determine the temporal and environmental significance of the external increments as seen by Mazzullo (1971) and the internal, primary shell increments as seen by MacKinnon and Williams (1974). Studies of *Terebratalia transversa* by Hughes and Rosenberg (1983) questioned the diurnal periodicity of brachiopod growth increments. They reported that intertidal specimens show growth increment patterns in the primary shell layer that narrow and widen cyclically (Figure 2) similar to tidal growth increment patterns reported in bivalved molluscs (Evans, 1972). Also noted were differences in the distinctness of growth lines and in the patterns in specimens collected from deep (40–65 feet) vs intertidal (−1.0 to −3.0 feet below the mean low low water) habitats, with the intertidal specimens showing the more distinct growth lines and cyclical patterns. At present, brachiopods collected, marked along the growing margin, and replaced into the intertidal marine environment, suggest that the growth patterns reflect the mixed semi-diurnal tidal cycles found in the study area near Anacortes, Washington, rather than day/night cycles. Controlled experiments in the laboratory are currently being used by Hughes and Rosenberg in an attempt to better understand the significance of brachiopod growth patterns which appear to be as complex as those found in bivalved molluscs.

2.2. NAUTILOIDS

Changes in the rotational history of the Earth-Moon system have been inferred from the growth patterns in fossil and Recent nautiloids (Kahn and Pompea, 1978; Kahn *et al.*, 1978). Kahn and Pompea (1978) proposed that each 'growth line' found on the surface of living and fossil nautiloid cephalopods forms in response to a solar daily periodicity. Kahn and Pompea further postulated that the internal chamber walls (septa) were formed under lunar monthly controls. Hence, they proposed that the number of days per lunar month could be determined by counting the number of external growth lines between adjacent septa. Examining Phanerozoic fossil nautiloids, they concluded that the lunar month during the Silurian period consisted of approximately nine days – a time considerably shorter than previously proposed (Pannella, 1972), and a conclusion which Professor Runcorn found inconsistent with current theories involving Earth-Moon dynamics (Runcorn, 1979; see also review by Hansen, 1982).

Hughes *et al.* (1980) tested Kahn and Pompea's (1978) conclusions with additional data from fossil nautiloids. The number of growth lines between consecutive septa were counted in many of the same genera of nautiloids that Kahn and Pompea examined. The number of growth lines per chamber was found to be far more variable than Kahn and Pompea acknowledged in their study. The variability was such that it rendered highly questionable Kahn and Pompea's assumption that external growth lines and septa in nautiloids are accreted with regular daily and monthly periodicity respectively. Consequently, their conclusions must be regarded as unfounded. The discussion below summarizes the Hughes *et al.* (1980) study.

Nautiloid collections at the British Museum (Natural History) and Geological Sciences Museum, London, were examined. Forty-three specimens which had distinct 'growth lines' were selected for study. The growth lines on these specimens could be seen clearly in the region where successive chamber walls intersect with the outer shell wall (Figure 3). Carbon paint applied to some specimens enhanced the distinctness of the growth lines. The junction of the chamber wall with the outer shell wall was marked with a paper pointer (see Figure 3), and external growth lines between successive septa were then counted four times (twice each by WWH and PJM independently) with a $10 \times$ dissection microscope using oblique illumination. A total of 127 chambers were thus examined. Chamber lengths were measured with vernier calipers in the region where growth lines were counted. (Note: Although the term 'growth line' is used to maintain consistency with Kahn and Pompea's (1978) study, in the author's opinion growth lines are better termed as lirae because of their morphology, and the presence of growth increments *within* the lirae.)

One additional specimen, *Dawsonoceras annulatum* (BM–C2866; Wenlockian, Upper Silurian) was selected for scanning electron microscopy because of the unusual distance that the growth lines extended out from the shell surface, and the possibility that smaller growth increments could be found within each of the lirae. It was prepared for SEM viewing by making a medial longitudinal section (see Figure 4A), and polishing with #440 and #600 wet/dry sand paper and finally with #1200 microgrit. The polished section was etched with 1.0% EDTA for 20 min, rinsed, air dried, and then gold coated with an ISI sputter coater for viewing with an ISI–40 scanning electron microscope.

Data for the 127 chambers examined on the 43 nautiloid specimens is presented in Table III. There was significant correlation ($P \leq 0.5$) between a chamber length and the number of growth lines per chamber, with the smaller chambers typically having fewer growth lines. For example, compare *'Orthoceras' chinense* (BM–C2575; Ordovician) with 64.3 growth lines per 16.4 mm chamber length to *Cyrtoceras corniculum* (BM–80298; Silurian) with 5.3 growth lines per 2.3 mm chamber length.

When multiple chambers were examined within single specimens there was usually

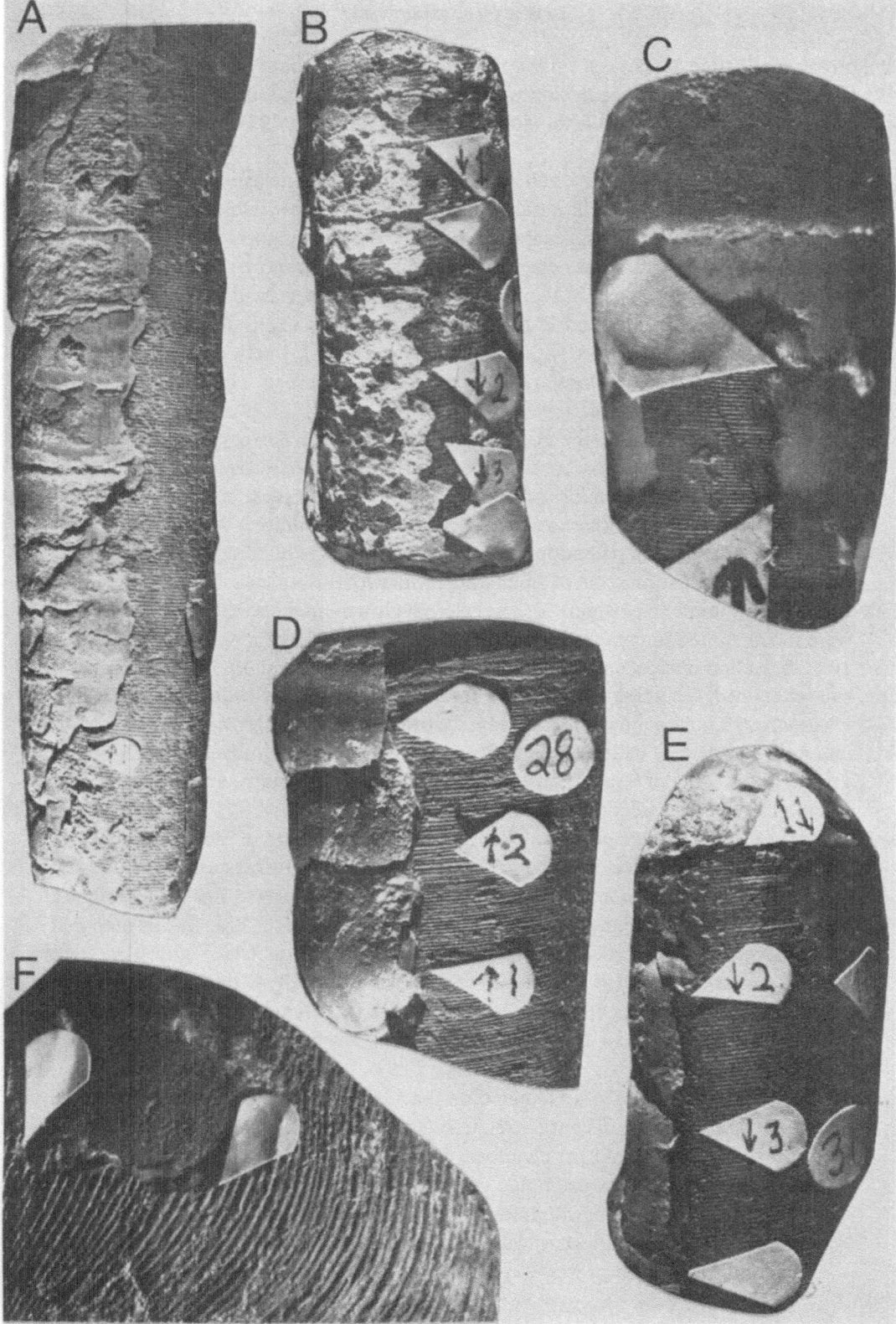

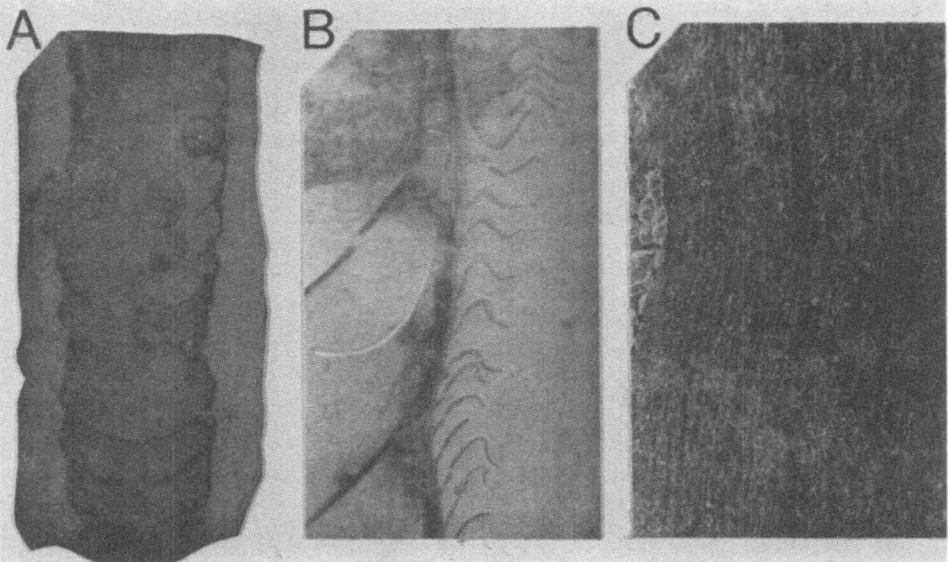

Fig. 4.   A. Medial longitudinal section through *Dawsonoceras annulatum*, BM–C2866, Silurian, X$\frac{1}{2}$; Note external 'growth lines' (lirae) and chamber walls, and the numerous cross-sectioned brachiopods within the nautiloid. B. Enlargement of upper right-hand side of 'A'. C. S.E.M. photomicrograph of external 'growth line' showing lamellae. Direction of growth is to the right. X200.

an increase in counts from the first-formed to the most recently formed chambers – probably linked to an ontogenic increase in chamber length. Examination of Silurian actinocerid cephalopods by Hewitt and Hurst (1983) showed that chamber volume and length increase with the ontogeny of single specimens.

The number of growth lines per chamber remained constant within single organisms in only four of 31 specimens. This is in contrast to Kahn and Pompea's (1978) claim that "the number of growth lines per chamber is constant for the various chambers in a single specimen". Nine chambers in '*Orthoceras*' *kinnekullense* (BM–C1951; Ordovician) gave the following counts per chamber length (mm): 24.5/9.8; 21.8/10.6; 18.0/10.6; 24.5/12.1; 23.5/12.4; 22.8/12.2; 22.5/12.5; 19.8/12.0; 15.5/11.9 (data given from posterior to anterior]. Gyroconic nautiloids were also found to have variable numbers of lines per chamber (e.g., *Gyroceras alatum* (BM–80281; Devonian): 16.0/2.8; 16.8/3.7; 22.8/4.1; 25.3/4.6; 23.5/5.0). Note also that in the latter example the number of growth lines per chamber increases with increased chamber lengths, whereas

←

Fig. 3.   Photomicrographs illustrating the external 'growth lines' (lirae) per chamber in fossil nautiloids from the British Museum (Natural History). A. '*Orthoceras*' *kinnekullense*, BM–C1951, Ordovician, X1; B. '*Orthoceras*' sp., BM–C4894, Silurian, X2; C. '*Orthoceras*' *victima*, BM–70984, Silurian, X4; D. '*Orthoceras*' sp., BM–C82059, Ordovician, X2; E. '*Orthoceras*' sp., BM–C358, Ordovician, X2; F. *Proclydonautilus spirolobus*, BM–C3101, Triassic, X2.

TABLE III

Data for fossil nautiloids from the British Museum (Natural History) and Geological Sciences Museum

System/ Series	Species	I.D. No.	Chambers counted	Mean chamber width (mm)	Growth lines per chamber		
					Range	Mean	S.D.
Miocene	*Aturia* sp.	BM–C73159	1	28.5	39-47	42.7	4.0
Eocene	*Nautilus regalis*	GSM–101137	2	10.8	11-22	15.4	3.8
Cretaceous	*Eutrephoceras* sp.	BM–C81350	2	4.1	24-28	26.1	1.8
	Eutrephoceras clementinum	GSM–114131	1	5.0	12-17	14.0	2.2
	Eutrephoceras clementinum	GSM–113132	2	5.4	11-14	12.5	1.7
Jurassic	*Cenoceras* sp.	GSM–3595	5	7.0	12-25	18.2	4.0
	Cenoceras polygonalis	GSM–1141289	1	15.1	16-22	19.5	3.0
	Cenoceras impendens	GSM–22392	2	4.8	9-11	9.9	0.8
	Cenoceras sp.	BM–37018	1	4.1	10-11	10.3	0.5
	Cenoceras polygonalis	BM–C4221	3	14.0	24-37	29.2	4.1
Triassic	*Syringoceras primoriense*	BM–C20308	1	8.9	50-55	52.8	2.7
	Proclydonautilus spirolobus	BM–C3101	1	15.6	21	21.0	0
Carboniferous	*'Orthoceras'* sp.	BM–C79064	3	3.1	31-44	38.0	4.3
	Metacoceras pulcher	BM–C18006	3	1.7	10-14	12.5	1.1
	Metacoceras pulcher	BM–C5277	3	2.2	20-28	22.1	2.1
	Metacoceras pulcher	BM–C18005	1	2.4	17-23	19.5	3.0
	'Orthoceras' cinctum	BM–C184	4	6.8	31-49	36.4	5.1
	Epidomatoceras maccoyi	BM–C69179	3	2.2	16-24	18.9	2.5
	'Orthoceras' cinctum	BM–C78373	1	6.2	72-85	77.8	6.3
Devonian	*Gyroceras alatum*	BM–80281	4	4.2	10-40	20.7	9.8
	Gyroceras alatum	BM–80281	5	4.0	8-38	20.9	4.2
Silurian	*'Orthoceras' argus*	GSM–104285	3	5.0	60-89	74.5	8.8
	Dawsonoceras cnnulatum	BM–C7423	3	9.1	30-39	34.3	2.9
	'Orthoceras' reticinctum	GSM–104290	2	6.4	20-23	21.1	1.1
	'Orthoceras' sp.	BM–C4894	3	5.9	14-19	16.3	1.6
	'Orthoceras' victima	BM–70984	2	4.9	25-45	36.4	6.0
	'Orthoceras' victima	BM–70984	2	5.2	30-34	32.0	1.2
	'Orthoceras' socium	BM–2005	2	6.6	18-28	22.7	3.5
	Cyrtoceras ambiguum	BM–80319	7	2.8	22-39	28.1	4.0
	'Orthoceras' transiens	BM–80026	5	4.7	10-14	12.3	1.2
	Cyrtoceras corniculum	BM–80316	2	6.0	17-29	24.4	4.3
	Cyrtoceras corniculum	BM–80298	4	2.3	4-6	5.3	0.7
	Cyrtoceras corniculum	BM–80298	3	2.3	15-23	17.5	2.9
Ordovician	*'Orthoceras' elongatocinctum*	GSM–103563	15	2.0	7-13	9.3	1.2
	'Orthoceras' subundulatum	GSM–103549	3	3.0	10-16	12.3	1.7
	'Orthoceras' kinnekullense	BM–C1951	9	11.6	15-24	21.1	2.8
	'Orthoceras' sp.	BM–C358	3	11.3	31-38	32.6	2.0
	'Orthoceras' sp.	BM–C82061	1	6.5	21-24	22.3	1.3
	'Orthoceras' sp.	BM–C82060	1	11.9	22-23	22.8	0.5
	'Orthoceras' sp.	BM–C85052	1	4.5	20-23	21.5	1.3
	'Orthoceras' chinense	BM–C2575	1	16.4	61-67	64.3	2.8
	'Orthoceras' avelini	GSM–104289	4	6.4	10-12	10.8	0.6
	'Orthoceras' sp.	BM–C82059	2	9.6	34-38	35.6	1.4

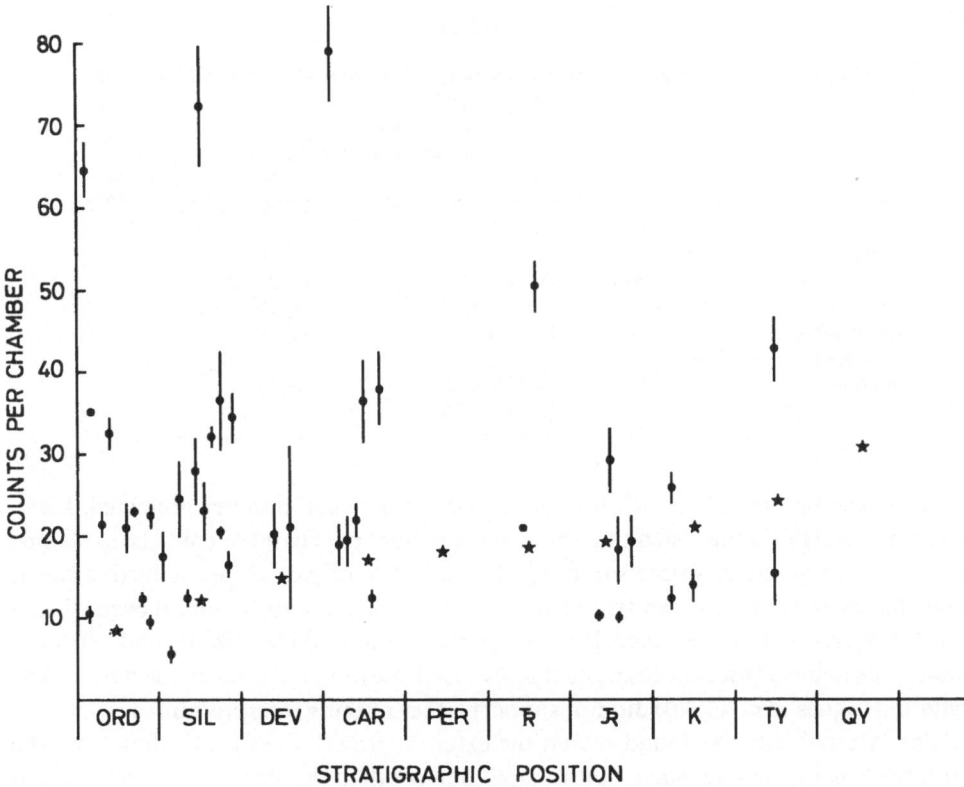

Fig. 5. Nautiloid 'growth line' counts per chamber for British Museum (Natural History) specimens. Points represent the mean of four counts; bars equal the standard deviation; and stars approximate the results of Kahn and Pompea (1978).

in the former example the overall trend is a decrease in counts per chamber with increasing length.

In *Dawsonoceras annulatum* (BM–C2866) the external growth lines (lirae) were seen to extend 5–12 mm out from the shell surface and were 0.1–0.3 mm wide (see Figure 3B). SEM photomicrographs of these cross-sectioned external 'growth lines' showed numerous lamellae (Figure 3C). The lamellae are similar in width (5–20 μm) to those found in the primary shell layer of bivalved molluscs, and these may represent actual daily or subdaily increments.

Figure 5 compares the results of Hughes *et al.* (1980) with previous results obtained by Kahn and Pompea (1978). Most Paleozoic nautiloids showed considerably *more* growth lines per chamber than reported by Kahn and Pompea (1978). A comparison of counts made in genera common to Hughes *et al.* (1980) and Kahn and Pompea (1978) (Table IV) shows a large variation in counts between the two studies. Kahn and Pompea's conclusion that the number of growth lines per chamber decreases with an increase in geologic time is not supported by the specimens examined by Hughes *et al.* (1980).

TABLE IV

Comparison of counts made on genera common to Hughes *et al.* (1980) and Kahn and
Pompea (1978)

System	Genus	Average 'growth lines' per chamber	
		Hughes *et al.* (1980)	Kahn and Pompea (1978)
Tertiary	*Aturia*	43	25, 26
Cretaceous	*Eutrephoceras*	13, 14, 26	22, 22, 23, 23, 24
Jurassic	*Nautilus*	10, 18, 20, 29	18, 18
Carboniferous	*Metacoceras*	13, 20, 22	15
Devonian	*Orthoceras*	- - -	12
Silurian	*Orthoceras*	12, 16, 21, 23	- - -
		32, 34, 36, 75	

When the number of growth lines exceeded the 'normal' number predicted, Kahn and Pompea (1978) suggested that the increase in lines per chamber could be explained as an 'integral multiple' due to sampling of 'inshore' nautiloids. Supposedly this means that the organism responded two or three times to each major temporal event (i.e., a lunar day), rather than just once. However, Hewitt and Watkins (1980) found 'offshore' nautiloids to have twice the lirae per chamber as those found in inshore specimens. The data in Hughes *et al.* (1980) did not support either of these interpretations.

The internal lamellae found within the external growth lines protruding from the surface of *Dawsonoceras annulatum* are similar in width to the growth increments found in the other molluscs with ornate shells, such as the gastropod *Pteropurpura* sp., and the bivalve *Venus lamellata* (Hughes, 1981). Although the 7–10 external growth lines per chamber (see Figure 3B) approximate Kahn and Pompea's (1978) results, Hughes (1981) questioned whether, in view of present growth rates for living *Nautilus macromphalus* (0.25 mm day^{-1}; Martin *et al.*, 1978), secretion of such structures could take place within the assumed daily growth period. Moreover, the internal growth lamellae found in bivalved molluscs, which have widths similar to those found in lirae of *D. annulatum*, are known to be influenced by numerous temporal and environmental parameters.

In addition to Hughes *et al.* (1980) study of fossil specimens, other biological, geophysical, and paleontological objections to Kahn and Pompea's (1978) study have been published, usually in the form of short critiques (Hughes, 1979; Jones and Thompson, 1979; Runcorn, 1979; Saunders and Ward, 1979). Experimental evidence from studies on living *Nautilus* fails to support Kahn and Pompea's initial assumptions that chamber wall formation is under lunar monthly control (Ward *et al.*, 1981; Cochran *et al.*, 1981) or that the formation of external growth lines results from vertical migration timed to solar days (Ward *et al.*, 1981).

The relationship between nautiloid skeletal patterns and environmental parameters may turn out to be as complex as those in bivalved molluscs and other taxa once thought simply to reflect solar daily rhythms.

When mistaken assumptions about the temporal significance of skeletal patterns are used (e.g., nautiloids and brachiopods) they can only lead to uncertain conclusions about the dynamical history of the Earth-Moon system.

3. Summary

Skeletal patterns in invertebrate taxa can no longer be viewed only as simple records of solar and lunar periodicities. Many environmental, temporal, behavioral and physiological parameters are now known to be encoded in invertebrate skeletons. Although considerable study has been given to some taxa, such as the bivalved molluscs, the method of formation and significance of growth patterns in other taxa is in need of further study (e.g., brachiopods, bryozoans, cephalopods, echinoderms, polychaetes).

To maximize the efficiency of research efforts, a survey on the skeletal patterns in all taxa with accretionary skeletons should be made. This might well be followed with concentrated studies on those taxa which show the most promise in providing data on the history of the Earth's rotation.

Although numerous techniques have been developed, it is not apparent from the literature that they are being consistently applied. The use of standardized techniques in the study of growth patterns in promising taxa is necessary so as to produce repeatable results among researchers.

In short, only when careful analyses replace tenuous assumptions regarding the temporal and environmental significance of skeletal growth patterns in living and fossil organisms, will we be able to use extant taxa as tools to assist in deciphering the history of the Earth's rotation.

Acknowledgements

Special thanks to Merilyn Christian, Juanita and Richard Ritland, and Gary Rosenberg for reviewing the entire manuscript. Bruce Saunders and Ida Thompson kindly reviewed the nautiloid section. Figures were prepared by Dorothy Cooper and Paul Denton. Travel assistance to attend 'Keithfest' was provided by R. W. Schwarz, A. U.; Geoscience Research Institute; and an A. U. faculty grant.

References

Barker, R. M.: 1964, 'Microtextural Variations in Pelecypod Shells', *Malacologia* 2, 69–86.
Berry, W. B. N. and Barker, R. M.: 1975, 'Growth Increments in Fossil and Modern Bivalves', in Rosenberg, G. D. and Runcorn, S. K. (eds.), *Growth Rhythms and the History of the Earth's Rotation*, John Wiley and Sons, New York, pp. 9–25.
Bonham, K.: 1965, 'Growth Rate of Giant Clam *Tridacna gigas* at Bikini Atoll as Revealed by Radioautography', *Science* 149, 300–302.
Bourget, E.: 1980, 'Barnacle Shell Growth and its Relationship to Environmental Factors', in Rhoads, D. C. and Lutz, R. A. (eds.), *Skeletal Growth of Aquatic Organisms*, Plenum Press, New York, pp. 469–491.
Clark, G. R.: 1968, 'Mollusk Shell: Daily Growth Lines', *Science* 161, 800–802.

Clark, G. R.: 1974, 'Growth Lines in Invertebrate Skeletons', *Annu. Rev. Earth Planet. Sci.* **2**, 77–99.

Clark, G. R.: 1980a, 'Study of Molluscan Shell Structure and Growth Lines Using Thin Sections', in Rhoads, D. C. and Lutz, R. A. (eds.), *Skeletal Growth of Aquatic Organisms*, Plenum Press, New York, pp. 603–606.

Clark, G. R.: 1980b, 'Techniques for Observing the Organic Matrix of Molluscan Shells', in Rhoads, D. C. and Lutz, R. A. (ed.), *Skeletal Growth of Aquatic Organisms*, Plenum Press, New York, pp. 607–612.

Cochran, J. K., Rye, D. M., and Landman, N. H.: 1981, 'Growth Rate and Habitat of *Nautilus pompilius* Inferred from Radioactive and Stable Isotope Studies', *Paleobiology* **7**, 469–480.

Crabtree, D. M., Clausen, C. D., and Roth, A. A.: 1980, 'Consistency in Growth Line Counts in Bivalve Specimens', *Palaeogeogr., Palaeoclimatol., Palaeoecol.* **29**, 323–340.

Craig, G. Y. and Hallam, A.: 1963, 'Size-Frequency and Growth Ring Analysis of *Mytilus edulis* and *Cardium edule*, and their Paleoecological Significance', *Palaeontology* **6**, 731–750.

Dillon, J. F. and Clark, G. R.: 1980, 'Growth-line Analysis as a Test for Contemporaneity in Populations', in Rhoads, D. C. and Lutz, R. A. (eds.), *Skeletal Growth of Aquatic Organisms*, Plenum Press, New York, pp. 395–415.

Dodge, R. E.: 1980, 'Preparation of Coral Skeletons for Growth Studies', in Rhoads, D. C. and Lutz, R. A. (ed.), *Skeletal Growth of Aquatic Organisms*, Plenum Press, New York, pp. 615–618.

Dolman, J.: 1975, 'A technique for the Extraction of Environmental and Geophysical Information from Growth Records in Invertebrates and Stromatolites', in Rosenberg, G. D. and Runcorn, S. K. (eds.), *Growth Rhythms and the History of the Earth's Rotation*, John Wiley and Sons, New York, pp. 191–222.

Ekaratne, S. U. K. and Crisp, D. J.: 1982, 'Tidal Micro-Growth Bands in Intertidal Gastropod Shells, with an Evaluation of Band-Dating Techniques', *Proc. R. Soc. Lond.* **B 214**, 305–323.

Emiliani, C., Hudson, J. H., Shinn, E. A., George, R. Y., and Lidz, B.: 1978, 'Oxygen and Carbon Isotopic Growth Record in a Reef Coral from the Florida Keys and Deep-Sea Coral from Blake Plateau', *Science* **202**, 627–629.

Evans, J. W.: 1972, 'Tidal Growth Increments in the cockle *Clinocardium nuttalli*' *Science* **176**, 416–417.

Gordon, J. and Carriker, M. R.: 1978, 'Growth Lines in a Bivalve Mollusk: Subdaily Patterns and Dissolution of the Shell', *Science* **202**, 519–521.

Hansen, K. S.: 1982, 'Secular Effects of Oceanic Tidal Dissipation on the Moon's Orbit and the Earth's Rotation', *Rev. Geophys. Space Phys.* **20**, 457–480.

Hewitt, R. A. and Watkins, B.: 1980, 'Cephalopod Ecology across a late Silurian Shelf Tract', *N. Jb. Geol. Palaont. Abh.* **160**, 96–117.

Hewitt, R. A. and Hurst, J. M.: 1983, 'Aspects of the Ecology of Actinocerid Cephalopods', *N. Jb. Geol. Palaont. Abh.* **165**, 362–377.

Hughes, W. W.: 1979, 'Matters Arising: Nautiloid Growth Rhythms and Lunar Dynamics', *Nature* **279**, 453–454.

Hughes, W. W.: 1981, 'Shell Ornamentation in Fossil Nautiloids: a Problem for Earth-Moon Dynamics' (abstr.), *Geol. Soc. Am. Abstr.* **13**(7), 477.

Hughes, W. W. and Clausen, C. D.: 1980, 'Variability in the Formation and Detection of Growth Increments in Bivalve Shells' *Paleobiology* **6**, 503–511.

Hughes, W. W. and Rosenberg, G. D.: 1983, 'Environmental Significance of Growth Increments in the Living Brachiopod *Terebratalia transversa* (abstr.)', *Geol. Soc. Am. Abstr.* **15**(4), 225.

Hughes, W. W., Maddigan, P. J., and Runcorn, S. K.: 1980, 'Nautiloid Shell Growth Rhythms?: Data from British Museum (Natural History) Specimens (abstr.)', *Geol. Soc. Am. Abstr.* **12**(7), 452.

Isley, F. B.: 1913, 'A Fifteen Year Growth record in Fresh-Water Mussels', *Ecology* **12**, 616–619.

Jones, D. S. and Thompson, I.: 1979, 'Matters Arising: Nautiloid Growth Rhythms and Lunar Dynamics', *Nature* **279**, 454–455.

Kahn, P. G. K. and Pompea, S. M.: 1978, 'Nautiloid Growth Rhythms and Dynamical Evolution of the Earth-Moon System', *Nature* **275**, 606–611.

Kahn, P. G. K., Pompea, S. M., and Culver, R. B.: 1978, 'Paleoastronomy', *The Astronomy Quarterly* **2**, 1–18.

Kennish, M. J., Lutz, R. A., and Rhoads, D. C.: 1980, 'Preparation of Acetate Peels and Fractured Sections for Observation of Growth Patterns within the Bivalve Shell', in Rhoads, D. C. and Lutz, R. A. (ed.), *Skeletal Growth of Aquatic Organisms*, Plenum Press, New York, pp. 597–601.

Knutson, D. W., Buddemeier, R. W., and Smith, S. V.: 1972, 'Coral Chronometers: Seasonal Growth Bands in Reef Corals', *Science* **177**, 270–272.

Lutz, R. A. and Rhoads, D. C.: 1977, 'Anaerobiosis and a Theory of Growth Line Formation', *Science* **198**, 1222–1227.

Lutz, R. A. and Rhoads, D. C.: 1980, 'Growth Patterns within the Molluscan Shell', in Rhoads, D. C. and Lutz, R. A. (eds.), *Skeletal Growth of Aquatic Organisms*, Plenum Press, New York, pp. 203–254.

Ma, T. Y. H.: 1934, 'On the Seasonal Change of Growth in a Reef Coral, *Favia speciosa* (Dana) and Water Temperature of the Japanese Seas during the Latest Geological Times', *Proc. Imp. Acad. Tokyo* **10**, 353–356.

MacKinnon, D. I. and Williams, A.: 1974, 'Shell Structure of Terebratulid Brachiopods', *Palaeontology* **17**, 179–202.

Martin, A. W., Catala-Stucki, I., and Ward, P. D.: 1978, 'The Growth Rate and Reproductive Behavior of *Nautilus macromphalus*', *N. Jb. Geol. Palaont. Abh.* **156**, 207–255.

Mazzullo, S. J.: 1971, 'Length of the Year during the Silurian and Devonian Periods: New Values', *Geol. Soc. Am. Bull.* **82**, 1085–1086. ′

Neville, A. C.: 1967, 'Daily Growth Layers in Animals and Plants', *Biol. Rev.* **42**, 421–441.

Olive, P. J. W.: 1980, 'Growth Lines in Polychaete Jaws (Teeth)', in Rhoads, D. C. and Lutz, R. A. (eds.), *Skeletal Growth of Aquatic Organisms*, Plenum Press, New York, pp. 561–592.

Orton, J. H.: 1923, 'On the Significance of "Rings" on the Shells of *Cardium* and other Molluscs', *Nature* **112**, 10.

Pannella, G.: 1972, 'Paleontological Evidence on the Earth's Rotational History Since Early Precambrian', *Astrophys. Space Sci.* **16**, 212–237.

Pannella, G. and MacClintock, C.: 1968, 'Biological and Environmental Rhythms Reflected in Molluscan Shell Growth', *J. Paleont., Mem.*, **42**, 64–80.

Pope, J.: 1976, 'Comparative Morphology and Shell Histology of the Ordovician Strophomenacea (Brachiopoda)', *Paleontogr. Am.* **8(49)**, 129–214.

Rhoads, D. C. and Lutz, R. A.: 1980, *Skeletal Growth of Aquatic Organisms*, Plenum Press, New York, 750 pp.

Rhoads, D. C. and Pannella, G.: 1970, 'The Use of Molluscan Shell Growth Patterns in Ecology and Paleoecology', *Lethaia* **3**, 143–161.

Rosenberg, G. D.: 1980, 'An Ontogenetic Approach to the Environmental Significance of Bivalve Shell Chemistry', in Rhoads, D. C. and Lutz, R. A. (eds.), *Skeletal Growth of Aquatic Organisms*, Plenum Press, New York, pp. 133–168.

Rosenberg, G. D. and Jones, C. B.: 1975, 'Approaches to Chemical Periodicities in Molluscs and Stromatolites', in Rosenberg, G. D. and Runcorn, S. K. (eds.), *Growth Rhythms and the History of the Earth's Rotation*, John Wiley and Sons, New York, pp. 223–242.

Rosenberg, G. D. and Runcorn, S. K.: 1975, *Growth Rhythms and the History of the Earth's Rotation*, John Wiley and Sons, New York, 559 pp.

Rosenberg, G. D., Ashton, M., Hewitt, R., and Simmons, D. J.: 1980, 'Application of Normalized Power Spectra to the Analysis of Chemical and Structural Growth Patterns', in Rhoads, D. C. and Lutz, R. A. (eds.), *Skeletal Growth of Aquatic Organisms*, Plenum Press, New York, pp. 675–686.

Runcorn, S. K.: 1964, 'Changes in the Earth's Moment of Inertia', *Nature* **204**, 823–825.

Runcorn, S. K.: 1966, 'Corals as Paleontological Clocks', *Scient. Am.* **215**, 26–33.

Runcorn, S. K.: 1979, 'Matters Arising: Nautiloid Growth Rhythms and Lunar Dynamics', *Nature* **279**, 452–453.

Saunders, W. B. and Ward, P. D.: 1979, 'Nautiloid Growth and Lunar Dynamics', *Lethaia* **12**, 172.

Scrutton, C. T.: 1965, 'Periodicity in Devonian Coral Growth', *Palaeontology* **7**, 552–558.

Scrutton, C. T.: 1978, 'Periodic Growth Features in Fossil Organisms and the Length of the Day and Month', in Broche, P. and Sundermann, J. (eds.), *Tidal Friction and the Earth's Rotation*, Springer-Verlag, Berlin, pp. 154–196.

Scrutton, C. T. and Hipkin, R. G.: 1973, 'Long-Term Changes in the Rotation Rate of the Earth', *Earth Science Reviews* **9**, 259–274.

Shuster, C. N., Jr.: 1951, 'On the Formation of Mid-Season Checks in the Shell of *Mya*', *Anat. Rec.* **111**, 127.

Tang, S. F.: 1941, 'The Breeding of the Scallop *Pecten maximus* (L.) with a Note on the Growth Rate', *Proc. Liverpool Biol. Soc.* **54**, 9–28.

Ursula, O.: 1981, 'Effect of Simulated Tidal Patterns on Growth and Growth Line Formation in the Little Neck Clam, *Protothaca staminea*', Thesis, Loma Linda University, California, 39 pp.

Ward, P., Greenwald, L., and Magnier, Y.: 1981, 'The Chamber Formation Cycle in *Nautilus macromphalus*', *Paleobiology* **7**, 481–493.

Weber, J. N.: 1969, 'Origin of Concentric Banding in the Spines of the Tropical Echinoid *Heterocentrotus*', *Pacific Science* **23**, 452–466.

Wells, J. W.: 1963, 'Coral Growth and Geochronometry', *Nature* **197**, 948–950.

Weymouth, F. W., McMillin, J. C., and Rich, W. H.: 1931, 'Latitude and Relative Growth in the Razor Clam, *Siliqua patula*', *J. Exp. Biol.* **8**, 228–249.

GROWTH RHYTHMS, EVOLUTION OF THE EARTH'S INTERIOR, AND ORIGIN OF THE METAZOA

GARY D. ROSENBERG

Dept. of Geology, Indiana/Purdue Univsersity, Indianapolis, IN 46202, U.S.A.

Abstract. Growth patterns preserved in the accretionary skeletons of fossils provide the only known method of directly measuring the rate of the Earth's rotation in the distant past. From seasonal and tidal growth patterns of fossils, one can determine the number of days per year and per month, respectively, in the distant past. Together, these values can be used to distinguish the effects of moment of inertia changes on the length of day from those of tidal friction. When the Metazoan accretionary skeleton originated in the Late Precambrian-Cambrian, the length of day determined from fossils was approximately 19 hr. This value requires that density differentiation of the Earth was essentially complete well before the end of the Precambrian. The growing length of day, as well as prior differentiation of oxygenated outer layers (atmosphere, hydrosphere, and crust) from the Earth's dense layers within, were prerequisites for the origin of the Metazoa. Circadian (= approximately 24 hr) rhythms in living Metazoa do not readily adapt to environmental cycles less than about 19 hr. Prokaryotes generally lack circadian rhythms because their generation times are less than a day; prokaryotes were well-adapted to Precambrian days less than 19 hr duration, as well as to oxygen-poor environments. As the length of day increased to 19 hr or more during the Late Precambrian, eukaryotes with life spans substantially longer than a day (and consequently with an ability to postpone energy usage beyond a day) evolved. During the Phanerozoic, moment of inertia changes were relatively small, so that lunar tidal friction became the most important cause of changing length of day. However, some researchers believe that even the former may have left an imprint on fossil growth patterns. This conclusion is difficult to confirm, given the uncertainties of growth pattern analyses. But facies-by-facies comparisons of growth patterns can help reduce this uncertainty; presumed tidal growth patterns should change systematically with depth of habitat, for example. Preliminary analyses for Late Ordovician brachiopods from Indiana suggest that this approach will be productive, and may help evaluate the suggestion that the Late Ordovician-Silurian was a time of unusual evolution of the Earth's moment of inertia during the Phanerozoic.

1. Introduction

What more could a student want than to find a teacher who inspires him? The study of growth rhythms is inspiring because several pioneers in the field have shown how to view nature in a new way. Imagine using fossils to measure the rate of the Earth's rotation and changes in he Earth's moment of inertia in the distant past. As you say in England, "Growth rhythms are a jolly good show".

Professor Runcorn's (1964) role was pivotal. He was the first to use fossil growth patterns to distinguish the causes of the changing length of day in the distant past. Geophysicists have long suspected that the Earth's rotation has been slowing during the Phanerozoic due to the retarding effects of lunar tidal friction. Early in Earth history, the Earth's rotation rate may have increased due to changing mass distribution (= changes in the moment of inertia following the growth of the core, for example). The relative importance to rotation of tidal friction and moment of inertia changes at different times during Earth history remained conjectural until Wells (1963) used Devonian corals to determine the number of days in the Devonian year, and Scrutton (1964) used Devonian corals to determine the number of days in the Devonian month. Wells counted the number of (presumed) daily growth increments within annulations (apparently seasonal patterns) and Scrutton counted the number of (presumably) daily growth increments within monthly (apparently tidal) patterns. Runcorn (1964) used Wells' and Scrutton's values, combined with a consideration of Kepler's Laws, to separate the effects on rotation of tidal friction from moment of inertia changes since the Devonian. That is,

$$(1 + \beta)(L_0 - L) = 2\pi I/t - 2\pi I_0/t_0$$

where

β = ratio between the solar and lunar tides
L_0, L = orbital angular momentum of Moon, present and past, respectively
I_0, I = Earth's moment of inertia, present and past, respectively
t_0, t = length of day, present and past, respectively

which gives the change in the Earth's moment of inertia as well as the loss of angular momentum to the Moon since any specified time in the past. These two values are important for they have implications for the problem of the origin of the Earth-Moon system, differentiation of the Earth's interior into layers of different density, and location of the occurrence of tidal friction (whether in the oceans or also within the mantle).

Runcorn (1964) maintained that Wells' and Scrutton's data required that fractionation of the Earth's interior (growth of the core) was substantially complete prior to the end of the Precambrian. Lambeck (1981) recently evaluated the data gathered from fossils and concurred that lunar tidal friction was of primary import in increasing the length of day, since the Precambrian.

Runcorn (1964) noted that the fossil data did not favor theories for an expanding Earth (increasing moment of inertia) unless G varies. If G were to be decreasing, then the Earth would be expanding, as gravity is the glue which holds the planet together. Creer and other researchers (fide Runcorn, 1964) had proposed that the Earth is expanding at rates up to $1.2\,\text{mm}\,\text{yr}^{-1}$ – but Runcorn noted that this value becomes untenable considering the results of fossil growth pattern analyses, as well as modern determinations of the bounds on changing G.

Nevertheless, Creer (1975) studied the fossil growth pattern data assembled to date and noted inflexions in the curves for days per year and per month through geologic history (Figure 1). The inflexions occur in both curves during the Ordovician-Silurian and Triassic and, because tidal friction can only retard the Earth's rotation, Creer suggested that apparent accelerations in rotation implied moment of inertia effects (variable heat flow within the mantle affecting the depth of the olivine/spinel transition zone). Uncertainties in Creer's inferences are discussed elsewhere (Rosenberg, 1982).

The purpose of this paper is to suggest that, not only do fossil growth patterns imply that the Earth's moment of inertia changed (i.e. density differentiation occurred) most rapidly long before the end of the Precambrian, but that the consequent establishment of relatively oxygen-rich outer layers (atmosphere, hydrosphere, and crust) and growing length of day were requirements for the origin of the Metazoa with accretionary skeletons. And, although Phanerozoic changes in the Earth's moment of

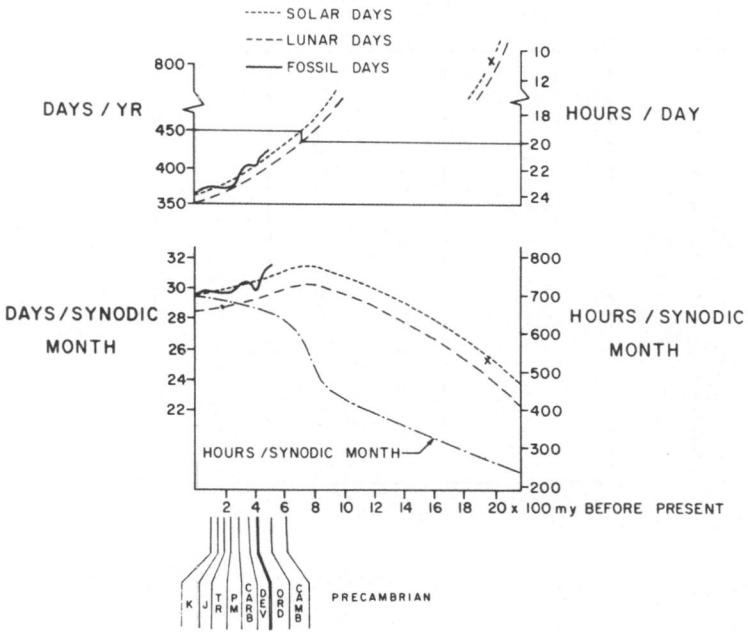

Fig. 1. Variation in number of days per year (top) and per synodic month (bottom) through geologic history. Dashed lines show number of solar days (...) and synodic days (– –) calculated for a 2 msec cy^{-1} increase in length of day. Solid lines (———) Creer's (1975) representation of the fossil data. The 'X' refers to Mohr's (1975) stromatolite data.

inertia influenced the length of day less than did tidal friction, accretionary skeletons may yet reveal changes within the Earth's interior during the Phanerozoic. This conclusion requires distinguishing the environmental causes of growth rhythm variability.

2. Origin of the Metazoa and Early Fractionation of the Earth

The differentiation of the Earth into zones according to density, melting point, and oxygen content is a widely accepted theory for the Earth's early evolution, but the rate and time of completion of differentiation are still conjectural. It is not certain whether evolution of the Earth's interior, changes in the Earth's radius, and consequent changes in the Earth's moment of inertia had ceased during the Precambrian or whether they continued at a significant rate through the Phanerozoic. However, the sudden appearance of the Metazoan accretionary skeleton at the end of the Precambrian and beginning of the Phanerozoic, the composition of the skeleton, the inherent characteristics of biological rhythms, and the Late Precambrian length of day extrapolated from fossil growth patterns themselves are consistent with substantial completion of fractionation prior to the Phanerozoic. The relevance of the first appearance of the Metazoa to this problem was first stated by Rubey in 1951, who believed that the Metazoa would have required prior origin of an ocean whose salinity was close to its

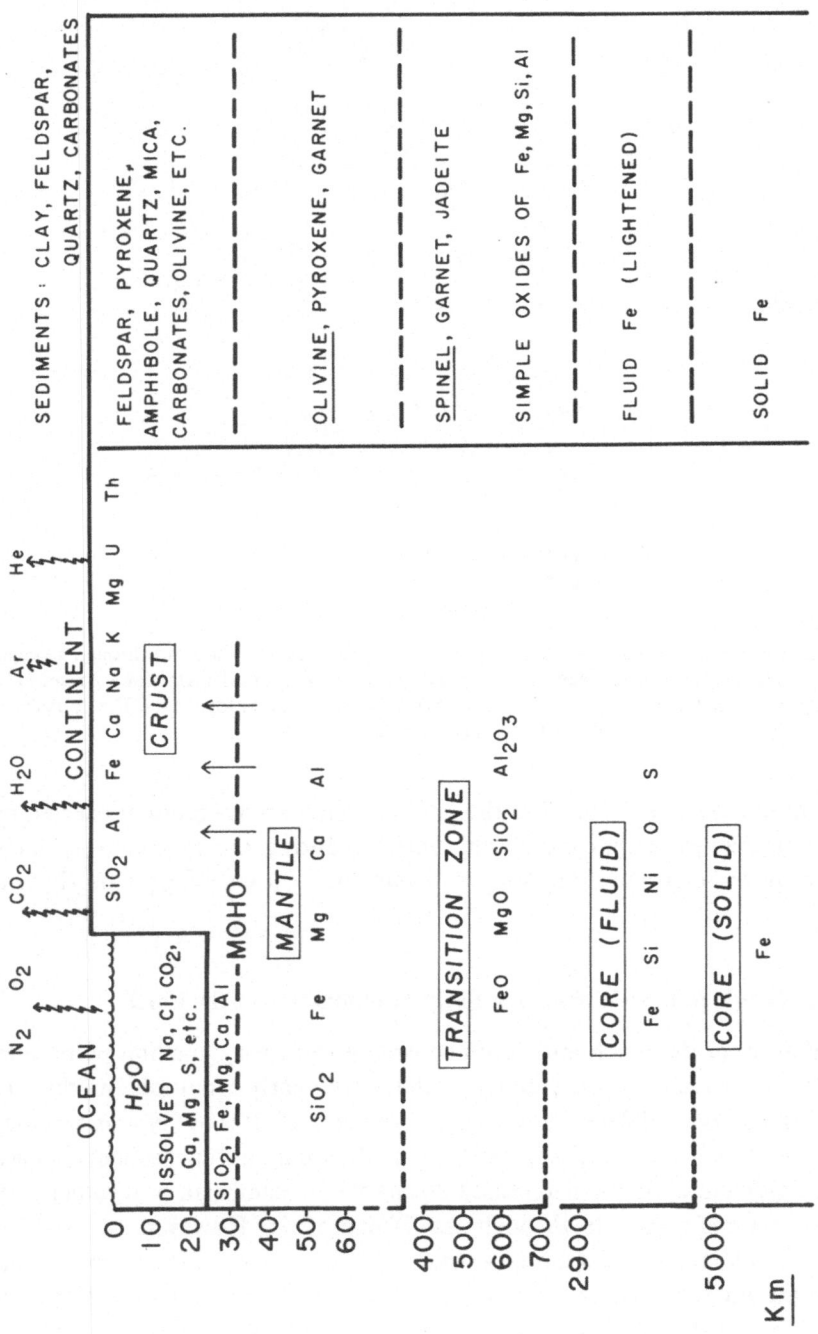

Fig. 2. Zonation of the Earth. The Earth's zonation developed during the Precambrian. This is a compositional zonation (left) with an iron-rich interior and oxygen-rich outer layers, and a mineralogical zonation (right). All sediments, including carbonates, are ultimately a product of the fractionation process. During the Phanerozoic, differential heat flow may have altered the depth of the olivine/spinel transition zone, hence Earth's moment of inertia. Adapted from Dott and Batten, 1981.

modern value. The bearing of the length of day determined from fossils on the evolution of the Earth's interior is explicit in Runcorn's (1964) paper. The point here is that these matters form a coherent picture of the early evolution of the Metazoa.

Figure 2 depicts the well-known density, compositional, and mineralogical zonation of the Earth. The inner layers of the Earth are relatively dense. The outer layers, including the hydrosphere and atmosphere are less so. The Earth's core and mantle are iron-rich and oxygen-poor. The Earth's hydrosphere and atmosphere are oxygen-rich. Even the crust should be considered oxygen-rich for it is 90 % oxygen by volume. Iron predominates in the core, simple oxides of iron predominate in the lower mantle, but increasingly polymerized silicates (with lesser amounts of iron) predominate in shallower layers of the crust. Weathering of the crust has produced sediments composed primarily of clay minerals, feldspar, and quartz. Both the atmosphere and hydrosphere were 'outgassed' from the Earth's interior during differentiation, although the modern oxidizing atmosphere is probably the product of the photosynthetic conversion of the primordial, reducing atmosphere (see Margulis, 1982 for a discussion of the timing of this event).

The oxygen and calcium-rich skeleton is ultimately a product of the above chemical evolution of the Earth in three ways. First, the carbonate is derived from (ultimately) outgassed and respired carbon dioxide and from outgassed water by the reaction, $CO_2 + H_2O \rightarrow H_2CO_3$. Second, the calcium is derived from sedimentary differentiation of pre-existing rocks (sea water, the source of calcium for the early skeleton is salty because calcium and sodium, among other dissolved ions, are volumetrically the most important products of weathering). Third, precipitated calcium carbonate is stable in an oxidizing environment, but not a reducing environment. The first appearance of extensive carbonate sediments in the Middle Precambrian and the origin of the Metazoan skeleton (all of which are accretionary) in the Late Precambrian and Early Cambrian thus testify to substantial completion of chemical evolution within the Earth, and indirectly to consequent changes in the moment of inertia.

Moreover, the growing length of day may have determined the time of origin of the Metazoa. Bonner (1965) provides the link between the history of the Earth's rotation, and the evolution of the Metazoa. (Figure 3 summarizes the following discussion.) Bonner showed that the generation time of organisms tends to increase throughout geologic history. This trend parallels the trend in phyletic size increase described for vertebrates and invertebrates by Cope and Newell, respectively (Newell, 1949). This applies to trees, horses, foraminifera, and molluscs among many other groups at several taxonomic levels. Generation time is the length of time from birth until reproduction, so it establishes a minimum period for the duration of the organism's life span. Protozoa commonly have generation times less than a day. Metazoa generally live longer than a day before reproduction. Not all big clams live longer than little species, and not all big mammals live longer than little mammals, but the phylogenies of many different species show parallel trends in size and life span – enough to constitute a major evolutionary trend. The organism's life span determines the lowest, fundamental frequency of its biological rhythms. Individuals living for an hour cannot display

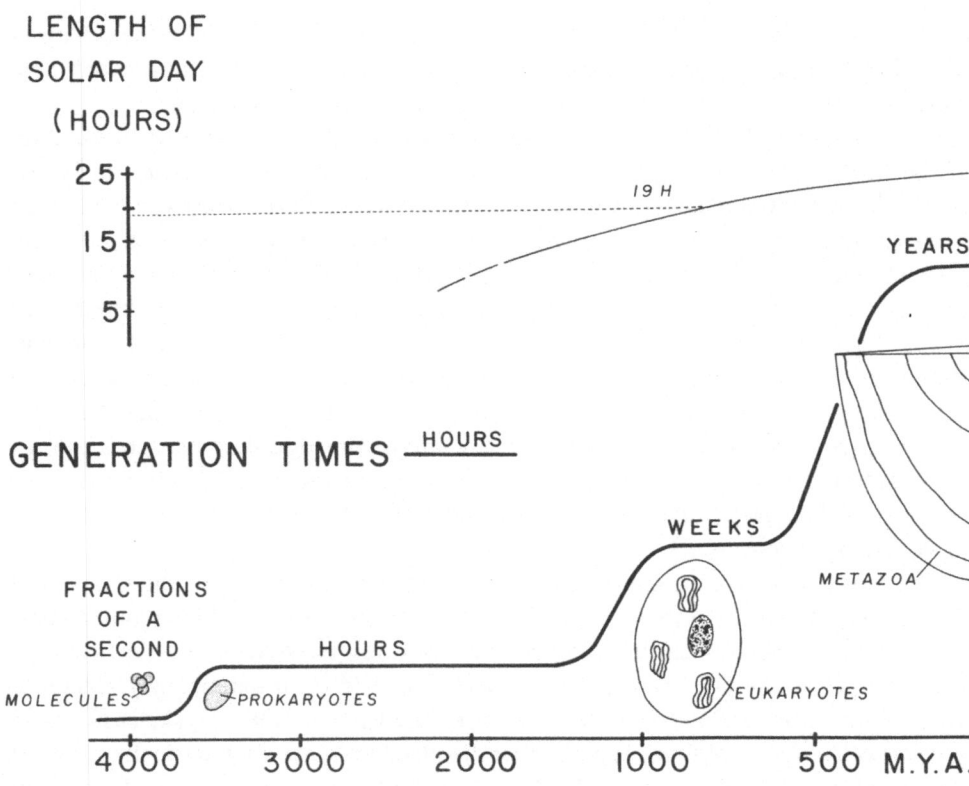

LENGTH OF
SOLAR DAY
(HOURS)

GENERATION TIMES ─HOURS─

Fig. 3. Major grades of biological evolution (molecules, prokaryotes, eukaryotes, and Metazoa) also mark quantum jumps in generation time (fractions of a second, hours, weeks, and years respectively). Generation times determine the fundamental frequencies of biological rhythms. Eukaryotes and Metazoa are the only organisms with circadian rhythms, and they did not originate until the length of day increased to about 19 hr.

rhythms with a 24-hr period, although such rhythms can occur at the population level. Individuals which live for a month do not have annual rhythms. Trees may have 11-yr growth cycles in phase with sunspot cycles, but an *E. coli* which divides hourly cannot on an individual basis be proven to have an 11-yr rhythm. Circadian rhythms "have not been demonstrated in prokaryotes ... a fundamental difficulty (being) that cell division times in prokaryotes tend to be much shorter than 24 hr ('ultradian') and this might obscure circadian phenomena" (Hastings, 1976).

In short, just as little bells naturally ring at higher frequencies than big bells, so the rhythms of little organisms have higher fundamental frequencies, determined by their short life span, than do those of larger organisms with longer life spans.

Tappan and Loeblich (1971) suggest that trends in size and life span could be related to changes in the environmental energy supply. They first recognize major grades of evolution (Figure 3): the molecular, the prokaryote, the eukaryote, and the higher plant and Metazoan levels. The molecular grade (inorganic chemical reactions) originated when the Earth originated about 4500 million years ago and predominated until the prokaryotes arose 3500 million years ago. Eukaryotes first appeared between

2000–900 million years ago, and the Metazoa arose about 700 million years ago, followed soon (600 million years ago) by Metazoa with accretionary skeletons. Land plants originated some 400 million years ago.

The generation times increased in quantum jumps through geologic time as each new grade of evolution was established (Figure 3). The characteristics of the energy utilized by the respective grades changed simultaneously. Thus, the molecular grade is characterized by inorganic reactions which had very short $(10^{-9}$ sec) generation rates with solar radiation and radioactivity as the energy sources. The generation times of prokaryotes are generally less than 24 hr, and these life forms used lower frequency ultraviolet, infrared, and visible light (for photosynthesis) and organic compounds (for heterotrophy) for energy sources. The origin of photosynthesis and the consequent release of oxygen into the atmosphere resulted in an atmosphere with oxygen levels similar to those at present perhaps by 2 billion years ago. The atmosphere could then filter out high frequency solar radiation which facilitates chemical evolution but also destroys organic compounds. The attainment of the eucellular grade brought generation times of up to weeks for life forms such as eucellular protists and multicellular algae, which exploit energy sources such as visible light and organic compounds for heterotrophy. The Metazoa and higher plants followed and were even more successful at delaying growth and energy expenditures, for their generation times increased to years. By this time, visible light and organic compounds had become the most important energy sources for higher life forms.

This brings us to the connection between the length of the day and the origin of the Metazoa. It is interesting (Figure 3) that, by the time of origin of the Metazoa in the Latest Precambrian, the length of the solar day had reached the approximate limit of entrainability (17–19 hr) of circadian rhythms in living organisms (Bunning, 1973; Brown *et al.*, 1970; Saunders, 1977). According to current chronobiological thinking, this means that circadian rhythms in living organisms can be shortened to periods substantially less than 17–19 hr only with difficulty. Circadian rhythms in living organisms exposed to environmental cycles substantially less than 17–19 hr will 'frequency demultiply' (return to their circadian periodicity). Or, at best, if their periods do decrease in response to the shortened environmental cycle, the rhythms will return to their usual period when the circadian environmental cycle is restored. This has prompted many chronobiologists to believe that there is within all Metazoa an inherited, internal, clock which times physiological processes to a 24 hr period and which is independent of environmental cycles. Because there was a rapid change in the length of the month in the Middle-Late Precambrian (Figure 1), it is tempting to speculate that the month had also grown to a duration crucial to Metazoan physiology (although limits on the entrainability of monthly rhythms are far less studied than those on circadian rhythms).

The point here is that prokaryotes were well-adapted to the short length of day that prevailed during most of the Precambrian. They had short life spans, so that their cycles of energy requirements matched environment cycles. There was little selective pressure for the development of longer life spans until the frequency of environmental energy supply declined to 19 hr. The ability of the Metazoa to delay growth and regeneration is also an ability to defer energy usage beyond the daily rate at which it is supplied. But why 19 hr should be the limiting value for the adaptation of circadian rhythms (and not 12, 15, or 22 hr) is a mystery.

3. Growth Patterns and Moment of Inertia Changes During the Phanerozoic

Moment of inertia changes during the Phanerozoic were small compared with those that occurred during the Precambrian. The length of day increased throughout the Phanerozoic at the average rate of about 2-2.5 msec cy^{-1}, a value that reflects solely the retarding effects of lunar tidal friction. The question here is whether one can confirm the minor Phanerozoic accelerations that Creer (1975) recognized and suggested were due to changes in the depth of the olivine-spinel transition zone (see Introduction). Given the variability of biological processes, inaccuracies inherent in measurements of growth patterns and the slight changes in Phanerozoic length of day due theoretically to moment of inertia effects (Figure 1), one would have great difficulty answering this question conclusively.

The point of this section is that the issue is not dead. After all, the evidence that fossil growth patterns are seasonal or tidal, and that constituent growth increments are daily, is largely circumstantial in the first place. The credibility of studies of growth patterns does not depend on whether researchers come up with the expected number of days per year or per month in the distant past as much as it depends on whether or not the growth patterns themselves vary consistently with paleo-facies or paleoenvironmental differences. For example, presumed seasonal growth patterns should vary with latitude as Hall (1975) has shown, and presumed tidal growth patterns should vary with depth as Dolman (1975) has shown. Herein lies the method for reducing the uncertainty of analyses of growth rhythms and, ultimately, confirming the utility of the equations that Runcorn (1964) derived, and the trends that Creer (1975) recognized.

Rosenberg (1982) described the growth patterns in *Rafinesquina alternata*, a Late Ordovician brachiopod from southeastern Indiana. Fine growth increments, presumed to be daily (Pope, 1976), appear in clusters, presumed by Rosenberg to be monthly or fortnightly. Figure 4 depicts both structural and compositional growth rhythms in *Rafinesquina alternata*. The top curve (from Rosenberg, 1982) shows approximately seven cycles in growth increment width (each delimited by dashed lines and each having approximately 21 increments) in a continuous series of growth increments beginning at the surface of the shell and proceeding in the direction of growth (to the right). This graph is a smoothed version of the original data and is derived from a recombination (summation) of all but the highest frequencies in the amplitude spectrum (Fourier analysis). The lower curve shows the distribution of calcium (weight percent) within the structural growth series shown above. This curve is also a frequency smoothed version of the original data (electron microprobe measurements of calcium concentration at 1 μm intervals with a 1 μm beam diameter thus yielding continuous traverses across the growth increments). Boundaries of calcium

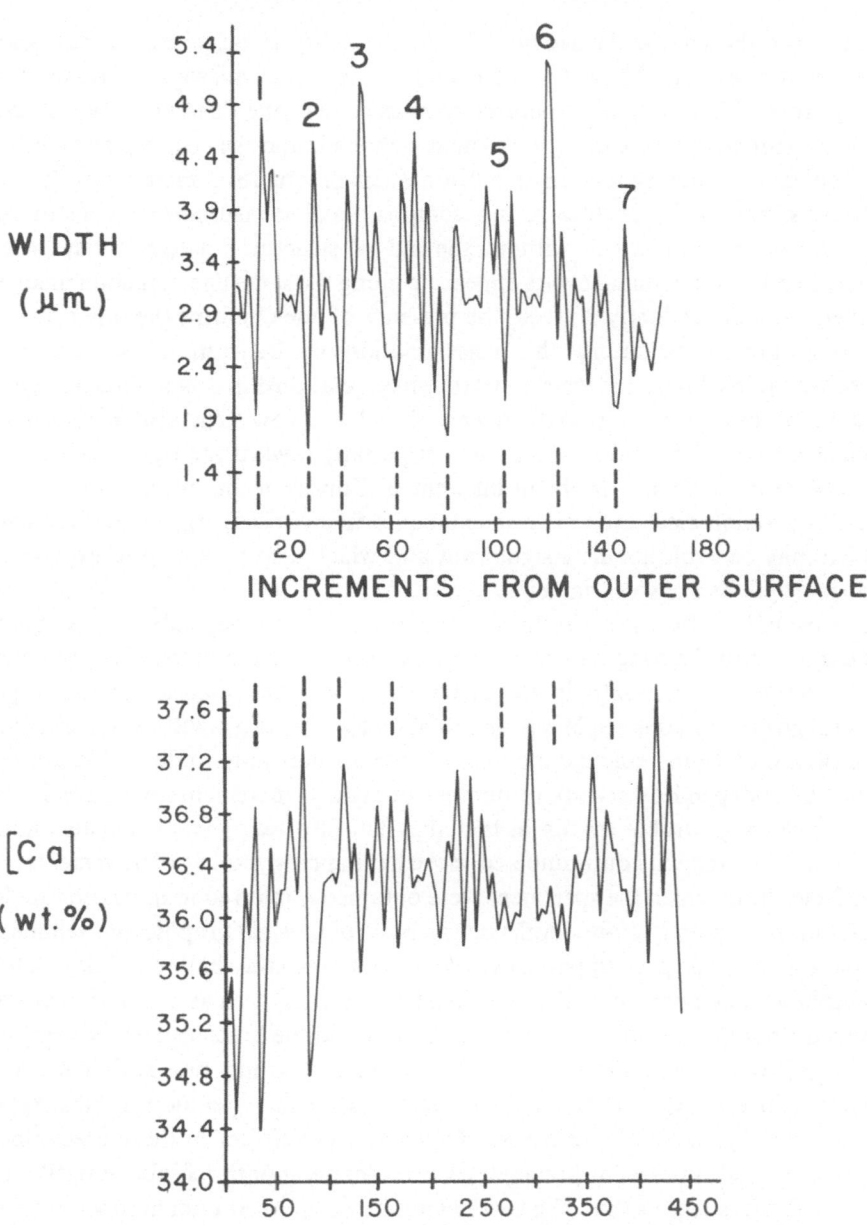

Fig. 4. Structural and compositional growth rhythms in the brachiopod *Rafinesquina alternata* (Late Ordovician, southern Indiana). The top curve shows approximately seven cycles of growth increment width, each averaging about 21 increments (from Rosenberg, 1982). The lower curve shows calcium distribution (electron microprobe analyses) across the structural growth series at top. Dashed lines denote boundaries of (apparent) tidal clusters of structural increments. Note that narrow increments appear relatively calcium-rich at left, but not at right. Direction of growth to right.

cycles are also shown with dashed lines. The replicability of such chemical traveses will be discussed elsewhere. Here it is interesting to note that microprobe traverses are initially more objective than structural measurements; one does not have to decide during measurement where one growth increment ends and the next begins, which can be a problem considering the range in distinctness of structural increments. It would not be possible to decide how many compositional measurements constituted a fortnightly or monthly cycle without the aid of structural patterns, but because structural and compositional cycles appearing along the same line of section through a specimen are independently derived, the presence of one confirms the other. It is not possible at present to specify the phase relationship between the structural and compositional rhythms if indeed a constant phase relationship exists. That is, Figure 4 shows that in some cases (especially toward the left) narrow structural increments are relatively calcium-rich, and in some cases (especially toward the right) they are not. This may be more than a problem inherent in Fourier techniques. It may present interesting physiological and environmental questions (Rosenberg, 1980), the answers to which may be evolutionarily significant and which also may help to explain why biological rhythms are so variable.

The variability of growth rhythms is one reason why independent techniques of measurement must be sought to ensure confidence in determinations of the number of days in the year and month in the distant past. Fourier analyses of the original structural growth series in *Rafinesquina* reveal that widths of growth increments varied with a period of 19 increments, modulated with a lower amplitude oscillation of 27 increments. Independent counts of number of growth increments within individual clusters yield comparable results in that they fall into two groups; clusters having between 8–17 increments outnumbered those having between 18–30 increments.

The facies from which the specimens were obtained is inferred to have been shallow subtidal (approximately 10-m depth) on the basis of biostratigraphic and sedimentologic parameters. The growth patterns are consistent with the shallow subtidal setting. The amplitude of environmental cycles timed to the tides, such as circulation of water and substrate shifts, decline with depth, and so should the brachiopods' perception of them. Shallow subtidal brachiopods thus probably would not add one growth increment with every tide, nor record the true number of days per month. This explains why there were generally fewer growth increments per cluster in these brachiopods, than the expected number of (lunar) days per synodic month (cf. above results with Figure 1). It also suggests that 30 growth increments, the maximum number of growth increments per cluster, would more closely approximate the true number of lunar days per Late Ordovician month than would the mean in the monthly range (23.6 increments).

Although this value supports Creer's (1975) suggestion that the Late Ordovician-Silurian was a time of evolution of the Earth's moment of inertia, the value is difficult to evaluate statistically. But, if the growth increments are lunar daily, if the clusters are fortnightly or monthly, and if the habitat from which the specimens were obtained was shallow subtidal, then the number of growth increments per cluster should change systematically with depth of facies deposition; the average number of growth increments per cluster and the relative power within the monthly vs fortnightly range (amplitude spectra) should decline with depth.

The cumulative percent of the power within the monthly-fortnightly range of increments is graphed in Figure 5 for the *Rafinesquina sp.* specimen (No. 24) pictured in Rosenberg (1982). The spectrum of this specimen is representative of the population from which it is drawn in that more than 17 % of the normalized power within the entire spectrum resides within periodicities between 7–19 increments vs less than 8 % between 21–33 increments. Cumulative percentage graphs should shift in the direction of the arrow with increasing depth, reflecting a trend towards a smaller percent of the power in the monthly range.

Preliminary data support this prediction. Figures 6 and 7 show the stratigraphic section of Late Ordovician sediments in southeastern Indiana from which additional strophomenid brachiopods have been obtained. The section includes calcareous shale and fossiliferous limestone. Several different facies type have been defined (Hay, 1977) on the basis of ratio of limestone to shale, thickness of beds, texture and composition, and weathering characteristics. Subdivision of the section into formations, members, and faunal zones is based on lithologic and faunal characteristics, but it is currently in a state of flux following Hay *et al.* (1981) revision. Depth interpretations are, of course, drawn from these same characteristics and Hay's depth inferences (Figures 6–7 right columns) are subject to re-interpretation.

The data will be presented elsewhere, but at present specimens from locality *T*, a shallow subtidal habitat (Figure 6), approximate the cumulative percentage curve for specimen No. 24 (Figure 5), whereas the populations from the H localities (slightly deeper water) have a smaller percent of the cumulative power in the monthly range (they plot to the right of the curve for specimen No. 24, Figure 5). Preliminary data from locality F, the shallowest facies represented in these strata (Figure 6) plot close to the line for specimen No. 24. This may not be anomalous; Dolman (1975) showed that in the shallowest habitats in which *Cardium* live, the amplitude of fortnightly growth rhythms increased relative to the monthly rhythms. This may reflect a declining frequency of submergence and increasing duration of emergence as the highest tide levels are approached. That is, brachiopods at shallow-most levels would add fewer growth increments than days per month for just the opposite reason that subtidal specimens add fewer growth increments than days per month. Finally, some of the specimens from shaly (deep water) facies (such as *W* and *X*, Figure 7) possess a greater than expected percentage of the cumulative power in the monthly range. Although this contradicts the prediction shown in Figure 5, it is interesting because the depth interpretations of the shaly units are subject to considerable debate. Some workers believe that a high percentage of clastics in these Late Ordovician units implies deposition in quiet water, offshore, whereas others believe that the same characteristic implies deposition in shallow water, close to the source of the sediments (see Rosenberg, 1982 for discussion). The data presented here would support the latter

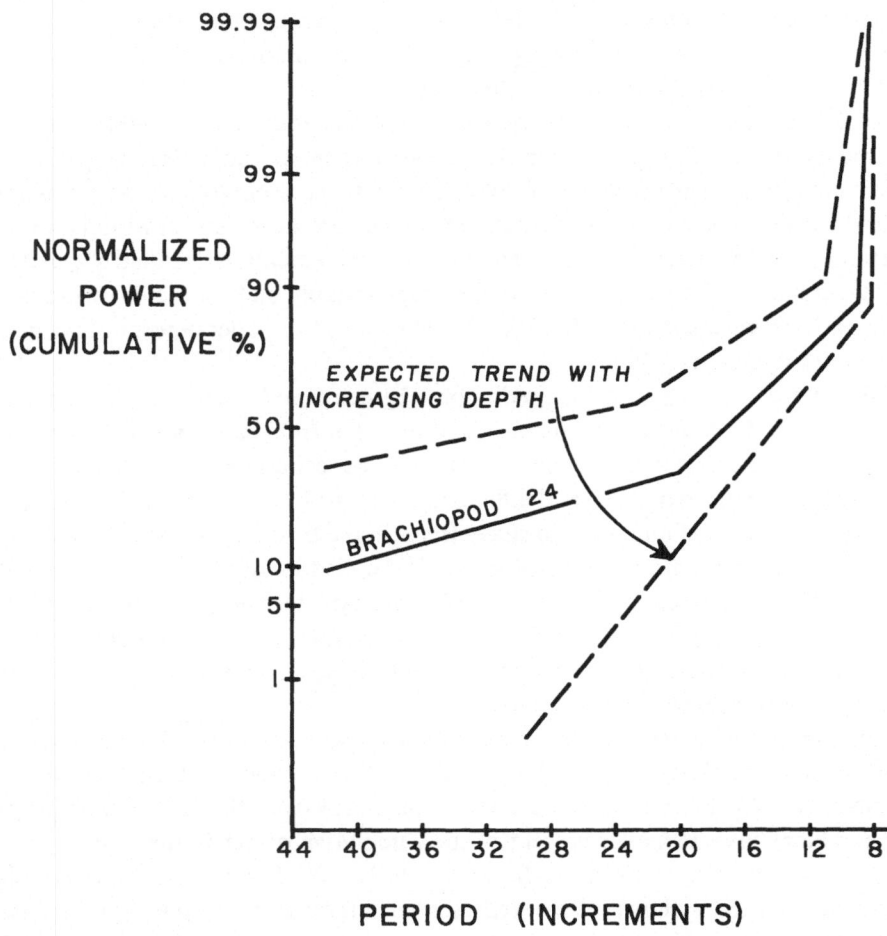

Fig. 5.   Cumulative percent of power (monthly-fortnightly range in power spectrum) in brachiopod skeletal rhythms. Brachiopod No. 24 figured in Rosenberg (1982) is shallow subtidal. Power in monthly range should decrease with depth, and cumulative curves move in direction of arrow.

hypothesis, so that the shaly limestone perhaps represents quiet mud flat environments nearer to sea level than previously supposed.

Additional data must be obtained before any firm conclusions can be drawn. But this writer submits that such comparisons of growth patterns by facies are good means to test assumptions concerning the temporal and environmental significance of fossil growth patterns. Such comparisons are necessary to test claims that the length of day has changed throughout geologic history. Ultimately, such comparisons may help limit the uncertainty of analyses of growth patterns and help to distinguish the effects of lunar tidal friction from those due to changes in the Earth's moment of inertia.

BROOKVILLE NORTH SECTION

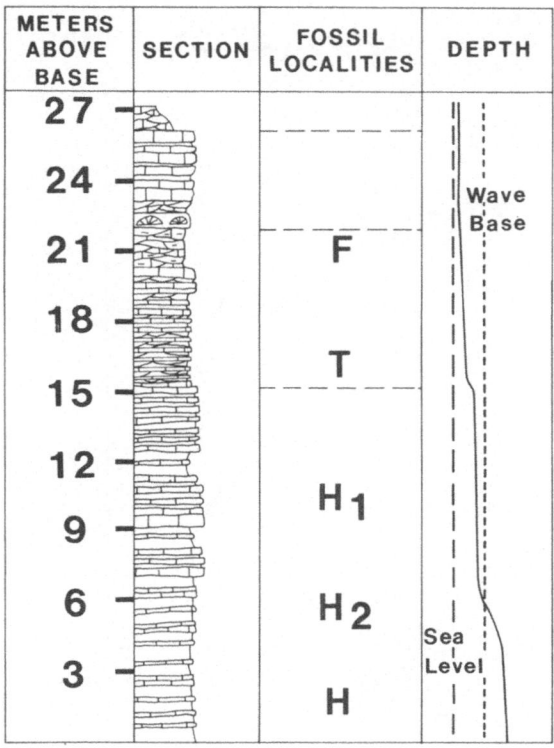

Fig. 6. Stratigraphic section of Late Ordovician strata in southeastern Indiana (Brookville North area, see Hay (1977)). Brachiopods obtained from localities shown for growth pattern analyses. Limestone (brick and nodule patterns) and shaly limestone (unpatterned) predominate. Depth interpretations from Hay (1977) and Hay *et al.* (1981).

4. Conclusions

Growth patterns preserved in the accretionary skeletons of fossils may yet confirm suggestions that the Earth's moment of inertia evolved during the Phanerozoic, even though Phanerozoic evolution of the Earth's moment of inertia was slight compared with the changes that occurred during the Precambrian. Conclusive distinction of the causes of rotational change requires limiting the uncertainty of the analyses of growth patterns. Facies-by-facies comparisons of growth patterns provide one approach to the problem. If growth patterns do change systematically with facies, as independent sedimentologic-stratigraphic analyses predict, then it will be possible to say with increasing conviction that the growth patterns have a determinate temporal signifi-cance that is useful for geophysical predictions. If the skeletal patterns do not change systematically with facies, then the evolutionary implications will be nevertheless

BON WELL HILL SECTION

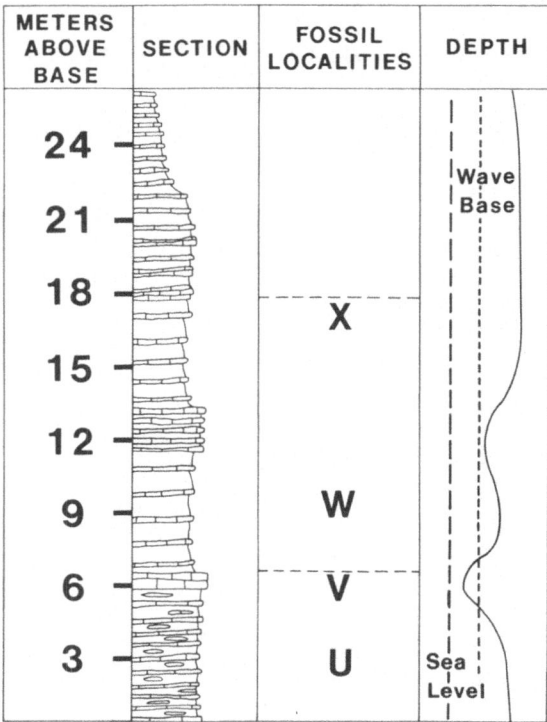

Fig. 7.   Similar to Figure 6, this portion of the section (Bon Well Hill area (Hay, 1977)), underlies that of
Figure 6.

important. In any case, it is worthwhile to remember that the consequences of
Precambrian changes in the moment of inertia on the origin of the Metazoa are
demonstrable. First, because the oxygen-rich accretionary skeleton is the biological
end result of density differentiation within the Earth, and, second, because the
lengthening day in the Late Precambrian, consequent to the effects of changes in the
value of tidal friction and moment of inertia, may have directly determined the time of
the first appearance of the Metazoa.

Acknowledgements

I thank Arthur Mirsky, Chairman of the Department of Geology, IUPUI for his
patient review of the manuscript and for his good suggestions. Helen Loeblich and
Norman Newell kindly encouraged an early version of the manuscript. Bill Davis
prepared the diagrams. Research supported by NSF Grant EAR 82–05824.

References

Bonner, J. T.: 1965, *Size and Cycle*, Princeton University Press, Princeton, New Jersey, 201 p.

Brown, F. A., Jr., Hastings, J. W., and Palmer, J. D.: 1970, *The Biological Clock: Two Views*, Academic Press, New York, 91 p.

Bunning, E.: 1973, *The Physiological Clock*, The English Universities Press, London, 239 p.

Creer, K. M.: 1975, 'On a Tentative Correlation between Changes in the Geomagnetic Polarity Bias and Reversal Frequency and the Earth's Rotation through Phanerozoic Time', in Rosenberg, G. D. and S. K. Runcorn (eds.), *Growth Rhythms and the History of the Earth's Rotation*, John Wiley, London, pp. 293–318.

Dolman, J. W.: 1975, 'A Technique for the Extraction of Environmental and Geophysical Information from Growth Records in Invertebrates and Stromatolites', in Rosenberg, G. D. and S. K. Runcorn (eds.), *Growth Rhythms and the History of the Earth's Rotation*, John Wiley, London, pp. 191–222.

Dott, R. H., Jr. and Batten, R. L.: 1981, *Evolution of the Earth*, McGraw-Hill, New York, 575 p.

Hall, C. A.: 1975, 'Latitudinal Variation in Shell Growth Patterns of Bivalve Molluscs: Implications and Problems, in Rosenberg, G. D. and S. K. Runcorn (eds.), *Growth Rhythms and the History of the Earth's Rotation*, John Wiley, London, pp. 163–175.

Hastings, J. W.: 1976, 'Group Report', in Hastings, J. W. and H. Schweiger (eds.), *The Molecular Basis of Circadian Rhythms*, Dahlem Konferenzen, Berlin, pp. 49–62.

Hay, H. B.: 1977, 'Cincinnatian Stratigraphy from Richmond to Aurora, Indiana', in Pope, J. K. and W. D. Martin (eds.), *Field Guidebook to the Biostratigraphy and Paleoenvironments of the Cincinnatian Series of Southeastern Indiana*, Seventh Annual Field Conference of the Great Lakes Section, SEPM. Miami University, Oxford, Ohio, pp. I–1 to I–34.

Hay, H. B., Pope, J. K., and Frey, R. C.: 1981, 'Lithostratigraphy, Cyclic Sedimentation and Paleoecology of the Cincinnatian Series in Southwestern Ohio and Southeastern Indiana', in Roberts, T. G. (ed.), *Geological Society of America, 1981, Field Trip Guidebook*, Vol. 1: Stratigraphy, sedimentology, pp. 73–86.

Lambeck, K.: 1981, *The Earth's Variable Rotation*, Cambridge University Press, 449 p.

Margulis, L.: 1982, *Early Life*, Science Books, International, Boston, 138 p.

Mohr, R. E.: 1975, 'Measured Periodicities of the Biwabik (Precambrian) Stromatolites and their Geophysical Significance' in Rosenberg, G. D. and S. K. Runcorn (eds.), *Growth Rhythms and the History of the Earth's Rotation*, John Wiley, London, pp. 43–56.

Newell, N. D.: 1949, 'Phyletic Size Increase: An Important Trend Illustrated by Fossil Invertebrates', *Evolution*, 3, 103–124.

Pope, J.: 1976, 'Comparative Morphology and Shell Histology of the Ordovician Strophomenacea (Brachiopods)', *Paleontogr. Am.* 8(49), 129–214.

Rosenberg, G. D.: 1980, 'An Ontogenetic Approach to the Environmental Significance of Bivalve Shell Chemistry', in Rhoads, D. C. and R. A. Lutz (eds.), *Skeletal Growth of Aquatic Organisms: Biological Records of Environmental Change*, Plenum, New York, pp. 133–168.

Rosenberg, G. D.: 1982, 'Growth Rhythms in the Brachiopod *Rafinesquina alternata* from the Late Ordovician of Southeastern Indiana', *Paleobiology* 8(4), 389–401.

Runcorn, S. K.: 1964, 'Changes in the Earth's Moment of Inertia', *Nature* 204, 823–825.

Saunders, D. S.: 1977, *An Introduction to Biological Rhythms*, Blackie and Son, Glasgow, 155 p.

Scrutton, C. T.: 1964, Periodicity in Devonian Coral Growth', *Paleontogy* 7, 552–558.

Tappan, H. and Loeblich, A. R.: 1971, 'Geobiologic Implications of Fossil Phytoplankton Evolution and Time-Space Distribution', in Kosanke, R. and A. T. Cross (eds.), *Palnology of the Late Cretaceous and Early Tertiary*, Geological Society of America, Special Paper 127, pp. 247–340.

Wells, J.: 1963, Coral Growth and Geochronometry', *Nature* 197, 948–950.

S. K. RUNCORN'S COMMENTARY

I recall first hearing of the paper of J. W. Wells on the existence of daily and annual growth lines on the epithica of Devonian corals when I paid a call on Sir Mark Oliphant in the Australian National University on my first visit in 1963: he had just received *Nature* and was intrigued by the idea. Later after lecturing in Oxford Professor L. R. Wager introduced me to a research student, C. T. Scrutton, now in the Geology Department of this University, who was discovering what turned out to be monthly growth rings in Devonian corals. I was excited though sceptical but I showed in a paper in 1964 in *Nature* that Wells' and Scrutton's counts of the number of days in the year and month respectively, which were independent observations, of course, were together consistent with the dynamical laws of the Earth-Moon system. Thus these biological growth increments were indeed the astronomical periods that Wells and Scrutton supposed – that two numbers were correct by chance was remote. I said afterwards that this was the first time I had seen advantage in the absence of mathematics and physics in the training of geologists.

These interesting contributions are from two scientists who came to the department as post-doctoral fellows following a number of Ph.D. students here who have worked in the field. All were keen to understand the biological mechanism by which it seemed that the great astronomical periods controlling the environment impressed themselves on marine life. Only a few enthusiasts are working in this field and Dr Hughes and Dr Rosenberg are making sure and steady progress – a field which hopefully will blossom forth with results of great geophysical value. Even now the observations, limited though they are, are sufficient to exclude the most extreme suggestions about the expansion or contraction of the Earth with the consequent large changes in its moment of inertia.

The dilemma of extrapolating back the history of the lunar orbit – which goes back to a paper by L. B. Slichter who showed, using the present lunar tidal friction values to calculate the values in the past, that as the Earth-Moon distance diminishes a 'catastropic' approach of the Moon to the Earth occurs not at 4500 Gyr but unacceptably late in the Pre-Cambrian. Although changes in the oceans, where tidal energies are dissipated, have been suggested as a way out, we must await the better determination of the rotation of the Earth and Moon from palaeontology, especially in the Pre-Cambrian by stromatolites, before real progress in the understanding of the history of the Earth-Moon system can be made.

NON-TIDAL CHANGES IN THE LENGTH OF THE DAY:
700 BC TO AD 1982

F. R. STEPHENSON

Dept. of Physics, University of Durham

and

L. V. MORRISON

Royal Greenwich Observatory

Abstract. Occultations and eclipses from ancient times down to the present are analysed to determine changes in the length of the day. By subtracting the expected tidal contribution from the observed changes, the non-tidal variations are obtained. The non-tidal variations are shown to occur on time-scales of decades and millennia.

1. Introduction

Changes in the rate of rotation of the Earth, usually referred to as changes in the length of the day (LOD), have been measured astronomically on timescales of weeks to millennia. Tidal braking by the Moon and Sun is expected to cause the rate of rotation to decrease almost linearly with time, thereby producing a steady increase in the LOD of about 2.5 ms cy^{-1}. When this tidal component is subtracted from the observed changes in the LOD, the non-tidal changes, which are of particular interest in geophysics, are obtained. These non-tidal variations provide data appertaining to the dynamical interaction between the atmosphere, mantle and core of the Earth. The nature of these interactions has been an abiding interest of Keith Runcorn, especially in the studies of electromagnetic coupling and paleontological clocks. His active interest in these areas is partly responsible for the progress in the astronomical determination of the non-tidal changes in the LOD reported in this paper.

The introduction in 1955 of atomic clocks as a comparative standard of time-keeping made possible the resolution of changes in the LOD of the order of 0.1 ms over a period of a few weeks. Changes in the LOD on a time-scale of a year or less have been shown conclusively to arise from the interchange of angular momentum between the atmosphere and mantle (Hide *et al.*, 1980). Before 1955, no man-made clock was stable enough to provide a standard of reference for the detection of changes in the LOD on a time-scale longer than a year. Only astronomical observations of the regular motions of the planetary system or phenomena associated with these were capable of providing the necessary standard. This paper is mainly concerned with the collection and analysis of the most accurate astronomical observations made before 1955 which are pertinent to the determination of changes in the LOD.

The astronomical observations analysed in this paper fall naturally into two periods: the pre-telescopic period from 700 BC to AD 1600 and the telescopic era since the latter

Geophysical Surveys 7 (1985) 201-210. 0046-5763/85.15.
© 1985 *by D. Reidel Publishing Company.*

date. The telescopic data reveal that the LOD fluctuated by up to ± 4 ms on a time-scale of some 30 y. This degree of resolution cannot be derived from the pre-telescopic data and for this period only long-term trends can be discerned. Between 700 BC and some undefined epoch in the interval from AD 1 to AD 950 the LOD increased at an average rate of 2.4 ms per century. However from AD 950 to AD 1800 the rate of increase was only 1.4 ms cy^{-1}. The increase in the LOD due to tidal braking is 2.4 ms cy^{-1}. Hence it is apparent that there was no significant non-tidal contribution before AD 1 but there was a non-tidal *decrease* in the LOD of 1.0 ms cy^{-1} after AD 950. The non-tidal contribution to the change of LOD is thus variable on a timescale of millennia. In looking for mechanisms to account for this behaviour, it should be noted that it is not possible to distinguish in the astronomical data between secular changes in the moment of inertia of the Earth and the redistribution of angular momentum within the Earth.

In this analysis, the Moon's orbital motion forms the basis of the uniform time-standard used in deriving the changes in the LOD. It is assumed that the tidal acceleration of the Moon has been constant over the last 2700 y and that conversely, tidal friction in the oceans and body of the Earth has also been constant during this time. This assumption is crucial to our method of analysis, so we begin by considering the determination of the tidal acceleration of the Moon and what evidence there is for it having been constant over the period in question.

2. Tidal Accelerations of the Moon and Earth

The realisation of a uniform time-standard from observations of the Moon's position is dependent upon having a complete description of the lunar motion. This knowledge has been somewhat deficient in the past because of the large uncertainty in the lunar secular acceleration ($\dot{n}$). The cause of this acceleration lies in the reciprocal action of the tidal bulges which the Moon raises on the Earth. Recent determinations of $\dot{n}$ using a variety of techniques have considerably reduced the uncertainty in this parameter as is shown by the results summarised in Table I.

TABLE I

Recent measurements of the lunar orbital acceleration

Method	Author(s)	$\dot{n}$ (arcsec cy^{-2})
Transits of Mercury	Morrison and Ward (1975)	$-26 \quad \pm 2$
Numerical Tidal Model	Lambeck (1980, p. 335)	$-30.6 \quad \pm 3.1$
Artificial Satellites	Cazenave (1982)	$-26.1 \quad \pm 2.9$
Lunar Laser Ranging	Dickey and Williams (1982)	$-25.1 \quad \pm 1.3$

In the above table we have not included determinations based on the type of material analysed in this paper (i.e. eclipses and occultations). In our opinion it is not possible to separate adequately the contributions from $\dot{n}$ and $\dot{\omega}$ (the Earth's rotational acceleration) from these observations – these unknowns are too highly correlated. An independent result for $\dot{n}$ is required to enable a solution for $\dot{\omega}$ and related quantities to be made.

In our analysis we have used the result derived by Morrison and Ward (1975) since this is based on data covering the longest time-span (some 250 y). This value is equal to the mean of all four determinations (-26.0 ± 1.0 arcsec cy^{-2}). If a substantially different value for $\dot{n}$ is accepted in the future, our results can be readily adjusted to conform to it.

Many of the eclipse observations discussed below date back more than two millennia, the earliest being made in Babylon around 700 BC. The constancy of $\dot{n}$ over this time-scale is dependent on the constancy of tidal friction. It is thus important to consider the stability of these quantities over the required period.

Tidal friction in the solid body of the Earth is generally calculated to be only about 5 or 10% of that in the oceans (e.g. Zschau, 1978; Cazenave, 1982). Dissipation in shallow seas was formerly regarded as much greater than that occurring in deeper water, but Lambeck (1980, pp. 330–334) argues against this. There seems to be no valid reason why tidal friction in either the solid body of the Earth or in the deep oceans should vary on a time-scale of millennia so that further discussion can be restricted to the continental shelves. Studies by Fairbridge (1961) and Mörner (1971) indicate that over the past three millennia the mean global sea-level was never more than 1 or 2 m above or below the present level. As the average depth of water on the sloping shelves is about 100 m, over the historical period changes in the total area covered by these stretches of relatively shallow water might be expected to be of the order of one per cent. Even if shallow seas provide the bulk of the tidal dissipation of the Earth's rotational energy, it seems reasonable to treat $\dot{n}$ as effectively constant over the period covered by our observational data.

Assuming a fixed value for $\dot{n}$, the rate of change in the length of the day due to tidal action may be calculated on the basis of conservation of angular momentum in the Earth–Moon system. Lambeck (1980, Equation 10.5.2) finds the tidal acceleration of the Earth's spin in terms of the lunar acceleration to be:

$$\dot{\omega}_T = 51\,\dot{n}\,\text{arcsec cy}^{-2}$$

i.e. $$(\dot{\omega}_T/\omega) = 1.07 \times 10^{-11}\,\dot{n}\,\text{y}^{-1}. \tag{1}$$

The above expressions include both lunar and solar tides (in the solid body of the Earth, oceans and atmosphere) as well as the contribution from the tidally-induced secular changes in the eccentricity and inclination of the lunar orbit.

With our adopted figure for $\dot{n}$ of -26 arcsec cy^{-2}, we obtain

$$(\dot{\omega}_T/\omega) = -27.8 \times 10^{-11}\,\text{y}^{-1}$$

which leads to a result for the rate of change in the LOD due to tides of $+2.40$ ms cy^{-1}. The eclipse and occultation observations analysed in this paper lead either to individual values or ranges of values for the change in LOD relative to the adopted standard (86 400 SI s). Hence at the epoch of each observation the appropriate non-tidal change in the LOD may be obtained by direct subtraction of the tidal term.

In the present investigation, lunar positions were computed using the analytical lunar ephemeris $j = 2$ (IAU, 1968). This ephemeris is based on a value for $\dot{n}$ of -22.44 arcsec cy^{-2}. In order to fully incorporate our choice of $\dot{n}$, it was necessary to apply the following additions to the mean lunar longitude of the ephemeris:

$$\Delta L = -1''.54 + 2''.33\, T - 1''.78\, T^2 \tag{2}$$

where T is measured in centuries from the epoch 1900.0 (Morrison, 1979b).

Our results are affected to a minor extent by inaccuracies in the adopted expression for the longitude of the lunar node. Van Flandern (1983) deduced a correction to the mean longitude of the node in $j = 2$ of

$$-1''.55 + 5''.87\, T - 0''.049\, T^2.$$

We have included this correction in our calculation of eclipses, but – as for occultations – its effect is negligible.

3. Observations

The observational material analysed in this paper is entirely in the form of occultations and eclipses. This covers the period from earliest times – in practice about 700 BC – to AD 1954. Each observation was initially analysed to give a result for ΔT, the difference between Dynamical Time (TDT) and Universal Time (UT) arising from variations in the Earth's rate of rotation. With the inception of Atomic Time (TAI) in 1955, observations of this kind were superseded. Beginning with that year, annual values of the difference TAI–UT1 (where UT1 is a precise measure of UT freed from the effects of polar motion) are published by the Bureau International de l'Heure, Paris. These values may be converted to ΔT by the relation:

$$\Delta T = 32^{s}.184 + (\text{TAI} - \text{UT1}). \tag{3}$$

3.1. THE TELESCOPIC PERIOD (c. AD 1600–1954)

In the telescopic period we have concentrated almost exclusively on timings of occultations of stars by the Moon. Unlike eclipse contacts (other than at the total phase), occultations are effectively instantaneous events. They have the advantage over other types of observation that no secondary measurements – such as aligning a star with a wire – are required. The observations investigated here are taken from the catalogues of Morrison et al. (1981) for the years 1623–1942 and Morrison (1978) for the years 1943–1954. In the very earliest period, relatively few occultations are preserved and the precision of timing is low. Hence before AD 1670, timings of fourth contacts for solar eclipses – taken from Morrison et al. (1981) – were used to supplement the occultation data.

The occultations were reduced to give values of ΔT as explained in Morrison (1979a). Corrections were applied for the lunar limb profile using the charts of Watts (1963). The standard deviation (σ) of ΔT and the number of observations within 3σ at various periods are given in Table II.

TABLE II

Standard deviation of ΔT and number of observations within 3σ

Period	σ	$N < 3\sigma$	Period	σ	$N < 3\sigma$
1620–1669	60 s	94	1820–1860	1.5 s	1 265
1670–1699	15 s	65	1861–1942	1.3 s	24 800
1700–1759	5 s	169	1943–1954	1.0 s	~ 10 000
1760–1819	2 s	313			

In the period 1620–1860 the individual values of ΔT were smoothed by cubic splines having 13 knots spaced at proportionately smaller intervals of time according to the increase in the number of observations. After 1860 the values of ΔT were grouped into annual means. These latter values are published in Table I of Morrison (1979a).

3.2. THE PRE-TELESCOPIC PERIOD (about 700 BC to AD 1600)

Before the invention of the telescope, timings of occultations are extremely rare. However, observations of eclipses of both Sun and Moon are fairly numerous and we have restricted our attention to these.

Useful eclipse observations are of three main kinds. These are: (i) timings of the contacts for both solar and lunar eclipses; (ii) observations of the occurrence of lunar eclipses near moonrise or moonset – without estimates of time; and (iii) untimed observations of the occurrence of total solar eclipses. Data in category (i) are from both Babylon (700 BC to 50 BC) and the Arab lands (AD 800 to 1000). Those in category (ii) are exclusively from Babylon while the most reliable observations of total solar eclipses are mainly from China (200 BC to AD 800) and Europe (AD 900 to 1600). There are few usable eclipse observations of any kind between about AD 100 and 800. In all we have analysed some 60 ancient eclipses and a similar number of medieval eclipses. A full list of data will be published elsewhere.

Of the above material, eclipse times were typically estimated to the nearest 4 minutes, either directly (as in Babylon) or by measuring altitudes (the usual technique in the Arab lands). A timed observation gives a result for ΔT, the uncertainty being estimated by considering factors such as clock drift. However, if only the occurrence of an eclipse is recorded – whether in categories (ii) or (iii) – a range of values of ΔT is indicated. Although this range has sharp limits, these are often widely spaced. In consequence, one limit is frequently redundant but in medieval times, when ΔT is comparatively small, it is not unusual for both limits to be valueless. This results in the unfortunate wastage of otherwise good data.

4. Non-tidal changes in the LOD

4.1. THE PERIOD SINCE AD 1600

Changes in the LOD over this period were calculated from the first derivative of ΔT as follows. Before 1861 the derivative of the spline curves was evaluated at yearly intervals. From 1861 onwards the derivatives were calculated annually by applying the following 5-point quadratic convolute to the annual mean values of ΔT listed in Table I of Morrison (1979a):

$$d(\Delta T)/dt = 1/10\{[f(+1) - f(-1)] + 2[f(+2) - f(-2)]\}. \tag{4}$$

This convolute transmits changes in LOD undiminished when they persist for 4 y or longer. Changes in LOD lasting for 2 y are reduced by half.

The changes in the LOD are measured relative to the standard mean solar day of 86 400 SI s. If we assume that underlying the decade fluctuations there is a secular linear change in the LOD, we find that the observed LOD is equal to the standard LOD of 86 400 SI s near the epoch AD 1800. This is not surprising because the standard LOD is defined in relation to the mean motion of the Sun in Newcomb's Tables, which was obtained by fitting the theory to observations of the Sun made in the 19th century. We therefore chose AD 1800 as the epoch from which long-term secular changes were to be calculated. In particular we measure the linear change in the LOD due to tidal friction from this epoch.

The non-tidal changes in the LOD were calculated from the expression

$$\Delta LOD_{NT} = \Delta LOD - 2.40 \, (T - 18.00) \tag{5}$$

where $+2.40 \, \text{ms cy}^{-1}$ is the secular increase in the LOD due to tidal friction, as outlined in Section 2, and T is the epoch of the observation in centuries. The results are displayed in Figure 1. On this diagram the annual values of ΔLOD_{NT} in the period AD 1780–1982 are joined by a continuous line. However, in the period AD 1620–1780, where the results are comparatively uncertain, a dashed line is used. For comparison, an estimate of the standard deviation (σ) is also plotted in Figure 1.

4.2. THE PERIOD FROM 700 BC TO AD 1600

On average there are only about 5 observations per century throughout this interval although the actual numbers are quite variable. The relative infrequency and low precision of the observations allow only general trends to be determined. The data form two distinct groups having mean epochs around AD 1000 (medieval) and 300 BC (ancient). As there is no useful material of intermediate date we have assumed constant rates of lengthening of the day between (i) the medieval period and AD 1800 and (ii) the ancient period and a selected mean epoch in medieval times.

(i) For the medieval data (covering the time-range from AD 702 to 1567) we have deduced ΔLOD values from the observed ΔT values using the expression:

$$\Delta LOD = \frac{2\Delta T}{36525(T - 18.00)} \, \text{s}. \tag{6}$$

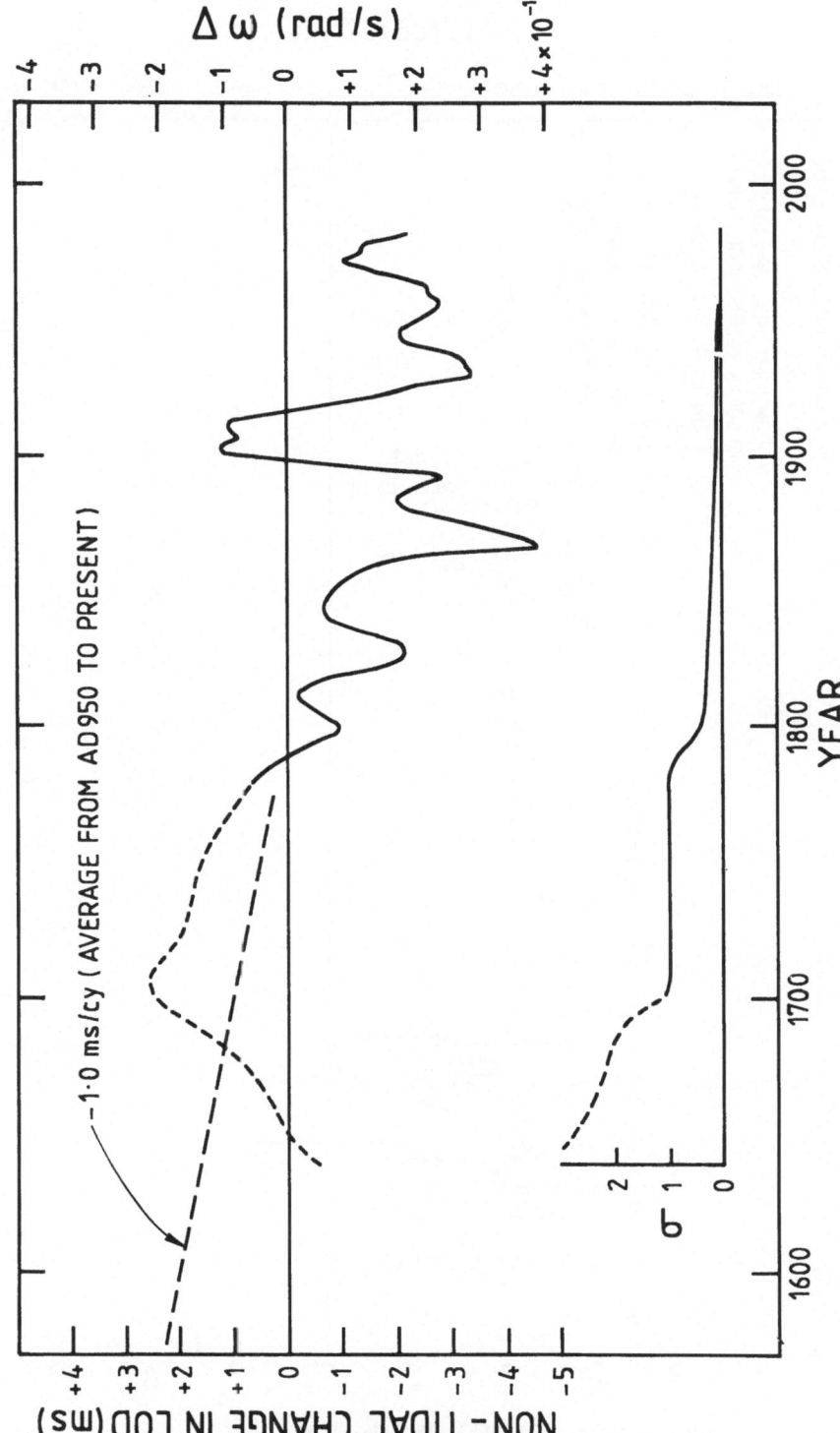

Fig. 1.　Non-tidal changes in LOD in the telescopic period.

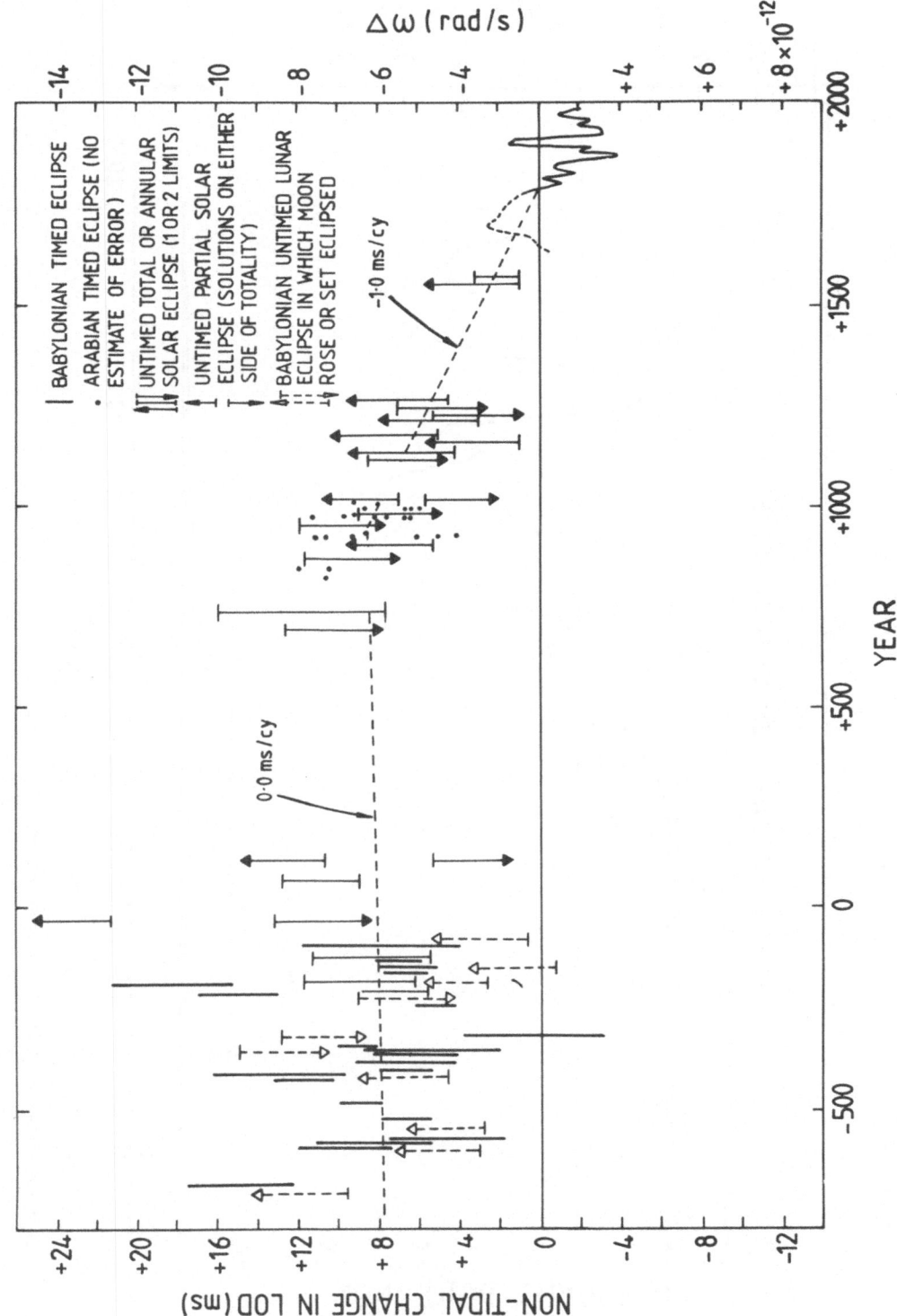

Fig. 2.   Non-tidal changes in the length of the day from ancient times to the present, as determined from eclipses and occultations.

The denominator of Equation (6) is the number of days which elapsed between any individual observation and AD 1800. In each case the non-tidal part of ΔLOD was calculated from Equation (5).

(ii) A useful medieval reference epoch from which to determine ΔLOD in ancient times is provided by the Arabian eclipse timings (23 in number covering the date range AD 829 to 1004). These give a mean value for ΔT of 1850 $\pm$ 80 s at the epoch AD 948; the corresponding ΔLOD value is -11.9 ± 0.5 ms. For the ancient data (695 BC to AD 120) we have assumed a constant rate of change in LOD down to AD 948. ΔLOD was calculated from the expression:

$$\Delta\text{LOD} = \frac{2(\Delta T - 1850)}{36525(T - 9.48)} + 0.0119 \, \text{s}. \tag{7}$$

As before, Equation (5) was used to calculate ΔLOD_{NT} figures.

The ancient and medieval results for ΔLOD_{NT} are plotted in Figure 2, which also shows the modern curve for comparison. In this diagram, solid vertical lines, without bars or arrow heads, represent individual means and standard deviations as deduced from the Babylonian eclipse timings (to avoid confusion due to overcrowding of data, only about half of the approximately 40 known observations are represented here). Dots indicate medieval Arabian timings of eclipses; in the absence of evidence to the contrary, the errors of measurement are assumed to be constant. Single vertical solid lines with bars and/or arrow heads denote ranges of ΔLOD_{NT} deduced from untimed sightings of total and annular eclipses; horizontal bars denote firm limits and arrow heads redundant limits. Occasionally a record expressly denies the occurrence of a central eclipse, asserting instead a very large partial obscuration of the Sun; in such cases a pair of ΔLOD_{NT} ranges is indicated. Finally, the series of Babylonian records which state that the Moon rose or set whilst eclipsed is represented by dashed vertical lines having open arrow heads.

The average epoch for the 40 or so Babylonian timed data is 390 BC. At this epoch the mean value of ΔT is 15600 $\pm$ 350 s, which leads to a result for ΔLOD of -44.4 ± 1.0 ms. The equivalent value of ΔLOD_{NT} is $+8.2$ ms. At AD 948, the mean epoch of the Arabian data, ΔLOD has a value of $+8.5$ ms. Hence the average rates of increase in the LOD due to non-tidal causes over the pre-telescopic period are (i) 0.0 ms cy^{-1} between 390 BC and AD 948 and (ii) -1.0 ms cy^{-1} (i.e. $\dot{\omega}_{NT} = +2.7 \times 10^{-22}$ rad s^{-2}) since AD 948. These rates of change are represented by dashed lines in Figure 2. It will be noted that very few of the untimed data are inconsistent with the general trends deduced from the Babylonian and Arabian timed observations, which provides useful confirmation of the validity of our interpretation.

5. Conclusion

The modern data since AD 1780 reveal fluctuations in the LOD of about ± 4 ms in amplitude. A power spectrum analysis indicates a broad peak around the 30 year period (Morrison, 1979a). It seems very likely that similar fluctuations occurred in the period AD 1620–1780, but here the resolution is inadequate for such changes to be detected. The variations observed since AD 1780 are equivalent to torques $\sim 10^{18}$ Nm acting on the mantle of the Earth. Around 1900, the non-tidal torque reached an amplitude of some 20 times the steady tidal component over a period of about 5 y. Such magnitude and temporal behaviour imposes fairly severe constraints on the possible mechanisms for core-mantle coupling.

The pre-telescopic data indicate the existence of long-term non-tidal variations in the LOD. Decade fluctuations may be present but they are too transient to be detected. The origin of the variations occurring on a time-scale of millennia is obscure. Possible causes are changes in the principal moment of inertia of the Earth–whether in the deep interior of the Earth or at the surface (e.g. sea-level changes) – or core-mantle coupling. Without a full discussion of the various causal mechanisms, it is perhaps inadvisable to attribute the observed variations to any particular geophysical process.

References

Cazenave, A.: 1982, in P. Brosche and J. Sundermann (eds.), *Tidal Friction and the Earth's Rotation, II*, Springer, Berlin, p. 4.
Dickey, J. O. and Williams, J. G.: 1982, *Trans. Amer. Geophys. Union* **163**, 301.
Fairbridge, R. W.: 1961, *Phys. Chem. Earth* **4**, 99.
Hide, R., Birch, N. T., Morrison, L. V., Shea, D. J., and White, A. A.: 1980, *Nature* **286**, 114.
IAU: 1968, *Trans. Int. Astr. Union*, Ser. B **13**, 48.
Lambeck, K.: 1980, *The Earth's Variable Rotation*, Cambridge Univ. Press. Cambridge.
Mörner, N.: 1971, *Palaeogeogr. Palaeoclimatol. Palaeoecol.* **9**, 153.
Morrison, L. V.: 1978, *R. Greenwich Obs. Bull.* **183**.
Morrison, L. V.: 1979a, *Geophys. J. Roy. Astron. Soc.* **58**, 349.
Morrison, L. V.: 1979b, *Monthly Notices Roy. Astron. Soc.* **187**, 41.
Morrison, L. V., Lukac, M. R., and Stephenson, F. R.: 1981, *R. Greenwich Obs. Bull.* **186**.
Morrison, L. V. and Ward, C. G.: 1975, *Monthly Notices Roy. Astron. Soc.* **173**, 183.
Van Flandern, T. C.: 1983, preprint.
Watts, C. B.: 1963, *Astr. Pap. Amer. Eph.* **17**.
Zschau, J.: 1978, in P. Brosche and J. Sundermann (eds.), *Tidal Friction and the Earth's Rotation*, Springer, Berlin, p. 62.

S. K. RUNCORN'S COMMENTARY

The potential of F. R. Stephenson was first spotted by the long retired Headmaster of the Royal Grammar School, Newcastle upon Tyne, Dr E. R. Thomas, who was working in our department pursuing a long time interest in the physics of singing sand not I am sorry to say by the elaborate examination mechanism or ritual on which university dons expend so much effort! Anyway I agreed to have him as a research student and, when he said he had an interest in astronomy, I suggested that it might be worth studying the ancient eclipse data. The older analysis of Fotheringham and de Sitter had given the result that tidal friction over the last 2000 yr was twice that derived from observatory data, as analysed by Sir Harold Spencer Jones, from whose fine lectures I derived my interest in the subject. At the time the acceptance of Stephenson as a research student was resisted in the Faculty of Science. I remembered this and reminded the Dean, then moved on to much higher things, when Sir Harold Jeffrey's report on his Ph.D. thesis contained the words "This is the best thesis I have ever read". But University administrators have other Papal attributes besides mantles of office, his response being "Oh, I never said that!" when I repeated his dismissive words.

Under the American system of postgraduate education, Stephenson would have had to have had courses in Laplace transforms, seismology etc. – and French and German – but here he started off his study by teaching himself Chinese. These discursive reflections, prompted by the contribution of Stephenson and Leslie Morrison, not one of our graduates, but one whose contribution here most happily symbolises the happy – and essential – collaboration we always have had with many other laboratories, show that we don't know much about education: "the wind bloweth where it listeth".

This contribution might be the last word in refining the historical data on the variations of the Earth's rotation. It demonstrates unambiguously two results challenging the geophysicists: the so called 'non tidal acceleration' and the 'decade' fluctuations in the length of the day. On the reasonable assumption, urged by Sir Harold Jeffreys, that the now well established value of tidal friction from modern data be assumed constant over the last 2500 yr, the existence of a long term increase in the Earth's rotation due to changes of the Earth's interior has been proved. Such a 'non-tidal acceleration' was first suggested by Urey (1952), but in this contribution it is shown to be only present from mediaeval time to the present. If this is really so it is hard to ascribe the effect to continuous changes as suggested by Lyttleton, in his Earth-contracting model, or by Peltier who, from the Lageos satellite orbit decay, deduces that the ellipticity of the Earth is diminishing and ascribes it to the melting of the ice caps.

The second is the decade fluctuations, and these are so large that they have never been plausibly explained except as interchange of angular momentum between the core and the mantle resulting from the varying electromagnetic torques between the core and mantle. I have remained sceptical as to whether smoothing, beloved of astronomers, has not removed some of the sharpness of the changes – and the recent discovery of the geomagnetic jerk of 1969 has reinforced my view. The time constant of course – whether it is 10 yr or 1 yr – is the parameter which, if known with some certainty, would probably give renewed impetus to understanding the basic physics of core-mantle coupling.

NEW PUBLICATIONS

Principles of Pleistocene Stratigraphy Applied to the Gulf of Mexico

Edited by
N. HEALY-WILLIAMS

1984, 224 pp.
Cloth Dfl. 130,–/£ 33.25   ISBN 0–934634–72–6
Available in North America from IHRDC

This study of one of the world's major oil provinces is an examination of advances made in the past decade in high resolution stratigraphy of Pleistocene marine sediments. There is a strong need for such a book which presents the very new techniques, not yet widely publicized, that can be used to analyze deep water marine and continental margin sequences. Topics covered include magnetostratigraphy, planktonic foraminiferal biostratigraphy, oxygen isotope stratigraphy, tephrochronology and a review and updating of terrestrial-marine correlations during the Pleistocene. The emphasis is on the Gulf of Mexico, but the techniques described can be applied to other marine sedimentary basins. This is a resource for petroleum biostratigraphers, sedimentologists, and others active in exploration and production, and it will also be of use to students of marine geology and continental margin stratigraphy in general.

Contents
Preface. Introduction. 1. **Michael T. Ledbetter:** Pleistocene Magnetostratigraphy. 2. **Robert C. Thunell:** Pleistocene Planktonic Foraminiferal Biostratigraphy and Paleoclimatology of the Gulf of Mexico. 3. **Douglas F. Williams:** Correlation of Pleistocene Marine Sediments of the Gulf of Mexico and Other Basins Using Oxygen Isotope Stratigraphy. 4. **Michael T. Ledbetter:** Late Pleistocene Tephrochronology in the Gulf of Mexico Region. 5. **Richard H. Fillon:** Continental Glacial Stratigraphy, Marine Evidence of Glaciation, and Insights into Continental-Marine Correlations. Summary. References. Index.

D. REIDEL PUBLISHING COMPANY

A Member of the Kluwer Academic Publishers Group

    P.C. Box 17, 3300 AA Dordrecht, The Netherlands
190 Old Derby Street, Hingham MA 02043, U.S.A.
Falcon House, Queen Square, Lancaster LA1 1RN, U.K.

First *Break*

A monthly geophysics magazine published with the
European Association of Exploration Geophysicists by
Blackwell Scientific Publications Ltd

Edited by Ian Williamson *Petroleum Geology Section, Department of Geology, Imperial College of Science and Technology, London*

Associate Editors
Andrew McBarnet and Steve Sasanow

Aimed at the practising applied geophysicist, *First Break* provides a mixture of short but authoritative articles on all aspects of seismic, gravitational and electromagnetic geophysical techniques, with current news and views about the industry. The journal is published in association with the European Association of Exploration Geophysicists, and its emphasis on rapid publication in a magazine format provides complementary reading to the EAEG's established research journal *Geophysical Prospecting*. High quality submitted material is published in addition to commissioned articles. Filling a gap between the trade publications and the research journals, *First Break* provides geophysicists with practical, accurate and relevant information about the geophysics industry.

Subscription Information
First Break is published monthly. Subscription rates for 1985 are, for individuals £34.50 (UK & overseas), $58.00 (USA & Canada), and for institutions £85.00 (UK & overseas), $145.00 (USA & Canada) post free.

Order Form

Please tick the appropriate box and return to **Blackwell Scientific Publications Ltd, P.O. Box 88, Oxford, England.**

☐ I would like to subscribe to *First Break*

☐ I wish to pay by cheque/money order *(delete as necessary)* and enclose the sum of ....

☐ I wish to pay by Access/Barclaycard/VISA/Mastercard *(delete as necessary)*

Please debit my credit card no. ⬚⬚⬚⬚⬚⬚⬚⬚⬚⬚⬚⬚⬚⬚⬚⬚

Expiry date........................... with the sum of ....

Signature.. Date ....

☐ Please send me a specimen copy of *First Break*

Name ....

Address ....

....

BLACKWELL SCIENTIFIC PUBLICATIONS LTD
P.O. Box 88, Oxford, England.

GEOPHYSICAL SURVEYS / *Vol. 7 No. 3 June 1985*

HEAT TRANSFER AND PLANETARY EVOLUTION

D. C. TOZER

University of Newcastle upon Tyne, England

'Clear your mind of cant'   (Johnson)

Abstract. The object of this account is to show how much one can interprete and predict about the present state of material forming planet size objects, despite the fact we do not and could never have the kind of exact or prior knowledge of initial conditions and in situ material behaviour that would make a formal mathematical analysis of the dynamical problems of planetary evolution an efficient or meaningful exercise. The interest and usefulness of results obtained within these limitations stem from the highly non linear nature of planetary scale heat transfer problems when posed in any physically plausible form. The non linearity arising from a strongly temperature dependent rheology assumed for in situ planetary material is particularly valuable in deriving results insensitive to such uncertainties. Qualitatively, the thermal evolution of a planet is quite unlike that given by heat conduction calculation below a very superficial layer, and much unnecessary argument and confusion results from a persistent failure to recognise that fact. At depths that are no greater on average than a few tens of kilometres in the case of Earth, the temperature distribution is determined by a convective flow regime inaccessible to the laboratory experimenter and to the numerical methods regularly employed to study convective movement. A central and guiding quantitative result is the creation in homogeneous planet size objects having surface temperatures less than about half the absolute melting temperature of their material, of internal states with horizontally averaged viscosity values $\sim 10^{21}$ poise. This happens in times short compared with the present Solar System age. The significance of this result for an understanding of such processes and features as isostasy, continental drift, a minimum in seismic S wave velocity in Earth's upper mantle, a uniformity of mantle viscosity values, the survival of liquid planetary cores and the differentiation of terrestrial planet material is examined. After a discussion and definition of 'lithospheric' material, it is concluded that endogenous tectonic activity only continues on Earth's surface on account of water enhancing the deformability of its rocks.

   Metal/silicate differentiation of terrestrial planet material is predicted to have been a global scale catastrophic process in the many objects it formed predating the existing planets, but intersilicate and volatile/silicate separations are necessarily protracted, quasi continuous processes arising from local shear instabilities in the convective flow of such a viscous material. In particular, these local magma producing instabilities require the involvement of 'lithospheric' planetary material in convective movements and it is shown how this unsteadiness accounts for the distribution and salient features of planetary seismicity and vulcanicity at the present time.

   The picture that emerges for the state of Earth's silicate shell material after more than four billion years of average viscosity regulation and shear instability is one of chemical and isotopic heterogeneity on a wide range of length scales. The larger length scales of this range are introduced by the pattern of heterogeneity remixing rather than its generation. For example, at the largest scale, the predicted heterogeneity is radial and a feature indirectly arising from properties conferred on the shell material by major mineral phase transitions at depths ~ 700 km. These increase the adiabatic temperature gradient and have the effect of a barrier adequate in strength to prevent wholesale mixing of the material above and below for at least a large fraction of the Earth's history in which radiogenic heat has been the dominant cause of large scale internal movements. That such a barrier actually marks a chemical and isotopic heterogeneity of the mantle is because only the convective movements above it are prone to the shear heating instabilities on which differentiation absolutely depends. Many millions of such instabilities in this shallower shell material would by now have created a three dimensional heterogeneity extending downward in length scale to ~ 1 km. However, only 10% of this shell material has yet experienced these highly localised shear heating instabilities and one would predict a continuing emission of primitive volatile phases and a widespread metasomatism even if the same convective movements had not recycled material from the hydrosphere. Such recycling is a further aspect of convective self regulation.

   The mesoscale and lateral heterogeneity of near surface material more familiarly referred to as continental crust and its underlying mantle is another cumulative feature of the remixing process – in this case the result of

Geophysical Surveys 7 (1985) 213-246. 0046-5763/85.15.

separated ultrabasic and less refractory fractions of the upper shell material from many shear heating events being able to form superficial blocks, whose net buoyancy and coherency make them immune to entrainment and remixing by the radiogenically driven flow. This partial but permanent concentration of lower melting point silicate and volatile phases near the external surface has in turn caused a gradual increase of the horizontally averaged temperatures associated with the self regulating convective state at upper mantle depths. This thermal evolution has strengthened the barrier to convective mixing of the whole silicate shell presented by its major phase transitions but it could explain a persistent small scale incorporation of more primitive, i.e. less differentiated shell material from the phase transition region into the upper shell convective circulation.

1. Introduction

The first well known contribution to this subject was made by Thomson (later Lord Kelvin) in the eighteen sixties. Although geologists have sometimes taken satisfaction in the fact that Kelvin's estimate of an age for the Earth was subsequently revised in a direction that pleased them, that overlooks the point that his discussion was the appropriate response of a physicist to a view of planetary evolution then expressed in the geological doctrine of uniformitarianism by Hutton, Playfair, Lyell and others. Perhaps in overreaction to an even earlier view that the Earth was only a few thousand years old, uniformitarianism had been taken by them to mean geological process as we observe it today, had, for all practical purposes, always occurred. Although the word 'uniformitarianism' is still used in geological circles, it now seems to mean no explicit time dependence of the form and parameter values of the physical laws governing geological processes.

The particular way in which Kelvin introduced the concept of planetary evolution by using the then newly formulated conservation of energy principle, was easily criticised once a natural radiogenic heating had been discovered and its significance for Earth's heat budget roughly evaluated. However, to my mind a much more serious consequence of his particular approach that one can still see amply demonstrated in modern textbooks of planetary science is that thermal studies became unhealthily detached from the business of interpreting observable phenomena – the whole point of any scientific theorising. Although an obvious fact, there is still a strong need to emphasise that the present internal temperature distribution, let alone those of the past that have led to it, is not itself an observable feature forming the basis of a test of planetary heat transport theory. If planetary heat transport theory were no more than a theory of internal temperature, little more need be said, but a major burden of this and several articles I have written in the past is to point out that a modern treatment of heat transfer can explain several observable things about the Earth and other planets without the need to commit oneself to any very precise views about their internal temperature distributions. However, instead of using this opportunity for theoretical development, many workers in the field seem to have been totally inhibited by a quite spurious validity conferred on a narrow range of upper mantle temperature values by numerous, highly elaborated versions of Kelvin's heat conduction calculations. Such timidity has allowed a whole nexus of notions about internal temperature and its evolution to survive virtually unscathed among those who have made a great display in recent years of rejecting the fundamental assumption on which they all rest – a static Earth model. Just how reluctant many Earth scientists have been to re-examine the foundations of their beliefs about planetary temperature and the extent to which they remain unintegrated with now widely held views about geodynamics has been shown by the piecemeal and obviously question begging way in which Plate Tectonicians try to explain their picture of large scale secular movements of Earth's superficial material. If the particular pattern of surface movements at the present time must for ever remain unpredicted by planetary heat transport theory – for the very simple reason that we do not have access to the 'initial' data that might make such a prediction possible c.f. the weather forecasting problem, the fact that there is some pattern of large surface movements with the inferred gross characteristics is far from being the theoretical mystery it has often been made out to be by Plate Tectonicians. My confident assertion stems from the fact that there are certain self regulatory characteristics to the solutions of very generally based versions of heat transfer theory for the Earth that make the observed type of geodynamic activity look natural-certainly more understandable than its absence would be!

Although there was a considerable effort prior to the nineteen sixties to connect the thermal and geodynamic problems through thermally induced changes in the overall planetary volume, the tendency to see them as distinct problems arose from the practise of treating deformation as a perturbation – something sufficiently small to be determined by but not determining the thermal problem. However, if there is any one thought distinguishing the modern approach to planetary thermal history from those based on heat conduction theory, it is the acknowledgement of a mutual dependence or closed loop character of the relationship between planetary material deformation and its thermal state. Another important underlying change has been the recognition that heat conduction and convection are not the easily distinguished and mutually exclusive heat transport processes they are usually made out to be in elementary textbooks of physics and geology. They should both be seen as regimes of solution to appropriately general formulations of heat transfer in a gravitational field that continuously merge into one another as material properties and/or boundary conditions are changed. Heat conduction or what I prefer to call 'state of rest' solutions to the heat transfer equations, can be the only acceptable solution to a planet's heat transfer problem, but that acceptability can only be established by arguments based on the physical properties assigned to its interior. It is *not* to be judged solely by the absence of any signs of endogenous deformation in its surface rocks. This change of outlook about the use of the more general convection equations stems from greater awareness of the enormous differences in rheology of superficial and deeply buried rock that can be induced by temperature and pressure variation with depth, one aspect of the mutual coupling of the thermal and dynamic problems referred to above*. Thermal convection is hardly a

---

* It may still be wishful thinking on my part that this is now recognised. In the nineteen seventies, there were many attempts to solve the thermal history problems of other terrestrial planets that showed all the old determination to use surface features as justification for the use of the heat conduction theory of heat transfer.

new idea in Earth science, but it is precisely because it has usually been seen in what I call an open loop relationship with thermal problems that it has generally appeared to be more a vehicle for self expression by a few enthusiasts than a quantitative theory. With a little thought, it could have long offered a serious threat to the validity of all those absurdly detailed heat conduction calculations of internal temperature in the nineteen fifties and sixties that now bedevil the subject.

Although it is obvious that internal movements at speeds exceeding some value based on the time constant of heat conduction in a system of given size are going to affect its internal temperatures, a more revealing way of showing the size of the departures from a heat conduction solution is to examine the consequences of introducing any one of a variety of temperature dependent functions chosen to represent the extremely strong temperature dependence of a rock's irrecoverable deformability. Even such a simple and broad assumption as a strong decrease with rising temperature of the effective viscosity for the 99.99...% of any planet's material that is always going to remain outside a laboratory will not appeal to those who think all questions of material behaviour have to be and can be unequivocally settled in advance of one's main problem by 'hard' data. However, this objective view of material properties is certainly not appropriate to Earth's heat transfer problem and in fact I incline to the view that the material properties of any substance have a subjective quality that is bound to change with one's perspective of the system it comprises. After all, an experimenter has no option but to *choose* a mathematical form to define the properties of his specimens and assign values to its free parameters on the basis of observations made entirely outside them. He would certainly not contemplate sampling his specimen simply to check the validity of such an assignment of properties – why should the Earth be treated differently?

I have made this point explicit because quite unnecessary controversy and needless expense is still being incurred by geophysicists who naively believe in the objective reality of the concepts of continuum mechanics*. We all know that scale models and laboratory measurement help engineers design and predict the workings of somewhat larger systems. However, when the discrepancy of size is as large as that between a planet and laboratory system it becomes a fundamental mistake to confuse the points of an abstract continuum on which similarly abstract 'properties of matter' are mapped with the points of physical space. I believe it can only do planetary science good for those who worry about what the properties of planetary interiors 'really are' to become more aware of the fact that no more or less abstraction and imagination is involved in quoting a Young's modulus or viscosity for a laboratory specimen than in developing a

---

* One would have thought the Plate Tectonics view of the Earth would have cured any objective view of material properties. Is the Hookean elastic model of Earth's interior, formerly believed by seismologists, to be rejected because no material behaves just like that in the laboratory or because it cannot possibly explain the generation of earthquakes? In view of its relevance to interpretations of the low velocity layer in the light of modern heat transport theory, one should note that even parameters like a 'melting' or 'solidus' temperature are subjective in the sense that they merely mark a temperature at which a continuously decreasing effective viscosity becomes apparent to the most casual of observers.

quantitative theory of planetary behaviour. In both cases, the objective is the interpretation and connection of observable phenomena outside a particular system; troubles only arise when any useful theoretical model is invested with the deadening stamp of 'reality'.

One of the paradoxes of the present situation is that one will frequently hear complaints of a lack of data but actually notice that only a trivial fraction of an already enormous body of information about rocks under stress has ever been incorporated into a quantitative planetary heat transfer theory. One groundless assumption frequently invoked to discard most of this data or restrict the breadth of one's choice of a rheological model for in situ Earth material is that its 'true' properties are only to be found by studying 'good' specimens of its principal mineral phases in isolation. One might justify that approach in talking of a rock's quasi elastic properties, but everyday observation of fractured or granular materials should be sufficient to convince anyone it can be wildly misleading about their irrecoverable deformability if the principle phases of such materials happen to be the least deformable. We shall see later that this kind of situation probably exists in the Earth, and that quite minor components are actually given a rheological importance by the nature of the heat transport process. I can only understand the continuing preoccupation of so many heat transfer theorists with the rheology of good quality magnesium rich olivine crystals as a reflection of the fact that by choosing the most refractory of all common minerals, subsequent calculations do least damage to their quite groundless preconceptions about internal temperature. The further and much more important consequence – that such creep resistant Earth models cannot explain the involvement of surface materials in the convective process, gets submerged in a morass of talk about 'rigid plates', whose most conspicuous property seems to be a propensity to bend and/or break. What good is a theory whose premises, taken in their literal meaning, deny the very phenomena one seeks to explain? I remain committed to the view that such olivines only set a physically plausible upper bound to the creep resistance of any terrestrial planet's material at a particular temperature and pressure (Tozer, 1965).

Another obvious reason for not pursuing a simple laboratory approach to Earth material rheology specification is that a convective heat transfer process will, if it occurs at all, impose shear stresses on an enormous length scale compared with any likely scale of phase heterogeneity in planetary material. There is a negligible chance of the most creep resistant phases ever forming a framework controlling deformation on the relevant length scales. What is much more clearcut in the geological field evidence is that the current pattern of deformation is itself determined by a pattern of 'defects' e.g. faults, created by previous deformation – another example of the way convective heat transfer is able to regulate the properties which sustain it. Laboratory experiment is only helpful here in suggesting that such 'defect' controlled deformability may be less geologically important if temperatures exceed about half an absolute solidus temperature, and pressure* exceeds the imposed deviatoric stress components.

---

* An effective pressure if a relatively mobile phase fills the defects of a material.

If such comments on rock deformability seem utterly trivial to the civil engineer or field geologist, their more theoretically minded colleagues often appear to be in need of a reminder when trying to dismiss superficial rocks as 'rigid'. Their desire to produce quantitative results seems to tempt them to neglect the complexity of the rheological situation near the external surface on the grounds that it is really a tiny fraction of the Earth. However, one has to keep in mind that it is an explanation of the deformation of superficial material and the cryptic signals travelling through such superficial material from deeper sources which are, or should be the raison d'être of planetary heat transfer theory.

The 'plate' idea originated as a way of describing the broad character of the geologically recent surface velocity field as regions of approximately quasi rigid body motion separated by comparatively narrow regions of positive and negative divergence. Unfortunately, use of the word 'rigid' in this kinematic picture was misconstrued and subsequent investment of 'plates' with a thickness has only made it harder to achieve the radical revision of rheological thinking needed to explain such a velocity, pattern at Earth's surface. Clearly, nothing more subtle than the idea of matter being either an elastic solid or viscous fluid lay behind the choice of the seismologist's 'low velocity layer' to mark the lower boundary of this mysterious plate rigidity, and long experience with heat conduction theory of all too easily reaching melting temperature at such depths would have reinforced it. However, just as the use of such a naive rheological outlook to interpret seismic signals thwarted intelligent discussion of Wegener's ideas, it has now deflected attention from a more subtle rheological transition whose depth is crucial in determining whether a planet's surface rocks move and deform irrecoverably under the influence of internal heat transfer (see below). Current understanding of the speed and effectiveness with which convective heat transfer regulates the horizontally averaged values of any plausible effective viscosity function at $\sim 10^{21}$ poise has totally changed the old views about internal melting. There is no longer any need to play with a distribution of radiogenic heat sources that balances the observed surface heat loss in order to avoid a seismologically unacceptable degree of 'melting' – more correctly seismic attenuation at depths ~ 100 km; no known rock has an effective viscosity as large as 10^{21} poise within hundreds of degrees of its solidus as usually defined. Solid state creep would still be an unknown phenomenon if that statement were wrong! The immediate implication of much lower horizontally averaged upper mantle temperatures than previously believed has stimulated a new analysis of magmatism (see later) and the fate of volatiles and their compounds with refractory silicates in the new thermal environment. As a result, one is now able to offer a quite different but quite unforced interpretation of the supposed evidence of extensive silicate melting in the upper mantle. A low velocity layer* at just the inferred depth range

---

* It is still far from clear whether this 'low velocity layer' even forms a closed surface that could, even in principle, be used to distinguish unequivocally 'plate' and deeper material. The fact that seismologists give even lower S velocities and rigidity modulus values to near surface material should have warned Plate tectonicians that seismic velocity distributions can make no *direct* contribution to their view of superficial Earth material as 'rigid'.

would have been sustained throughout Earth history by an intergranular liquid aqueous phase created by the dissociation of some of Earth materials' more stable hydrous phases, the amphiboles (Tozer, 1973, 1981). The most stable hydrated phases collectively provide a mechanism by which Earth material, firmly held in a self regulating, very viscous average state for at least 4 billion years, has been able to retain at least some of its primitive water content, and to remix with the interior at least some of this primitive water that has managed to escape and lie as oceans on its external surface*. However, the amphiboles are of special interest in that at pressures of less than a few kilobars they are thermodynamically stable right up to a solidus temperature, i.e. where an effective viscosity is $\leq 10^{14}$ poise, but they dissociate at significantly lower temperatures as pressure is raised (see Wyllie, 1971; Green, 1972). As a result, oceanic water can be subducted in the form of amphibole well below those depths at which cracks provide a significant permeability for any free water phase to reach the surface and it is then released primarily by the effect of an increasing pressure**. This kind of behaviour, combined with the negligible diffusivity of water through such high viscosity material (see later) makes it effectively impossible for any planet having basaltic components and water in its starting composition ever to rid itself of all water in its subsequent evolution. The free water content of material at the depths of the low velocity layer would be regulated at a value that just ensured a full hydration when convectively raised to shallower depths; any larger amount would have been lost quite early in Earth material accumulation by the porosity and permeability its own presence would have created in denser silicate phases***. Although the regulated free water content at depths ~ 100 km in Earth is probably less than 1% by weight, we now know from laboratory experiments that deformational wave speeds in rocks and particularly their damping are sensitive to a free water content. (Pandit and Tozer, 1970; Tittman, 1977).

---

* I shall later explain that ocean water probably plays a crucial role in creating the rheological conditions needed for a subduction of rehydrated surface material to the deep interiors.

** I am here disregarding the possibility of material being heated by its deformation well above the horizontally averaged thermal state (see later). If, as seems logical, one regards water of hydration as just another rock forming component, one could still say that the seismic low velocity layer is due to a partial, if highly incongruent 'melting'. The solidus is then recognised as one that decreases with increasing depth when pressures are in a tens of kilobar range; it will only revert to its normal upward trend when pressures $\gtrsim 200$ kilobars make the water freezing temperature higher than the dehydration temperature. It may be more generally remarked that a large decrease of 'solidus' temperature with increasing depth by at least 40% of its absolute value is a necessary condition for the survival of nominally liquid 'cores' in terrestrial planets for periods $> 10^8$ yr.

*** No record for such early volatile expulsion would be retained for objects orbiting the Sun at terrestrial planet like distances, but more distant objects would still retain such 'excess' volatiles as ices on their surfaces, e.g. Europa and Callisto.

2. A Minimal Rheological Model to Account for Endogenous Tectonic Activity

The presence of a low velocity layer is connected with the inferred existence of large displacements of Earth's surface rocks to the extent that water modifies two different aspects of Earth material rheology controlling the existence of such phenomena. To clarify this connection requires a rheological model for in situ Earth material at least capable of showing it: (1) Transmitting energy as P and S type deformation waves, (2) Deforming as an effectively viscous medium with increasing facility as its temperature is raised, (3) Cracking and slipping along internal fault surfaces in the manner familiar to every field geologist. Just why this third requirement is an essential feature even in a minimal model will only be apparent after the concept of 'lithospheric' material has been described with reference to heat transfer process. For the moment note that it contradicts the description of near surface material as 'rigid'.

It will be evident from my above remarks that there is no unique way of devising a rheological model meeting these requirements and I certainly would not wish to get involved in an argument over what I believe is a matter of judgement rather than fact. One is not trying to assert an objective truth so much as introduce a better description of in situ planetary material, allowing a wider variety of planetary deformation phenomena to be interpreted and discussed in a self consistent way. For purely hydrostatic stresses, there does not yet seem any need to change the traditional account of Earth material as a compressible elastic medium whose behaviour is summed up in a compressibility modulus dependent on local values of temperature and pressure. However, for shear stresses σ less than some value σ_c (see below), it is appropriate that it should now be described as a viscoelastic medium, characterised by an effective shear modulus and shear viscosity also dependent on local temperature and pressure values. The quantity σ_c is introduced in an attempt to mark the onset of an unignorable non linearity in the rheological behaviour*. The most important cause of non linearity for the heat transfer theorist is a potentiality for cracking, and the fact that subsequent rheological behaviour will be influenced by such cracking until sufficient time has elapsed for the material to become 'annealed'. Since both the initiation and annealing of cracks or faults depends on environmental conditions created by the heat transport process, one notices immediately that there are further possibilities here of convective heat transfer both regulating and being regulated by the rheological behaviour it induces. This additional mode of self regulation will eventually be seen to be the crucial factor in understanding the involvement of the most superficial Earth material in large horizontal movements.

In the 'low' shear stress range ($\sigma < \sigma_c$), one can assume the response is represented by a Maxwell type viscoelasticity with a viscous component that decreases with temperature T approximately like $A \exp(BT_s/T)$; A, B, and T_s are parameters generally only weakly dependent on pressure. If T_s is identified with a solidus temperature as

---

* Below σ_c any dependence of shear modulus or viscosity on shear stress values is to be allowed for in a linear viscoelastic theory by using 'effective' values of these parameters.

conventionally defined, B is typically ~ 20 and $A = 10^6$ poise. However, I deliberately leave the quantification of this temperature dependence vague in order to emphasise the predictive value of a convective regulation of viscosity values in terrestrial planet studies; one should note that regulated viscosity values $\sim 10^{21}$ poise correspond to temperatures $0.6\,T_s$ and that near the surface of Earth and other terrestrial planets $T < 0.3\,T_s$; this effective viscosity is then $> 10^{35}$ poise.

The 'elastic' part of this Maxwell type shear viscoelasticity, together with the compressibility, are assumed to be quantifiable from P and S type deformation wave velocities in the seismic frequency band. If any reader should feel strongly that such a Maxwell type viscoelasticity cannot possibly interpret observations of a seismic wave attenuation, I would have no objection to the introduction of any more complicated viscoelastic model of the low shear stress behaviour, providing it also showed the same kind of quasi viscous behaviour under an indefinitely sustained shear stress (however small) as is shown by a Maxwell type viscoelastic material. This important restriction follows from the assertion that Earth material can be 'annealed'; there are very general proofs to the effect that any material having a finite and fading memory of its previous deformation will behave like a Newtonian material in the limit of 'slow' deformation (see for example Truesdell and Noll, 1965). I do not believe it is worth pursuing this seismic wave damping problem any further here other than to point out that understanding of the planetary heat transfer problem following this attempt to give in situ Earth material a more accurate rheological description, has created a new interest in how much of the observed attenuation of seismic waves should be attributed to a scattering rather than a direct damping process (see concluding section).

The parameter σ_c, marking an upper shear stress limit for the use of any linear viscoelastic model, should also be taken as a function of local temperature and pressure and will have a smaller value in a 'cracked' than 'annealed' state. On both experimental and theoretical grounds one might expect any value given σ_c to increase with pressure and in general to depend on all the deviatoric stress components existing on fault surfaces at the time of interest. However, I shall avoid the horrendous complications of the general case by anticipating that our basic problem of understanding the occurrence of large horizontal movements of superficial Earth material is that of understanding the rheological behaviour under simple uniaxial tension or compression. This simplification looks possible because in any plausible quantified version of the low stress viscoelasticity the thickness of what I call 'lithospheric' material (definition below) is much too great to show a flexural instability in compression that would inevitibly lead to a more complicated stress distribution*. Given the automatic presence of sea water or groundwater as a stress corrosion agent in any surface crack that might open in Earth material as a result of horizontal tensional stress, I shall assume that downward crack propagation will occur so long as the tensional stress in 'lithospheric' material below its advancing tip is $> P_e$, where P_e is an effective pressure

---

* The compressive 'folding' frequently seen in the field is that of a water weakened material lying above the 'lithospheric' material.

(The lithostatic pressure minus water pressure at the crack tip). This equation of σ_c with P_e for tensional failure of the low stress shear viscoelasticity implies that surface cracks can be initiated with even infinitesimal tensional stresses if the water table is at or above the Earth's rock surface, but their propagation downward becomes increasingly difficult due to a rising P_e value resulting from the difference in rock and pore water densities.

Such an automatic presence of water as a weakening agent is not so obviously a factor in the failure of a linear viscoelastic model of surface material subject to horizontal compression. If it were uncracked, one would estimate a σ_c value $\sim (\sigma_0 + P)$, where σ_0 is a zero pressure compressive 'strength' estimated to be $\sim$ a kilobar and P the total pressure in the material – clearly a much more severe failure criterion to satisfy if subduction is to take place than the tensional one. However, at this point it is important to put the problem of defining a rheology of in situ planetary material into the context of a heat transfer problem. One anticipates that any superficial lithospheric material subjected to horizontal compression over a general region of convergence and downward movement in an underlying convective flow will normally have reached this position from another part of the surface in which it was subjected to horizontal tensional stress above lines of divergence in the convective flow. As such, its compressive failure would be conditioned by a much smaller value of σ_c than $\sim (\sigma_0 + P)$, reflecting a criterion for slip along established fault surfaces surviving* from its previous episode of tensional stress, and probably 'lubricated' by water introduced as a free phase or highly unstable hydrous minerals at that time. The question of whether the actual tensional and compressional stresses actually reach these proposed σ_c values for failure of the low stress viscoelasticity in uniaxial tension and compression can only be discussed after concepts of 'lithospheric' and 'asthenospheric' material implicit in this proposed rheological model have been defined.

Providing one chose the same dependence on temperature of the viscosity parameter of a Maxwell type viscoelasticity and that of the viscosity in a Newtonian Earth model – the type traditionally used by heat transfer theorists, there would be no significant difference in one's calculation of the *horizontally averaged* effective viscosities (and hence the horizontally averaged temperatures) associated with their respective self regulating convective states. If that seems at first sight to justify the simpler Newtonian models, there is a difference in the solutions crucial to any consistent treatment of the chemical/thermal evolution of the interior – the storage of a shear strain energy density $\sim 10 \, \mathrm{J \, m^{-3}}$ throughout the convecting material in the viscoelastic case. Without such a reservoir of strain energy, there is *no* way in which the primary radiogenic energy would ever create the inviscid conditions needed for planetary material differentiation. Even if it is still true that the decay of radiogenic heat sources is the ultimate determinant of terrestrial planet thermal evolution, it is the realisation that so far in Earth history its chemical evolution has depended on a succession of very local and geologically

---

* My criterion of what constitutes 'lithospheric' material is directly based on the idea that faults remain unannealed in it for at least the time it takes to travel from spreading to subduction zones (see later).

transient heating events, sometimes involving the radiation of waves in the seismic frequency band (see below), that provides the greatest contrast with traditional attempts to calculate planetary evolution. The intriguing paradox of the new account is that it is only by keeping the effective viscosity of an overwhelming proportion of the interior extremely large that planetary heat transfer ever achieves any magmatic activity!

This important evolutionary role played by stored shear strain energy means that no part of Earth's interior can be dismissed by the heat transfer theorist simply as a Newtonian medium – the notion lurking behind a common conception of 'asthenospheric' material. However, one can still indirectly make a valuable distinction between superficial and deeply buried Earth material based on a comparison of a locally definable stress relaxation time τ and some characteristic time scale of the deformation produced by heat transfer. For a minimal Maxwell model of the low shear stress behaviour, τ would be identified with the ratio of the two quantities η and μ defining it; for more general rheological models of low stress behaviour one can directly use the ratio of an effective viscosity regulated by the convective heat transport process and a seismologically determined shear modulus. Convective regulation throughout $>90\%$ of Earth history has prevented a horizontally average value of τ from decreasing with depth much below 10^3 yr – the exception to this now presented by the so called 'outer core' can only be understood as the effect of a change in constitution with depth that produces a very large decrease in an effective viscosity without any increase in temperature*. With hindsight, this convective constraint on horizontally averaged τ values (because μ can be treated as a constant to the accuracy needed here) can be seen as the reason why seismologists have sometimes thought a Hookean elastic model was an adequate statement of in situ Earth material behaviour – at least so long as they remained indifferent about why there were wave pulses to be timed through Earth's interior! More positively, this regulation of viscosity (and τ values) shows there are many possible quantifications of a low shear stress viscoelasticity all capable of predicting the salient rheological results about Earth's deep interior inferred from studies of the isostatic adjustment process following ice loading or sea level changes (Peltier and Andrews, 1976). However, a common feature of all heat transfer models incorporating plausible versions of the viscoelasticity is that above a depth never greater than a few tens of kilometres, horizontally averaged τ values increase extremely rapidly. Ignoring any sedimentary veneer, its superficial value exceeds the longest time scale of deformation in a self regulating convective state – an overturn time $\sim 10^8$ yr or its mean reciprocal strain rate, by an enormous factor $>10^{12}$.

In my view and to avoid further proliferation of poorly defined tectonic terms, the designation 'rigid plate' should now be dropped as patently self contradictory and the words 'lithosphere' and 'asthenosphere' reserved for regions of a solution to planetary heat transfer problems in which τ values respectively exceed or are less than the *longest*

---

* What could only be described in a Hookean formalism as the vanishing of a shear modulus in the 'outer core' is more plausibly interpreted as a fall in τ values to below the period of seismic body waves.

time scales of deformation produced by the heat transport. Although at first sight rather different from any previous definition, as far as one is able to judge this would be in accord with the intention of those who first used such terms. However, it is significantly different from recent attempts to locate a 'lithosphere'/'asthenosphere' boundary at the seismologist's low velocity layer because that links their definition with rheological behaviour at what one now regards as the shortest rather than longest time scales of deformation induced by heat transfer. Perhaps the ideal solution would be for everyone to recognise that the location of a lithosphere/asthenosphere boundary will vary with the problem one is dealing with, but there would possibly be objections from those who like to see everything in objectively real terms – e.g. how many would accept a seismologist's 'lithosphere' extending downward to the core/mantle boundary??? In any case, the significance of my definition of a region of 'lithospheric' material is that the ratio of the horizontal length scale of convective movements to its thickness shows the degree to which deviatoric stress can become enhanced in superficial material if it did not move horizontally in response to the stresses such movements impose on its lower surface. The question of whether large surface movements actually occur can now be seen as one of whether these enhanced stresses exceed the σ_c values for failure of the low shear stress viscoelasticity in *both* tension and compression, i.e. whether crack propagation and fault slippage can compensate for the absence of significant 'creep' in what I call 'lithospheric' material.

There is some complication arising from the fact that as a tensional crack propagates downward, a further concentration of the tensional stress occurs ahead of its tip. Furthermore if surface movement does result, this lithospheric thickness is altered from what it would otherwise have been by the upward and downward displacement of isothermal surfaces by the concomitant radial motion. Such lithospheric thickness changes associated with an established pattern of convective movements may well be a cause of unsteadiness in the pattern of lithospheric movements, in addition to any arising from an unsteadiness of the convective movements at asthenospheric depths*. This is because the compressive failure of lithospheric material is the more difficult condition to satisfy for surface movement to occur, and unlike the tensional failure condition, is actually made more difficult by the thickening of lithospheric material accompanying surface movement. This difficulty, arising from the fact that any enhancement of the heat flow above spreading centres has to be compensated by a deficiency elsewhere, would make it easier for surface tectonic movements to be sustained for a longer period of planetary history by unsteady asthenospheric flow.

A more immediate reason for our interest in these motionally induced changes of a lithospheric thickness is that they offer the possibility of an important observational test. My definition of lithospheric material puts its lower boundary at the depth at

---

* In fact, an asthenospheric unsteadiness is strongly suspected, even if there were no evidence of Continental Drift to support a picture of unsteadiness near the external surface – Holmes and Vening Meinesz noted long ago that merely to predict convective movement is not enough to understand geologically recent continental drift. Asthenospheric unsteadiness is suspected because self regulation creates an effective Rayleigh number $> 10^6$, a value at which no laboratory flow in a layer much wider than its depth is ever steady.

which τ has fallen to a value $\sim 10^{15}$ sec. With the usual seismologically assigned values for a shear modulus, this associates the boundary with the $\sim 10^{27}$ poise viscosity surface. The prospect of finding observational support for any prediction one might make for its position looks remote unless one reasonably assumes that τ is also about the period of time it would take any structural changes induced by previous deformation to anneal. As the depth to the 10^{27} poise viscosity surface is increased by cooling along the upper limb of any convective circulation, we have just the situation in which to expect the quenching of a strain induced anisotropy of properties. For any particular geographical location, the depth range in which to expect an anisotropy is both wider and shallower than that in which it would be being currently generated because any plausible effective viscosity function is very sensitive to temperature change when its values are as large as $\sim 10^{27}$ poise. This extreme temperature sensitivity and the fact that lithospheric material lies wholly within the upper thermal boundary layer of the convective process where radial temperature gradients are at least some tens of $^{\circ}C \, km^{-1}$ has the effect of keeping any reasonable estimate of a global average lithosphere thickness within the range 10–25 km*. A seismic anisotropy of sub oceanic rocks was detected by Raitt *et al.* (1969) at depths of only 10–20 km, but compelling evidence that seismic anisotropy has a cause like that pictured here has come from studies showing the 'fast' and 'slow' axes of the horizontal wave speed correlated with a local 'spreading' direction inferred from the azimuth of magnetic anomaly lineations (Forsyth, 1975).

These estimates of a lithospheric thickness dramatically reduce the difficulties one has always had in understanding the movement of tectonic 'plates' 100 km thick if plate 'rigidity' was intended to convey the same meaning as what I term 'lithospheric' behaviour. The deviatoric stress concentration factor in the superficial material is increased and the stress required for downward crack propagation or fault slippage decreased by factors > 5, a modification of great significance when satellite gravity observations and the theory of self regulating convective states agree that deviatoric stresses at asthenospheric depths are no more than a few tens of bars. Using the above condition $\sigma > P_e$ for tensional failure, it can be readily shown that this part of the failure condition for surface 'lithospheric' rocks to engage in convective movements is that their thickness t satisfy the inequality:

$$t < \sqrt{\frac{4\sigma L}{\delta \rho g}}$$

where σ is the horizontal shear stress imposed at the lithosphere/asthenosphere boundary by underlying convective movements of horizontal length scale L, $\delta \rho$ the difference in density of lithospheric rock and pore fluid, and g the gravitational acceleration. Note that the dependence of this and an associated condition for compressive failure on L has the effect of a low pass spatial frequency filter, biassing the

---

* There are grounds for thinking its average thickness beneath continents is greater than a corresponding figure for oceanic regions by at least a factor of three.

direct view we get of heat transfer movements towards the largest horizontal length scales present in an underlying flow of asthenospheric material. This bias can help to resolve the uneasiness sometimes felt at attributing the inferred velocity pattern of lithospheric Earth material to convective movements extending downwards only as far as a region of major phase transition at ~ 700 km (see below). Laboratory experiments have indicated how the combination of radiogenic heating throughout upper mantle material and its coupling to a self regulating state of convective flow with larger length scales in the lower mantle might create a range of ratios of horizontal to vertical length scales for the upper mantle convective velocity field extending upward to ~ 10. If we take $L = 3 \times 10^8$ cm as a representative figure for the current lithospheric velocity pattern and $\sigma = 10$ bars ($= 10^7$ dynes cm^{-2}), the above condition for tensional lithospheric failure is that its thickness $t < 24$ km. Estimates of a t value from a wide variety of laboratory rock creep data and near surface radial temperature gradient measurements readily satisfy this condition.

The associated condition for lithospheric movement of compressive failure is much more problematical and in my view, insoluble without invoking the activity of water as a lubricant of lithospheric fault surfaces. The inference that mountains have only been built on Earth's surface since the Pre Cambrian - perhaps for as much as half the Earth's present age, solely on account of the weakening effect of water on its 'lithospheric' rocks, may come as a major surprise for those who see water principally as an agent of erosion. If its rheological role in early stages of sedimentary rock formation is universally recognised, this wider orogenic significance has probably been overlooked because the problem of connecting surface movement and deformation with internal heat transfer is still only being 'solved' with Plate Tectonic slogans rather than quantitative analysis. It will be interesting if subsequent efforts to grapple with the 'messy' theoretical problem of 'lithospheric' deformation by heat transfer movements eventually lead others to a similar conclusion. Incidentally, in situ electrical conductivity determination is increasingly indicating water is distributed in even deep crustal layers in a way that is likely to make it rheologically very important as a 'lubricant' of fault planes.

The planet Venus may eventually provide us with the most conclusive evidence of what the Earth would have been like had water not been included in its composition in rheologically significant amounts. There is no elongated topography on the Venusian surface comparable to Earth's oceanic ridge system, a point given further interest by the demonstration of a surface temperature approaching 500 °C – the effect of infra red opacity in a dense CO_2 atmosphere. Any thought that such a high surface temperature makes the surface rock viscosity too small to support a perceptible ridge type topography is immediately discounted by the several kilometre elevation of large areas like Ishtar above the surrounding plains. The measurement of only ~ 10 cm of precipitable water above the Venusion surface, as compared with a > 3 km average ocean depth on Earth, may have pinpointed the reason for a Venusion lithosphere being able to withstand the probably very comparable shear stresses that a self regulating state of convective heat transfer imposes on its undersurface – a very remarkable testament to the lubricating effects of a pore fluid if confirmed. It is an astounding piece of good fortune that nature has provided such a good testing ground for ideas about the tectonic significance of water in another near Earth sized but dry object. Everything should be done to ensure that the instrumentation of any future Venus orbiter be designed to throw as much light as possible on what now appears to be a fundamental rheological question for terrestrial geology.

The high temperatures on the Venusian surface are themselves the result of no water being present as a medium extracting atmospheric CO_2 by carbonate formation and sedimentation; it is certainly a graphic illustration of how the thermal state of planetary rocks can depend on a minor volatile component. More subtly, the expulsion of carbon dioxide is probably connected with an absence of water in the deep Venusian interior, in that the internal temperatures associated with its self regulating convecting state are much more likely to be above those at which important carbonate phases dissociate. In the case of Earth, the entrainment and convective remixing of superficial water – only possible for such a long period of Earth history because it is rheologically active in surface rocks, has been a major factor controlling the average thermal state of its interior. One can see here a further twist to the idea of convective self regulation; the heat transfer process is not only controlling the horizontally averaged values of a highly temperature dependent viscosity function, but also that functional dependence. As a result, I would not be surprised if horizontally averaged temperatures in Earth to depths of several hundred kilometres, at least in the sub oceanic mantle, are little more than 500 °C – an untestable result that nevertheless provokes strong criticism whenever it has been mentioned.

The dependence of the conditions for failure of lithospheric material in tension and compression on the horizontal length scale L of asthenospheric movements may have perceptible and therefore more interesting consequences for the endogenous evolution of Earth's surface. The decline in radiogenic heat sources by a factor ~ 5 since a self regulating convective state was established over 4×10^9 yr ago will have meant an increase in average lithospheric thickness by a comparable factor. As a result, the lower length scale limit imposed on the surface velocity field by the lithospheric failure conditions is predicted to have increased perhaps tenfold throughout this period. Such a large change could be reasonably expected to have significantly changed the way in which a self regulating convective flow regime, having a wide range of horizontal length scales, would be expressed at the planetary surface.

I believe there is some geological field evidence for a smaller length scale of Pre Cambrian lithospheric flows, preserved in the not surprisingly small fraction of Earth's present surface rocks dating from several billion years i.e. more than ten predicted convective overturn times ago. Postponing for the moment consideration of why any rocks should have remained near the Earth's surface for so long, one can make some remarks about the future evolution of Earth's surface. I would expect the convective movement and deformation of surface rocks to cease before a continuing decline in radiogenic heat production has the effect of further doubling the average lithosphere thickness. This will take about 2.5×10^9 yr, but convective movements will continue in ever deepening and thinning shells of the upper and lower mantle for much longer. Such tectonically 'dead' convection can be sustained in an Earth sized chondritic planet by just the longer lived thorium content of it's rocks. Solar evolution to a much more luminous 'giant' stage of its evolution will probably put an end to the entire earth before traditional heat conduction type calculations become an appropriate form of solution to its internal heat transfer problem. It will be clear from these remarks why the smaller terrestrial planets and asteroids, having a smaller surface heat flow and thicker lithosphere on account of their larger ratio of surface area to volume and unable to retain any surface water as a lubricant of superficial deformation, show little or no sign of an endogenous surface evolution since their accumulation.

3. Heat Transfer and the Planetary Formation Process

There are no good reasons for believing terrestrial planet accumulation was directly capable of introducing the kind of large scale heterogeneity of metallic and silicate phases referred to as a core/mantle structure. I use the word 'directly' here in an attempt to distinguish a chemically unselective, gravitationally controlled accumulation of particles after they have reached, very approximately, a millimetre in size, from a possible contemporaneous process of internal differentiation that can be stimulated by an accumulation process if it succeeds in creating large enough objects (Tozer, 1978 and below). The other important type of differentiation of circumsolar matter that was eventually to establish the suspected differences in a metal/silicate ratio for the various terrestrial planets and their even bigger contrast in composition with the Jovian planets is attributable to heat and other energy transfers from a proto Sun while planetary material was still in a sub-millimetre stage of accumulation. It is then most vulnerable to non-gravitational interactions. This 'external' differentiation of terrestrial planet material could have arisen from a systematic difference in the rate of growth of silicate and metallic dust particles through the micron size range, the critical size range for radial heliocentric displacement of particles by photon and corpuscular radiation from a proto Sun. One should remember that the Jovian planets probably contain a much greater mass of these refractory phases than the terrestrial planets, and one could say that the early stages of accumulation of all planets was chemically very similar. I have already mentioned the expulsion of 'excess' volatiles from the interstices of a refractory accumulation by rising internal pressures; this is an example of a large scale differentiation directly stimulated by accumulation, but it is easy to see that a continuing gravitational association of these expelled volatiles and any fresh volatiles arriving at the surface of a refractory accumulation is still possible though its duration is very sensitive to the surface temperature maintained by solar radiation. The present Jovian/terrestrial planet chemical differences can be attributed to the fact that only a more helicentrically distant fraction of the refractory particles orbitting a proto Sun at $\geqq 5\,\mathrm{AU}$ was eventually able to form objects sufficiently massive to bind the lightest and most abundant volatiles purely gravitationally. The present atmospheres and ocean of the terrestrial planets are a minor volatile fraction that remained in association throughout early stages of accumulation mainly because it was chemically bound to the refractory phases. Where surface temperatures were less than about 250 K, the refractory accumulations would have served as 'nucleii' for an ever more indiscriminate addition of particulate icy and uncondensible matter. This essential nucleating role played by the gravitational attraction of refractory phase accumulations is a possible way of understanding the higher than cosmic abundance of these elements in the Jovian planets. Without sufficient refractory material to form nucleii in a few hundred million years, the volatile material now in the Jovian planets would have been forced outward by solar radiation and added to cometary material at heliocentric distances where particle densities would be too low for planet size objects to form even in the several billion years that have now elapsed. Even at the heliocentric distances of Uranus and Neptune the time required for the formation of refractory nucleating bodies could well have exceeded the time required for the collection of ambient volatiles by proto Jovian and Saturnian bodies or their repulsion to greater heliocentric distance – the reason for the higher abundance of condensible material in Uranus and Neptune?

One should be on one's guard against thinking the kind of compositional variation during accumulation specifically visualised above for the Jovian protoplanets, would 'directly' produce a core/mantle type heterogeneity. This is because the number of nucleating proto objects would have greatly exceeded the number of planets now existing. There have been attempts to produce the core/mantle structure of the terrestrial planets 'directly', but I believe they can all be dismissed as purely ad hoc on the grounds of appealing to a most improbable path of evolution for a cloud of refractory particles. Like most efforts to quantify planet formation, they visualise one particle as the embryo of an existing planet and all other particles as tiny bodies striking its surface. Of course, no one particle has any better claim than another to be seen as a planetary embryo – at least until accumulation is nearly over, and this ambiguity about which particle grows as the result of a collision must be considered a strong argument for accumulation being a mixing process. Features like Mars Imbrium, the Caloris basin and Hellas can only date from the terminal stage of terrestrial planet formation, but their length scale is as good evidence as one might now hope to find in support of a more symmetrical i.e. less embryo oriented view of planetary accumulation.

The absence of any unique centre to planetary accumulation might still be thought compatible with the idea of 'direct' or inhomogeneous planetary accretion if one could assume that one particle cloud was formed into a single body before a new cloud of different composition was created in its vicinity. The notion of a circumsolar gaseous nebula, in local thermodynamic equilibrium and cooling to produce particles of differing composition, has enjoyed a wide vogue in the last twenty years but observational evidence and theoretical argument is against it as a factor in terrestrial planet formation (Clayton, 1982). Infra red astronomy, in particular, has been supplying ever increasing evidence for refractory dust in the envelopes of giant stars, possibly approaching an explosive phase of their evolution, or in the interstellar debris from stellar explosions. This is the kind of material from which the Solar System is thought to have formed. Such indications of a prompt refractory dust condensation fit in well with the now numerous observations of cosmochemists that some refractory phases in meteorites must have condensed while a variety of radioactive isotopes, some having a half life of only a few years, were still significantly decaying. It may be shown theoretically that micron size dust particles are likely to be far colder than any solar nebula gas – on account of it's optical thinness, and on these grounds I believe one should regard both metallic and silicate dust as simultaneously condensing prior to the $\sim 10^8$ yr timescale it takes the terrestrial planets to accumulate from dust size particles.

This timescale (see for example Wetherill, 1980) has never been seriously questioned on dynamic grounds, although it has been indirectly by those who have been led to believe a powerful heat source is needed to explain the early differentiation of lunar highland material, or a lunar magnetisation dating from more than four billion years ago*. In my view, this raises more problems than it solves and is merely compounding the error of visualising planetary accumulation as an embryo growth. This is because the need for the heat source has arisen through the assumption of relatively small bodies striking an embryo surface dissipating their kinetic energy locally. This grotesquely overestimates the rate at which the gravitational energy that has to be dissipated during any accumulation, is lost from the material. In contrast, the very large collisions that are an inherent feature of the picture of a mutual particle growth, result in a much more efficient storage of this dissipated gravitational energy throughout all of an accumulation. The Moon is an interesting test case for theories of planetary formation because the amount of gravitational energy to be dissipated is only just sufficient to melt its material. The density of all terrestrial planets being the same to within a factor of two, one can say the amount of energy/unit volume to be dissipated is proportional to the square of their radius. In the cases of Earth and Venus, it amounts to about ten times the quantity of heat subsequently introduced by radioactivity.

Consideration of this gravitational energy input carries the implication that all protoplanetary accumulations reaching about 1200 km in radius would have become convectively unstable and developed internal heat transfer movements. One should view any subsequent general rise in internal temperature or chemical differentiation as evolution achieved in spite of two major features of heat transfer on such a large scale: (1) It's ability to maintain a state of near constant and an extremely large horizontally averaged viscosity value over a wide range of heat inputs. (2) It's tendency to remix planetary material even while it is extremely viscous. The first feature is easily able to prevent a general state of melting from occuring during a 10^8 yr period of accumulation – even in such an energetic case as the Earth. This temperature control is clearly an important factor to be considered in deciding whether any or a variety of volatile components can still be in the terrestrial planet interiors. I would cite the recent observation of a He^3 evolution from Earth's interior (Craig and Lupton, 1976) as a vivid confirmation of the ability of convective heat transfer to limit internal temperatures to sub solidus values when the mean rate of heat input to Earth material was hundreds of times greater than it's present radiogenic value.

* It has to be attributed to a radioactive source with less than a 10^8 yr half life in order to be undetectable in existing rocks.

These features of heat transfer, working together, constitute a major obstacle for any potential mode of differentiation and as such have fixed the timing and pattern of planetary evolution. This obstacle can be stated as the problem of getting the length scale of the density differences produced by a particular type of differentiation large enough for the two fractions to separate faster than they are remixed by a convective heat transfer process. Clearly, this obstacle is most severe for those differentiation processes producing the smallest density contrasts and ensures that the most coarse density separations of a polyphase starting material have to be more or less completed before more subtle ones can start. Making the reasonable assumption that all phases in the newly accumulated planetary material were mixed down to a length scale of centimetres or less, this obstacle would have been insuperable without a third feature of a convective heat transport process, growing in importance with its length scale, which offers a limited way of defeating the other two. This is the heat produced by the material deformation involved in a convective heat transfer. This heating is utterly negligible on a laboratory scale and of course, it is smaller in total quantity than the heat source primarily responsible for the material deformation. It does not augment the mean global rate of heat loss from a planet, and would have had no observable consequence for the planets if its release had not become highly localised in both space and time. I defer discussion of the cause of this behaviour and treat the large local viscosity reductions it can transiently produce as a way of tapping another potential source of internal heating – the gravitational energy to be dissipated in making a stable density stratification of the accumulated material.

The importance of this energy source as a driving force for further convective deformation can be expressed by a parameter β (Tozer, 1978), basically the product of the density difference produced by a particular mode of differentiation, the abundances of the two material fractions it involves, the square of the accumulation radius and reciprocal of the temperature rise needed to make their separation velocity greater than that of the concomitant convection. I show that if β exceeds a certain critical value β_c, deformational heat due to convection in the starting material can serve as a 'trigger' for a runaway process in which the rate of gravitational energy dissipation can grow to dominate all other heat sources. It will only start to decline when reorganisation of the accumulated material has made $\beta < \beta_c$ – essentially through a fall in the value of the abundance product for the two fractions in the source material. If $\beta < \beta_c$ for every possible differentiation of the starting material, the efficiency with which a purely thermally driven convection can produce deformational heat sets an upper bound to the mean rate of any differentiation and it must take the form of a sequence of purely local events rather than a catastrophic global process.

My studies indicate that only the metal and silicate content of a terrestrial planetary material accumulation could have created a $\beta > \beta_c$ situation in which the powerful negative feedback exerted by convection over horizontally averaged viscosity values can be temporily overwhelmed by what I term 'autocatalytic' differentiation. Even for this differentiation its autocatalytic nature only arises after an accumulation has reached a radius in the range 1500–2000 km, the precise value depending on the metal/silicate ratio of the material. Nevertheless, this result suggests that with the possible exception of the Moon, the immediate precursors of existing cores in the terrestrial planets were cores that had separated in several of their respective protoplanets and which would have promptly joined one another after collision of these objects. There is a certain amount of indirect evidence supporting such an early metal/silicate separation, of which the most critical now seems to be the inference of a mantle heterogeneity of xenon isotopic ratios (Allegre et al., 1983). As with the earlier problem of internal temperature, there appears to be some danger that at least the lay audience to complex geochemical arguments will be lulled into a belief that one can establish a particular pattern of trace element heterogeneity in mantle rocks from a study of surface rocks, when in fact all one can do with such data is test views about the extent of planetary material mixing throughout its post accumulation history that are based on a dynamical theory of some sort. Rather than resort to model building from the observations, I think one should try to use the geochemical data more directly. Using the fact that a primary metal/silicate fractionation or the joining of cores following the collision of protoplanets would be an intensely mixing process for the components constituting each fraction, one does not need any heterogeneity model to deduce that a metal/silicate separation must have predated the effective end of any radioactivity producing Xenon isotopes. Any evidence of some activity of I^{129} in the mantle, the parent nucleus of Xe^{129} with half life $\sim 1.6 \times 10^7$ yr, is therefore very decisively against core formation being spread over a significant fraction of Earth history.

Another aspect of the prediction of a multistage separation of metal and silicate phases in several generations of protoplanetary objects is that it is effectively accomplished at much lower pressures (tens of kilobars) than the megabars of the conventional single stage picture of its occurrence in Earth material. By 'effectively', I mean that the scale of mixing has been made too large at much lower pressures for atomic diffusion to have any significance in bringing the two fractions into chemical contact. As a result, it is neither necessary or relevant to speculate about chemical reactions which might have occurred between silicate and metallic phases had they been intimately mixed at megabar pressures, in order to decide what might be the alloying component(s) reducing the density of Earth's core below that expected of pure iron. It now appears that one of the other essential requirements of an alloying element in Earth's core not achieved by nickel is a substantial reduction of its final freezing temperature below that expected of iron at the same pressures. Sulphur still looks the most plausible geochemical choice to sustain the inferred liquidity of an outer core*. It

* It was this problem of maintaining any part of Earth's core in a liquid state, given that the lower mantle would have been in a self regulating, highly viscous state for billions of years, that motivated the original suggestion of the core material having become inhomogeneous in the course of freezing. The rejection of the alloying component to a residual liquid phase made the temperature range over which freezing took place both wider and lower than that predicted for a pure iron core. Later embroidery of this idea to make latent heat (a specific heat anomaly rather than heat source) the driving energy of core motions is not compatible with the existence of a self regulating state in the lower mantle. Some type of radioactive heating of the core material is necessary in these circumstances – K^{40} looks the most likely heat source.

is the very great difficulty in maintaining a sufficiently large volume or indeed any liquid core material in the other terrestrial planets, either or both on account of their smaller external radii and lower abundance of this moderately volatile element, that has made the Earth unique in being able to maintain a $\sim 1\,G$ surface magnetic field quasi continuously for billions of years. No iron/sulphur alloy could have delayed the freezing of lunar, Mercurian and Martian cores billions of years ago through convective heat transfer in their mantles. The less predictable observation of no significant Venusian field I would attribute to the complete or near complete solidity of its core, due to a much lower mean sulphur/iron ratio than that of the Earth's. Like its very much smaller abundance of surface water, this contrast of the two planets has to date from the 'external' differentiation of circumsolar material by solar heat transfer at a very early 'dust' stage of its accumulation. With the exception of the Venusian case, these predictions were firmly made before the advent of space probes, and it is amusing to reflect that the most unequivocal and successful forecasts about planetary magnetism have come from the study of mantle rather than core dynamics. Unlike the infinitely flexible post facto rationalisations of dynamo theorists, these predictions of core solidity do offer the necessary prospect of refutation by foreseeable, if expensive observation – planetary seismology or a long term plan to detect secular variation of an internally generated magnetic field from closely orbiting satellites.

4. Post Accumulation Planetary Evolution

A self regulating state of convective heat transfer would have been effectively established within half a billion years of the end of any significant accumulation or core/mantle separation. Since that time the internal thermal state could only have appreciably changed through a secular change in the temperature dependence of those material properties that define a self regulating state. Given the principle of uniformitarianism (see Introduction), the only way to achieve such secular change rests on a persistent occurrence of local viscosity reducing events that will facilitate a further differentiation of the silicates. For me, one of the most attractive and compelling features of the discovery of a regulation of extremely large averaged viscosity values is that it simultaneously creates the conditions necessary for large downward fluctuations of a local viscosity value.

The fact of an Earth's interior simultaneously able to transmit roughly one second period shear waves, yet vent a material, 'magma', with a very much shorter relaxation time, has been known for decades. This first major discovery of seismology, destroying an earlier generation's observational 'proof' of the interior being magmatic below a superficial crust, has only ever taken on the appearance of paradox through insistence on an internal temperature distribution that is very smooth, nearly spherically symmetrical and only capable of changing at the typical 1 °C/million year rate of direct radiogenic heating. Despite the enormous publicity given in recent years to the highly non random nature of the distributions of active vulcanism and deep seismicity, it is evidently still very widely thought that magma temperatures or the frozen phase equilibria of intruded material are representative of average thermal conditions at some depth of origin. One should be clear that our real problem is one of reconciling a prediction, confirmed by isostatic rebound studies, of an average effective viscosity in Earth's mantle $\gtrsim 10^{20}$ poise with the 10^{14} times smaller viscosity of a typical magma. If one rejects as hopelessly inconsistent the view that magma is free to flow from a material whose overall viscosity is constrained at $\sim 10^{20}$ poise, this rheological problem is never going to be solved by juggling with temperature values. It is my suggestion of horizontally averaged temperatures $\sim 500\,°C$ to depths of even some hundreds of kilometres that has drawn dismissive criticism, but one that I feel both free and justified in making on the grounds of water being an important and necessary lubricant of in situ Earth material accounting for the involvement of surface material in the convective heat transfer process. I have no doubt that when other theoreticians realise the essential uselessness of describing any, let only the uppermost 100 km of Earth material as a rigid, heat conducting shell, they will soon come to the conclusion that average temperature estimates have to be substantially reduced.

The regulation of a very large effective viscosity in what is more accurately described as a viscoelastic planetary material is tantamount to the regulation of a very large negative temperature coefficient for that viscosity*. It is the combination of a stored strain energy density and this large negative temperature coefficient that can give solutions to terrestrial planet heat transfer problems one of their most intriguing characteristics – local temperature and strain rate of fluctuations. The following remarks about their existence and nature are largely based on numerical studies of simple planar shear in a range of temperature dependent viscoelastic materials forming slabs. Their thickness and boundary conditions were chosen to interpret a mass of observations made on materials subjected in the laboratory to much larger shear stresses than would occur in any convection experiment, and to evaluate the significance of these observations for a much larger scale deformation process like planetary heat transfer. This roundabout approach is necessary because any direct demonstration of the existence of the suspected fluctuations in a numerical solution of the heat transfer equations for a viscoelastic planet would require one to follow the evolution of a solution over a vast range of length and time scales**. These planar shear studies (Mahboobi, 1981) have shown that the peak temperatures and strain rates reached in fluctuations actually decline in value if the general or pre fluctuation temperature in a slab is given a higher value. The fluctuation phenomenon disappears altogether in a slab of any relevant size if its pre fluctuation temperature is raised above ~ 0.4 of a material's absolute solidus temperature, a result that is supported by laboratory observations of the so called 'stick/slip' phenomenon. This result may be roughly translated into the context of terrestrial planet heat transfer problems with the perhaps

---

* For the usual type of exponential temperature dependence of the creep rate, the magnitude of this coefficient varies at least as fast as the viscosity value.

** Even using a finite element technique whose spatial resolution is thousands of times too large to show the suspected fluctuations, it is still necessary to constrain the solutions to convection equations in various ways, e.g. make them two dimensional, in order to store them inside the largest and fastest modern computers.

surprising remark that melting temperatures are only likely to be locally and transiently reached in those parts, if any, of a purely thermally driven convective circulation in which temperatures are on average well below half the absolute solidus temperature.

This last comment exploits the fact that the suspected fluctuations are so localised in a direction normal to the direction of shearing and so short lived that one can look upon them as a feature of different regions of a quasi steady state solution to a planet's heat transfer problem. However, one should be on one's guard against seeing them as a purely local property of the material, like a strength, because the sudden deformation is inseparable from the storage of strain energy throughout a volume of viscoelastic material. How extensive that volume is for a particular slip event or fluctuation is determined by variations in the properties of the material in the direction of shearing.

Clearly, the calculation of a peak fluctuation temperature and the possibility of a radiation of wave energy is relevant to an interpretation of the phenomena of magmatism and seismicity, but we can first take a global view of the whole fluctuation phenomenon to make an estimate of the state planetary evolution could have reached some four billion years after metallic core/silicate shell separation and the establishment of a self regulating convective state. By treating the convective process as a heat engine, it may be shown that an upper limit to the total amount of energy dissipated in fluctuations is a fraction $\sim g\alpha L/C_p$ of the total heat transferred, where α is the expansion coefficient, C_p the specific heat, L the length scale of convective movements and g a representative gravitational acceleration. This efficiency of heat energy conversion to mechanical energy is independent of the viscosity because the self regulation process keeps a high Rayleigh number at all relevant times, but its dependence on L makes any planetary scale convection quite unlike any convective regime studied in the laboratory. Even so, it is only a few per cent for convective movements limited to the upper mantle by the properties of the phase transition zone (see below) and would only lead to some minor and quite inconsequential redistribution of the surface heat flow if it were not expressed in large local fluctuations of temperature. This possibility makes it more appropriate to notice that this upper bound to the rate of energy release in fluctuations is of much the same order as estimates of the mean power involved in volcanic and seismic phenomena*.

From this estimate one can readily show that $<10\%$ of the upper mantle material has ever been heated by fluctuations throughout the past four billion years the necessary $500\,°C$ or more needed to make it molten and differentiate. This result is immediately helpful in explaining why even the chemically inert and non radiogenic atoms of He^3 can still be expelled from earth's interior (Graig and Lupton, 1976)–no atomic species can be expected to be significantly mobil while the overall viscosity is held at 10^{21} poise. Incidentally, this observation has made an excellent case for believing larger and more chemically reactive volatile molecules are still present in the

---

* I am using estimates of $\sim 10^{18}\,\mathrm{j\,yr^{-1}}$ that were made for vulcanism before it became fashionable to say that $>90\%$ of Earth's magmatism is in the vicinity of ocean ridge crests. In my opinion, these new estimates are wild overestimates based on a naive picture of the formation of new 'plate' 100 km thick.

interior at much the same concentration as existed at the end of planetary accumulation. Much the same point is made by the inference that the present rate of He^4 outgassing does not balance its estimated rate of production by members of the uranium and thorium decay series (O'Nions and Oxburgh, 1983).

If one wishes to say anything about the characteristics of individual fluctuations within the convection pattern with the present method of analysis, there is no option but to take the length scale of the fluctuation in the direction of shearing as a free parameter l. This freedom of choice is meant to indicate the possibility of sudden slip over areas of a slip zone l^2 up to a maximal value L^2 fixed by the length scale of convective movements. The strain energy density U stored in a volume $\sim l^3$, is dissipated in a slip zone of thickness $\sim \sqrt{Kt}$, where K is the thermal diffusivity and t the transit time of the stored energy to the slip zone. Only the more detailed calculations (Mahboobi, 1981) can show that most of the stored energy can be delivered to a slip zone in a time $\sim l/v_s$ (v_s the shear wave velocity $\sqrt{\mu/p}$) if its 'initial' temperature is ≤ 0.4 of the absolute solidus temperature, but in such a situation the maximum temperature rise ΔT in the slip zone is given by:

$$\Delta T = \frac{U}{\rho C_p} \left(\frac{l v_s}{K}\right)^{1/2}.$$

Taking $U = 10 \, \mathrm{j \, m^{-3}}$, corresponding to the average state regulated in Earth's interior, and conventional values for the other parameters, we see that l has to be ≥ 200 km for ΔT to exceed the four or five hundred degrees necessary to convert this average convecting material to a magma. In fact this calculation for planetary material having the average $\sim 10^{21}$ poise viscosity of its self regulating convective is unrealistic to the extent that this material has an initial temperature too high (0.5–0.6 of its absolute solidus temperature) for any appreciable fraction of its stored shear strain energy to be delivered to a slip zone at the assumed speed v_s. That is the reason why magma and seismic source regions do not occur much more uniformly throughout the region of convective flow. A better calculation would use the stored energy density U in the vicinity of any much cooler material comprising a descending thermal boundary layer for the convective flow and also take some account of the previously mentioned stress concentration that occurs in lithospheric material in such places. This makes a case for considering U as a function decreasing by a factor of at least a hundred as one's choice of an l value increases and only being $\sim 10 \, \mathrm{J \, m^{-3}}$ when $l \sim L$, the largest length scale in the convective flow pattern. This U variation may somewhat reduce one's estimate of a minimum area of slip zone l^2 needed to produce magma, but does not alter the prediction that magma and seismic sources should be strongly concentrated in the coolest parts of a planetary convective heat transfer*. This curious fact was once raised

---

* The small amount of earth's seismicity ($\sim 5\%$ of the seismic energy release) at depths of a few kilometres beneath ridge crests can be attributed to unstable downward crack propagation in lithospheric material induced by tensional stress concentration. The seismic characteristics of these ridge sources are predicted to differ systematically from those defining transform faults and Benioff zones.

by Beloussov as an argument against convection, but of course it could never be properly answered with the kinematic models of Plate Tectonicians. If it had been, more Earth scientists would have been prepared to accept my view of magma temperatures being quite unrepresentative of the horizontally averaged states of the upper mantle.

Just as we have predicted fluctuations not leading to magma production, one also expects some to occur without shear wave radiation. The detailed numerical calculations of viscoelastic slabs driven at a fixed mean strain rate always show a period of accelerating strain rate within a part of the slab before a stage of wave generation might be reached. This precursory aseismic part of a given fluctuation forms an increasing fraction of the total displacement associated with it as the chosen initial temperature of the slab is raised. In fact, the upper limit to the temperatures at which fluctuations would be expected can be seen as the result of this aseismic displacement rather rapidly growing from a small fraction to all the displacement as temperature is raised from ~ 0.3 to ~ 0.4 of a solidus temperature**. Unlike the case of magmatism, there is probably no minimum value of shear zone dimension l associated with the radiation of shear waves – at least for a zone near a planet's external surface where effective pressures tend to zero and fault gouge material can remain unannealed. At the other end of the scale of fluctuations we can make some useful remarks about the observable characteristics of the fluctuations and their consequences for planetary evolution. If we take as a maximum of $l \sim 500$ km, roughly corresponding to the vertical scale of convective flow above the phase transition zone of the mantle, it follows that up to $\sim 10^{18}$ J of wave energy and/or enough heat to melt a few cubic kilometres of rock in a slip zone could be associated with individual fluctuations. These maximal events would be associated with permanent displacements of ~ 10 m and could not recur in the same locality more frequently than about once a century. It is the transit time of the stored shear energy density to a slip zone that basically determines the dominant period of the wave packet radiated from a particular fluctuation – hundreds of seconds for the largest and scaling roughly as the (energy of the wave packet)$^{1/3}$. The general spatial association of vulcanicity and seismicity is explained and I suspect that the clustering of seismic foci at particular depths in descending material is due to the stimulation of slip instability by specific dehydration reactions that raise the magnitude of the temperature derivative of an effective viscosity in various intervals of the sub solidus temperature range.

I regard the general correspondence of these theoretical predictions with terrestrial observation as encouraging support for the new way of looking at planetary heat transfer – after all, the old methods made no substantial predictions about anything observable after free parameters had been deliberately chosen to avoid such obvious

** A solidus temperature for slab material is *defined* by the exponential relation chosen to represent the dependence of its effective viscosity on temperature. If the material was given a different functional dependence of its sub solidus viscosity, such as probably occurs in Earth material through the liberation of water from hydrated phases, the quoted upper limit for an initial slab temperature for seismicity and vulcanism to occur would be changed (see later).

defects as a generally molten mantle. One can foresee the other planets providing good tests of this view of magmatism and seismicity as an intrinsic local unsteadiness of the convective heat transfer process in large self gravitating and viscoelastic systems if they have a big enough negative temperature coefficient of effective viscosity in some part of the flow. The efficiency $g\alpha L/C_p$ with which a convective flow converts a primary radiogenic heat to deformational heat is not greatly different for all the terrestrial planets, and if it could have all been expressed in magma production and extrusion on their surfaces, their cratered terrains dating from accumulation would have long ago been completely inundated and obscured. However, as already explained the temperature coefficient of viscosity is not large enough negative for any magma producing fluctuations to occur in any part of a self-regulating convective state unless lithospheric material is disrupted and entrained by the heat transport process – that is the reason why ancient terrains survive as well as they do. More positively, I predict that all objects having extremely old surfaces will be found to be essentially aseismic in comparison with the Earth and would naturally suggest that is why the Moon, despite having a self regulating convective state in its interior, was found to be so grossly inefficient at producing seismicity from its heat flow. I have suggested elsewhere that the little endogenous seismicity that was recorded is probably due to tidally induced movements of an aqueous pore fluid, permanently trapped in the deep interior of the Moon for just the same reasons that a low Q zone now exists in the Earth's upper mantle, rather than any shear zone instability of its convective process. An analogous seismicity from sources in Earth's upper mantle would be below the tectonic or meteorologically induced microseism level. It is also no coincidence to me that a system like Mars, showing some signs of an endogenous volcanic activity, also has signs of surface modification by involvement in large scale convective movement – probably some billions of years ago.

If the natural seismicity of the heat transfer process in the Earth has been the most valuable probe of the current state of its evolution, it is the cumulative effects of hundreds of millions of magma producing fluctuations over the last 4×10^9 yr that has basically determined the present complexity of that evolutionary state. An inescapable feature of this model of magma production by strain rate fluctuations is its initial distribution in inclined zones perhaps hundreds of kilometres in extent but at most only tens of centimetres thick. It would all resolidify in situ within at most a few hours were there not immediately present large stresses arising from its density contrast with surrounding much cooler rock that tend to force it upward and into a more compact shape less vulnerable to rapid cooling. The dissipation involved in such upward flow could be an important factor in maintaining the liquidity of newly created magma, but there is no reliable way of calculating what fraction of it might form or join a magma reservoir surviving from earlier magma producing fluctuations or even reach the external surface. However, it seems true to say that any significant chemical differentiation of magma depends on it forming or reaching a pre-existing reservoir; if its flow from slip zone to reservoir were not so vigorous as to prevent differentiation it would cool and develop a viscosity too large for it to occur. Clearly, what could only be a succession of very discrete injections of undifferentiated magma into a cooling and partly fractionated magma chamber is one very direct way of producing the distinctive small scale chemical heterogeneity or compositional banding one sees in some intrusions. However, the key question that decides whether the formation of reservoirs is itself an intermediate stage in the cumulative production of silicate differentiation on a much larger planetary scale, is the degree and period for which one or more fractions of the cooling reservoir maintain a finite density contrast with the surrounding rocks. Elementary analysis indicates that below a certain volume of material that depends on its density contrast with the surroundings, convective movements will entrain and eventually remix it with the general body of convecting material. A relevant and surprising result of experimental petrology up to pressures of ~ 30 kilobars on silicate systems thought to resemble the starting composition of Earth's silicate shell is that separation into a more creep resistant, volatile free, ultrabasic fraction and a less dense sialic crustal fraction can result in a decrease in its mean density, *both* fractions being less dense than the starting material under the particular P, T conditions each fraction experiences after separation. The fractional decrease in mean density, mainly due to a high compressibility of the sialic fraction caused by phase changes, is only 10^{-2} but the relevant comparison is with the much smaller ($< 10^{-4}$) thermally induced fractional density differences within undifferentiated material that one predicts as a feature of a self regulating convective state and which are of just the right order to explain the observed departures of the geoid from a shape predicted by hydrostatic theory. The ratio of these thermally and chemically induced fractional density differences make it plausible to say that once a primitive mantle material is differentiated into sialic and ultrabasic fractions together forming masses with a length scale 1% of the heat transfer movements i.e. 5–10 km, they would be permanently immune to entrainment in the convective flow so long as they retain coherency. The coherency and continuing spatial association of the two fractions in the face of tectonic, i.e. convectively imposed stresses would have been greatly aided by the refractory ultrabasic fraction, even more creep resistant than the starting material by virtue of its smaller volatile and sialic mineral content, being emplaced under the less dense sialic phases. In other words, the two fractions end up in a thermal environment that optimises the coherency of the single region they comprise. I attribute the generation and maintenance of a continental 'freeboard' to this coherency and the 1% reduction in mean density entailed by differentiation. I have noted (Tozer, 1977) that this continuing association of both fractions, despite the large horizontal translations referred to as 'continental drift', is probably the reason why the mean heat flow in continental and oceanic regions are so similar. The somewhat lower values over shield areas can be attributed to the loss of coherency implied by aeons of sub aerial erosion. Clearly, the present horizontal differences in seismic properties to depths of perhaps hundreds of kilometres that are referred to as 'the deep structure of continents' would be very difficult to reconcile with current conception of continental drift unless there was a buoyant and relatively creep resistant foundation to the continental crust. The non hydrostatic stresses generated within such a slab of differentiated material by the surrounding heat transfer movements would grow with its horizontal dimensions, particularly if these became comparable or greater than the dominant length scales of these movements. While such tectonic stresses would not disrupt the slab on a small enough scale to pose any threat of subduction, they do impose an upper limit on the area of continental slab material likely to survive intact for something like the period of time it would take the pattern of convective movements to change radically – probably of the same order as the overturn time, $\sim 10^8$ yr for a flow which has its effective Rayleigh number regulated at 10^6 to 10^7. Quite apart from the idea of continental disruption and dispersal, such stresses are the most probable cause of the many deep fault surfaces revealed by recent crustal seismic reflection studies – incidentally, hardly convincing evidence for the 'rigidity of plates'!

The piecemeal and progressive removal of volatiles and the more fusible silicate fraction of Earth's primitive shell material to form buoyant crustal slabs would in turn have reacted on the horizontally averaged temperatures associated with the self regulating convective state. As already mentioned, the concomitant upward drift and temporal decline of the incompatible heat source elements, uranium and thorium, is of secondary importance to the evolution of mantle temperatures. The only further point stems from the total dependence of this drift on local magma producing fluctuations. Since these *only* occur if the mean convection velocity exceeds some finite value, it is impossible for heat source differentiation alone to produce a convectively stable planetary interior. Self regulating convective states can only be finally destroyed through a temporal decline of the heat sources by a factor that grows with planetary radius. I have already noted it would take $\gtrsim 10^{10}$ yr for any of the terrestrial planets.

So far in Earth history, the increasingly refractory nature of its interior produced by differentiation would have caused the horizontally averaged temperature in the upper mantle to rise by a few tens of degrees – mostly due to a rise of temperature under the continents. This would have increased the planetary volume by a few tenths of a percent and quite sufficient to produce tensional cracking in a surface shell. However, one has the observational difficulty that all planets which might exhibit this expansion cracking, due to differentiation, in their surfaces, are also those in which surface rocks have been subducted by the heat transport process. It is the youth of Earth's surface which explains the absence of any clear evidence of global expansion, but the enormous Valles Marineris feature on Mars, an object just large enough to show some endogenous renewal of a very ancient surface, may be such an expansion feature. The principle causative agent there is more likely to have been the declining rheological effect of water that is widely believed to have been expelled from the Martian interior, rather than a silicate differentiation. Since one is in effect saying that thermal expansion more than compensates for the volume of water expelled, this would be as dramatic a confirmation of the sensitivity of rock rheology to a water content as the absence of sea floor spreading on Venus it if could be proven. The cracked appearance of the surfaces of Europe and Ganymede, where temperatures are too low to permit the escape of any expelled water to interplanetary space may also be an illustration of the same effect.

The increase of Earth's upper mantle horizontally averaged temperatures may, however, be being revealed to us in a much more indirect way by the isotopic data currently interpreted as a sign of large scale heterogeneity in the present mantle. If the heat transfer theorist has to select a depth range marking the site of some kind of barrier to convective mixing of the homogeneous silicate shell created by metal/silicate separation, a region of first order phase change in major minerals would be the obvious first choice on account of the enhanced adiabatic temperature gradient generated by latent heat and volume changes (Tozer, 1959). This gradient could easily be ten times the 'normal' value regularly calculated for Earth's mantle ($\sim 0.3\,°\mathrm{K\,km^{-1}}$) and could therefore exceed the value needed to carry heat purely by conduction through its so called 'transition zone'. The only modification I would make to the original analysis is that the lower of two transitions now marking the transition zone at a depth $\sim 650\,\mathrm{km}$, is a more severe obstacle to convective mixing than the olivine/spinel transition now thought to occur at a depth $\sim 400\,\mathrm{km}$. Although we know even less of the temperature dependence of an effective viscosity function for the lower mantle than the upper mantle, convective self regulation can be relied upon to create convective movement and a very similar viscosity value to the upper mantle. Viscous homogeneity of the whole mantle is now supported by isostatic studies (Peltier and Andrews, 1976) and I would only say that this would be a quite extraordinary coincidence for a system showing a $\sim 40\%$ radial increase in density were there not a strong control being exerted over its viscosity values. Even if convective velocities are predicted to be much the same few centimetres/year in upper and lower mantles, one can also predict a major difference in their character – no magma producing local temperature fluctuations in the lower mantle because none of its material would have a large enough negative temperature coefficient. Hence, one does not expect any differentiation to have taken place and its horizontally averaged temperatures would have changed even less than those of the upper mantle. The relative change in horizontally averaged temperatures is such as to enhance the obstacle provided by the convective stability of the phase transition zone to convective mixing, but there is an interesting secondary effect. The rise in upper mantle temperatures would have increased the depth to the phase transition zone by several kilometres throughout Earth history because dp/dT is positive for the suspected phase transitions. This would imply that a comparable thickness of previously undifferentiated material has been gradually incorporated into the upper mantle circulation, its contrasting composition probably giving it such a large density contrast with its new surroundings in comparison to the latter's thermally induced density differences as to lead to rapid rise and local expression on the external surface – the 'plume' phenomenon?

5. Conclusions

This paper has attempted to trace the influence of heat transfer on the currently observable state of planetary material from its very earliest stages of accumulation. Although one might say that the very earliest stages in this history had the most profound influence in determining the way it is now divided among the planets of the

Solar System, it is the idea of convective self regulation which came into play when at least some of its more refractory components formed self gravitating objects exceeding about 1000–1500 km in radius, that is central in producing predictions that can be tested against new observations. Rather in the way Eddington was able to make predictions about stars without any knowledge of thermonuclear energy production beyond the knowledge that it would be self regulating, appreciation of a regulation in the transfer of heat from fixed heat sources incorporated into the interiors of planet sized objects has enabled us to bypass and redefine a number of problems that looked intractable or solved by ad hoc methods in previous accounts of planetary thermal history. In my view, the scientific gains greatly outweigh any imagined loss, planetary heat transfer theory having been converted from a rich source of untestable predictions about internal temperature into something able to connect and in some cases quantitatively explain the characteristics of observable phenomena. The new ideas are not precise in their predictions about internal temperature because that would require a knowledge of the viscosity dependence on temperature only fools would claim to possess, but they are entirely unambiguous in predicting horizontally averaged temperatures much less than those calculated with conduction theory. Unfortunately, temperature prediction with conduction theory has acquired a quite unjustified aura of respectability through decades of timid repetition and the new ideas have been attacked mainly because they do not conform to them. It is a graphic example of the compartmentalisation of current thinking that views derived from the era of static Earth model building should be so fiercely held at a time of such widespread devotion to the dogma of Plate Tectonics. Plate Tectonics is, in more senses than one, a very superficial approach to the question of planetary evolution. It may have marked a very important change in geological thinking, but its infinitely accommodating kinematic pictures disguise the fact that without a coherent theory of planetary dynamics its explanations can only be circular and its predictions self fulfilling. Its own confusion and double thinking over rheological questions, expressed in the absurdity of describing an obviously faulted and folded near surface material as 'rigid', will have to be discarded or very carefully explained if any theory is ever to account for the brute facts of tectonic change.

   Near-surface material is different rheologically from that at depth primarily because its intrinsic shear relaxation time due to creep processes is longer than the longest time scale associated with convective movements of a self regulating state of Earth's mantle. This inequality forms the basis of my definition of 'lithospheric' material and there is no need for any further concept like 'plate' based on Earth's structure at seismic frequencies of deformation. Involvement of surface material in the convective movements of a planet is then seen to be a question of the extent to which crack formation and propagation in any 'lithospheric' shell it may possess can compensate for its negligible thermally activated creep deformability. It is symptomatic of the confusion about rheological questions and what heat transfer theory is trying to achieve that heat transfer theorists have given much more attention to the weak non linearity of thermally activated creep, something having no perceptible effect on the planetary heat transfer process, than this crack induced non linearity of the rheology on which virtually all visible deformation of Earth material directly or indirectly depends. Disruption and translation of Earth's lithospheric material by its heat transfer process is not understandable unless it extends downward for much less than the depth to the seismologists' low velocity layer. Prediction and seismic anisotropy interpretation give it a thickness no more than a few kilometres under spreading centres and perhaps a few tens of kilometres near subduction zones. The weakness of Earth's 'lithospheric' material and indeed the fact that there are any large horizontal movements and deformation of it at this stage of Earth history are probably the result of a downward percolation of surface water or water derived from hydrates on previously formed and unannealed crack surfaces. This kind of interaction is a further ramification of the idea of heat transfer as a self regulating process that makes it look naive to regard tectonic activity as anything other than a function of the whole system Earth material comprises.

These problems of explaining the dynamical status quo of Earth material are meant to serve as an introduction to a discussion of how that state has evolved. It is suggested that the refractory components of circumsolar material accumulated homogeneously down to a very small length scale, the gross differences in composition between Jovian and terrestrial planets being attributable to the effects of a solar heat transfer while the material as a whole was in a very finely divided or uncondensed state. Unlike the situation at terrestrial planet – like heliocentric distances, there was enough refractory material in accumulations orbiting at several A.U. to act as 'seed' objects for a gravitationally unselective accumulation of all components in the circumsolar material. Primary differentiation of the refractory accumulations into a metallic core/ silicate shell structure dates from the time at which they formed objects roughly 1500 km in radius – a size at which the gravitational potential energy to be dissipated in a stable rearrangement of these phases is enough to make it 'autocatalytic'. Since that time, which in many cases considerably predates the formation of the few familiar objects in non intersecting orbits we call the planets, any further evolution of the density structure of the refractory accumulations has been determined by the characteristics of the internal heat transfer process. The establishment of a self regulating convective state within 10 % of the present age of the Solar System resulted in the freezing and cessation of magneto hydrodynamic activity in all the smaller terrestrial planet cores. Differentiation of their silicate shells since the establishment of such self regulating states has been determined by the rate at which local heating events in the convecting shell material, occurring on a timescale of minutes rather than millions of years, are able to circumvent the constraint on horizontally averaged viscosity values. This is necessary to produce a medium in which different silicate phases can separate faster than they are remixed by convective movement. The occurrence of such heating events, identified with and explaining the current pattern of vulcanism and seismicity in the Earth, has been predicted from a study of the consequences of an effective viscosity being regulated at such large values for a planetary scale system – in particular its automatic implication of a large negative temperature coefficient for that effective viscosity and the storage of shear strain energy throughout the convecting material if it is given any plausible viscoelastic properties capable of integrating the tectonician's and seismologist's views of Earth material.

The resulting picture firmly supports a qualitative view held by many geologists that the formation of continents has been a piecemeal process – a collective result of literally hundreds of millions of discrete magma producing events spread over the whole of past Earth history and only declining in their frequency of occurrence throughout this period as a result of a decline in the heat source density. Where modern heat transfer theory would add to this traditional view would be in connecting the occurrence of the magma producing events with an involvement of Earth's superficial material in a convective process. When radioactivity has declined further to the point at which the lithosphere has become too thick to be disrupted by heat transfer movements, magmatism and continental growth will abruptly cease. Indeed, that is the situation existing for most, if not all the post accumulation histories of the smaller and/or dryer terrestrial planets.

Such heating events have as yet only affected a small proportion of upper mantle material and are not expected to have occurred at all in a convecting lower mantle. This partial removal of volatiles and lower melting point silicate components from the interior to form a hydrosphere and the superficial slabs we call 'continents' has resulted in a slow upward drift of the horizontally averaged temperatures associated with a self regulating state. It is not expected to be more than a few tens of degrees and it would never be directly detectable, but it may be the underlying cause of so called 'plume' phenomena if one understands by that a gradual incorporation of undifferentiated and more volatile rich material from the phase transition zone of the mantle into an upper mantle convective circulation.

What new observations are suggested by the new view of planetary heat transfer? It is expected, although not proveable that because magma is only produced discretely in quantities of no more than a few cubic kilometres and in sheets initially very vulnerable to rapid cooling, a significant fraction of it may get remixed rather than incorporated into continental blocks. The control exerted over average viscosity values by heat transfer imply negligible rates of atomic diffusion and one can imagine that convective upper mantle material will now present an extremely complicated picture of locally differentiated material in which radiogenic isotopic ratios have been reset at a great variety of different dates. I would predict that sub oceanic mantle is not the isotopically simple substance often visualised, although having invoked the concept of crustal contamination to explain the scatter of their results, isotopic chemists may already have made it impossible to demonstrate such mantle heterogeneity. I find some recent evidence of Bougault *et al.* (1983) that sea floor basalts are isotopically heterogeneous on the length scale of a single drill hole very indicative of such a complex mantle heterogeneity, as well as supporting my view that no extensive and continually existing magma reservoir can be formed simply by decompression of the rising material beneath ocean ridges. Another consequence of remixing the material involved in individual heating events is that it may have had time to differentiate enough to be a source of seismic wave scattering. Unfortunately, one would not expect scattering to be very important at frequencies less than a few hertz, though the attempt to find it and possibly demonstrate that upper and lower mantles were different in this respect would greatly help to sort out the extent of convective mixing. It is certainly tempting to believe that a body wave Q of at least 10^3 often quoted for the lower mantle is only compatible with no mixing of the kind of scattering centres one might expect to be generated by individual heating events in the upper mantle.

I would like to end with a few remarks on the broader consequences of convective heat transfer for the Earth and planetary sciences. A great deal of time and money has been spent in the last twenty years in computing velocity patterns for a mantle model and it is still possible to find theorists who take such a deterministic view of planetary heat transfer. My experience of laboratory scale convection quickly convinced me that the determinism assumed by the modellers was often a pure illusion born of an arrogant presumption that the diversity of velocity patterns seen in many laboratory systems with nominally the same boundary conditions was due to the incompetence of the experimenter rather than something both real and important. That presumption, apart from leading a wider audience to believe that theoretical convectionists cannot make up their minds about anything, has merely delayed acceptance of the fact that there are both 'chaotic' and deterministic aspects of a solution to Earth's heat transport problem. Other subjects like meteorology and astrophysics have learnt to accept limits to the level of achievable understanding and this is clearly reflected in the way they are studied. Papers are not written on this afternoon's weather pattern simply because we live long enough to see many very different patterns and have learnt that today's may be entirely determined by extremely inauspicious events only a fortnight ago. Argument by analogy with a system in a very different dynamical regime can be very deceptive, but in the absence of any proof to the contrary and with what little one has been able to find out about self regulating planetary convection, there is no reason to doubt that most of what we observe of Earth material has been decided by unknown and unknowable details of an 'initial' state – retrodiction is every bit as illusory as prediction for a 'chaotic' dynamical system. One senses that our perspective of Earth's problems would radically change if we also lived long enough to see all the changes that accompany even small amounts of continental drift. What now seems to be required is a view of Earth material's present state as one drawn from an extremely large ensemble of distinguishable geological states, any one of which could just as easily existed had 'initial' conditions even only a few million years ago been trivially different. It is, of course, an unproveable and possibly over pessimistic assumption that any physical process, like geological change, conforms to the mathematician's idea of a 'chaotic' dynamical process, but one doesn't need to go that far to sense that a lot of observation is still motivated by the older static view of Earth as a unique work of creation worthy of mapping in every detail. In particular, 'the whole is the sum of its parts' reductionist position, motivating the collection of something called 'material property data' as a kind of prelude to the formulation of the 'true' theory of planetary dynamics, should be examined much more critically or even discarded as a naive misinterpretation of the formalism of continuum mechanics. We already know far more than enough about the behaviour of matter to realise that heat transfer equations in any plausible form for a planet can neither be rigorously solved nor length scaled to test their essential content in a laboratory experiment. Such problems raise fundamental questions about what we are to mean by scientific understanding in planetary science. On a more practical point, they lead me to believe the study of other planets, seen as systems differing by less than a factor of 100 in mass, are more likely to reveal key factors in earth's evolutionary process (and vice versa) than any amount of experience with systems having $<10^{-20}$ of their masses. It may be of no use in answering a host of pressing geological questions, but I believe we are greatly in need of a genuinely, if only modestly predictive theory of internal dynamics for planetary objects, transcending the details of individual bodies and rather like that which already exists for the larger material accumulations we call 'stars'. For me, the discovery of self regulative behaviour in the solutions to terrestrial planet heat transport problems is an important first step in the creation of such a general theory of planets. I can only hope this survey of its consequences convinces others of the value of taking a much more general approach to the bewildering complexity of planetary phenomena.

References

Allegre, C. J., Staudacher, T., Sarda, P., and Kurz, M.: 1983, *Nature* **303**, 762–766.

Bougault, H., Cande, S. C., Schilling, J-G., Turcotte, D. L., and Olson, P.: 1983, *Nature* **305**, 278–279.

Clayton, D. D.: 1982, *Quart. J. Roy. Astron. Soc.* **23**, 174–212.

Craig, H. and Lupton, J.: 1976, *Earth and Planetary Sci. Letters* **31**, 369–385.

Forsyth, D. W.: 1975, *Geophys. J. Roy. Astron. Soc.* **43**, 102–162.

Green, D. H.: 1972, *Tectonophysics* **13**, 47–71.

Mahboobi, N.: 1981, 'Viscoelastic Instabilities and the Generation of Earthquakes', Thesis, Univ. of Newcastle upon Tyne.

O'Nions, R. K. and Oxburgh, E. R.: 1983, *Nature* **306**, 429–431.

Pandit, B. I. and Tozer, D. C.: 1970, *Nature* **226**, 335–336.

Peltier, W. R. and Andrews, J. T.: 1976, *Geophys. J. Roy. Astron. Soc.* **46**, 605–646.

Raitt, R. W., Shor, G. G., Francis, T. J. G., and Morris, G. B.: 1969, *J. Geophys. Res.* **74**, 3095–3109.

Tittman, B. R.: 1977, *Phil. Trans. Roy. Soc. London. Ser. A.* **285**, 475–479.

Tozer, D. C.: 1959, *Physics and Chemistry of the Earth*, Pergamon, London, 3, pp. 414–434.

Tozer, D. C.: 1965, *Phil. Trans. Roy. Soc. London. Ser. A.* **258**, 252–262.

Tozer, D. C.: 1973, *Geofisica International Mexico* **13**, 363–388.

Tozer, D. C.: 1977, *Science Progress*, Oxford **64**, pp. 1–28.

Tozer, D. C.: 1978, *The Origin of the Solar System*, S. F. Dermott (ed.), Wiley, New York, pp. 432–462.

Tozer, D. C.: 1981, *Phys. Earth Plan. Int.* **25**, 280–296.

Truesdell, D. and Noll, W.: 1965, *Handbuch der Physik* 3, No. III, Springer, pp. 113–117.

Wetherill, G. W.: 1980, *Ann. Rev. Astron. and Astrophys.* **18**, 77–113.

Wyllie, P. J.: 1971, *The Dynamic Earth*. Wiley, New York, pp. 167–176.

S. K. RUNCORN'S COMMENTARY

The implications of solid state physics for the Earth and planetary interiors were very slow in being realised. Perhaps this is not surprising. Most of the brilliant minds going into geophysics went into seismology, because it was par excellence the happy mileau for sophisticated mathematics, and the advances in technique and data acquisition were challenging. As seismological observations of the Earth's interior could then – and even now – be entirely interpreted on the simple basis of classical elasticity and – for the core – a fluid, the longer term behaviour of the mantle was assumed to be that of a solid, or, when Jeffreys undertook his search for evidence for non elastic behaviour in planets and satellites, that of a Lomnitz solid, really the transient creep regime. I early came under the influence of Sir Nevill Mott, attending two of his summer schools at Bristol, and David Tozer in his undergraduate years at Bristol University would have come to accept the modern view of solids. Recently Sir Brian Pippard was asking me why geophysicists had been so slow in taking the evidence for continental drift seriously, and I replied that it was because they were all classical physicists and unacquainted with solid state creep. He replied that I shouldn't be too critical as when he returned to Cambridge in October 1945 after working at RRDE Malvern, no-one in the Cavendish Laboratory believed in the reality of dislocations and it was because of this that Bragg invented his bubble raft as a working model. [I had always supposed that this was because Sir Lawrence liked to lecture to children.]

David Tozer, using the sharp dependence of creep on temperature, resolved the old dilemmas on the thermal history of the Earth and terrestrial planets based on the old assumption of heat transport by conduction, and cleared away many obstacles to accommodating continental drift. Like many great advances, it all seems so obvious now and it is difficult to convince anyone entering geophysics today "what all the fuss was about".

GEOPHYSICS 2001

J. A. JACOBS

Department of Earth Sciences, University of Cambridge

Abstract. An attempt is made to try and predict those fields of research in geophysics which are most likely to prove rewarding in the next one or two decades. Topics covered are the early history of the Earth, satellite geodesy, mantle convection, geomagnetism, aeronomy and space research.

1. Introduction

It is not easy to look ahead and decide where we should put most of our effort in the years to come. It is easy to make a list of questions for which we do not know the answer – such as what is the origin of the solar system, what is the origin of the Earth's magnetic field, is it possible to predict earthquakes? We will probably never know for certain the answers to such fundamental issues, although we shall undoubtedly learn much more about them. A list of such problems would not be very helpful – it is far better to try and pinpoint those areas which hold out some hope of real progress being made by a more concerted effort and those in which little work has been done for some time but in which even a small increase in knowledge would be of value.

This meeting is concerned only with magnetism, rotation and convection in the solar system and I will confine my remarks to them. However, it should not be forgotten that seismology has the highest resolving power of all geophysical methods that are used to study the structure of the Earth and is the main tool for determining many parameters that are critically important to our understanding of the dynamic behaviour of the Earth. It is particularly difficult to determine variations of some parameters such as conductivity or viscosity and we may be forced to study related phenomena such as anelasticity or attenuation. Mechanisms of attenuation are very sensitive to thermal conditions and it is difficult to determine their radial and lateral variations in as much detail as the elastic properties. However they can offer basic information on thermodynamic conditions in the mantle. To understand attenuation in the core and solid inner core, it would be interesting to consider new physical mechanisms e.g. the effects of a magnetic field.

2. Early history of the Earth

I look forward to increased knowledge of the early history of the Earth – particularly on the question of differentiation during accretion and partial or complete melting. The greatest breakthrough in this area may come through isotopic comparative studies. Isotope geochemistry and cosmo-chemistry provide essential data for constraining speculations on the way in which planets are put together, differentiate, and evolve. Extinct radioactivities constrain the time scale of the formation of the solar system, lead

Geophysical Surveys 7 (1985) 249-255. 0046-5763/85.15.

isotope studies constrain the rapidity of core formation, and studies of the samarium-neodymium system provide tantalizing glimpses of the way in which the Earth's mantle may be layered. We should also be able to settle once and for all the question of whether the anomalous iridium-rich layer at the Cretaceous-Tertiary boundary is the result of a collision of a large asteroid with the Earth as proposed by Alvarez et al. (1980). In this connection, recent analyses by Rampino and Reynolds (1983) of the clay mineralogy of samples from the Cretaceous Tertiary boundary layer at 4 localities have shown that the boundary clay is neither mineralogically exotic nor distinct from locally derived clays above and below the boundary.

We can also hope to learn much more about the Earth from the next generation of space missions. The U.S. Committee on Planetary and Lunar Exploration gave the study of comets one of the highest priorities of future planetary research. Comets are the only obtainable source of the primitive material from which the solar system evolved, and are believed to contain a record of the physical and chemical conditions of the interstellar medium and the primordial solar nebula. Furthermore, comets may have contributed appreciable amounts of volatile material to the present atmospheres of the terrestrial planets. It has also been suggested that cometary material was important to the evolution of life on Earth.

My first interests in geophysics were concerned with the thermodynamics of the Earth's interior, and I look forward one day to a more confident history of the thermal evolution of the Earth. I am not too hopeful, however, of seeing any major developments – this would necessitate a significant breakthrough in the theory of melting which would be important in many other branches of physics and metallurgy. Improvements in our understanding of the behaviour of materials at high pressure can however be anticipated. Static and shock wave experiments provide the capability for a variety of property measurements over a range of temperatures and pressures including conditions in the Earth's core. Phase diagrams can be delineated and the partitioning of constituents determined (either directly in a diamond cell or indirectly by application of thermochemical arguments). Transport properties such as electrical conductivity are also measurable. The diamond cell anvil technique in particular, is still largely untapped and in the next decade could provide a wealth of thermodynamic data at high pressures.

3. Satellite Geodesy

The last 10–20 yr have seen tremendous developments in spatial geodesy – VLB1 and the use of lunar and satellite laser – ranging techniques. However the geophysical potential of such technological advances has not yet been fully exploited. The deployment of a network of permanent VLB1 type instruments and laser-ranging stations could lead to greatly increased improvement in the precision of measurements of polar motion. Most people believe that the Chandler wobble is a single damped free oscillation of the Earth maintained by some irregular excitation process – the recent work by Okubo (1982) confirms this conclusion. However, I don't think that we shall learn what excites the Chandler wobble, or what causes the damping of its motion although it should prove possible to settle once and for all the question of a possible relationship between earthquakes and the Chandler wobble. Here, as in the decade variations in the l.o.d., the mechanism of core mantle coupling has still not been satisfactorily resolved and deserves further attention. Again, using recent developments in spatial geodesy with stations strategically placed on different plates, it should become possible to determine whether present day plate motions agree with the average motion inferred from magnetic anomalies. It should also be possible to establish whether the motion is smooth (on a time scale of a few years) or more erratic. About 50 % of the Doppler observations made on a Navy Navigation Satellite over a 9-yr period have been analyzed to determine the motion of 8 sites on the North American plate and 12 sites on 7 other plates (Anderle and Malyevac, 1983). The computed plate motions were not statistically significant compared with the standard errors of measurement of 1 to 5 cm yr^{-1} except for the Australian, European, and Pacific plates. The measured motions of these plates were about twice those inferred from geologic records, but were in the proper direction.

An area that has been pursued at Newcastle in the past and in which I would like to see more work done is the use of palaeontological data for direct evidence of the length of the day, month and year in the geologic past. Growth increments on the shells of marine organisms are clearly observed, but can we always be sure that they really indicate diurnal, monthly and annual periods with all time increments faithfully recorded? More work on control specimens is needed and I believe would repay the effort involved.

Finally, have we yet settled the question of whether the gravitational constant G decreases with time as postulated by Dirac leading to an expansion of the Earth? Recent work by Canuto (1981) has suggested that in some theories of gravity the presently available data actually favour an increase in G with time.

4. Convection

A question which I anticipate will be resolved is whether there is whole or two-stage convection in the mantle. If the mantle is convecting throughout its volume (at the high Rayleigh number necessary to account for the observed mean surface plate speed), there should be a substantial heat flow across the core-mantle boundary. There should then be a well-developed thermal boundary layer at the base of the mantle. Seismic evidence for this is quite convincing, but further investigations are highly desirable. The existence of a thermal boundary layer at the base of the mantle could help in understanding another important characteristic of the convective circulation. Because of the combination of a sharp temperature gradient with low viscosity, this boundary layer would be convectively unstable and serve as an efficient source of small-scale thermal inhomogeneities which would rise plumelike to the Earth's surface at rates sufficiently rapid that their ascent would be quasi-adiabatic. Such thermal events, perhaps accompanied by partial melting at the base of the lithosphere, provide an attractive explanation of intraplate volcanism. The cold thermal boundary layer at the Earth's surface (the lithosphere) is affected in precisely the opposite fashion by the temperature dependence of viscosity. Here the effective viscosity is so high, due to the low surface temperature, that the lithosphere is able to withstand the large temperature contrasts across it without suffering disruption through buoyant instability.

Although the model of whole mantle convection is appealing, it does not necessarily correspond to reality. If the Earth initially accreted with a stable density stratification or if irreversible processes such as melting and partial differentiation led to subsequent stratification, the large-scale flow associated with plate motions could not penetrate throughout the mantle. The cessation of seismicity at 700 km, the compressive state of stress in some subducted slabs and the apparent segregation of the mantle into at least two geochemical reservoirs are readily explained by a stratified model. The question of whether the mantle is stratified, with each layer forming a closed convecting system, is one of the most interesting, and as yet, unresolved problems.

5. Geomagnetism

The theory of geomagnetism has developed comparatively recently. Twenty five years ago, it was shown rigorously that the dynamo mechanism can provide a feasible means for the generation of magnetic fields in electrically conducting liquid bodies, such as in the outer core of the Earth. So far there has been little comparison with laboratory data – experimental work is highly desirable for the understanding of non-linear phenomena. Work is proceeding on trying to determine the details of core motions – I believe, however, that more progress will come from a consideration of the possible driving mechanism for the geodynamo. Different mechanisms have in the past been proposed – precession, thermal convection and the one that at the moment enjoys the greatest support, gravitational differentiation of material in the core to form the solid inner core. I would hope that more definitive conclusions can be reached in the future. Experimental and theoretical work on compositionally layered, convecting systems should prove valuable in this respect.

I think that another fruitful area would be a detailed analysis of archaeomagnetic data to study intensity variations of the Earth's magnetic field. Can the intensity change by up to 50 % or more in a hundred years as some recent work has indicated? Apart from such comparatively rapid changes, it is rather surprising that the intensity of the geomagnetic field has apparently remained remarkably constant (within a factor of 2 or 3) over geologic time.

There are some very special features of the geomagnetic field that need an explanation and which deserve special study e.g.

(1) Reversals of the field – and particularly the so-called excursions or aborted reversals. Can we dismiss this problem just by saying that it is a common feature of non-linear systems?

(2) The persistent tilt of the dipole axis.

The exploration of planetary magnetic fields by space probes has given a great impetus to developing a theory of planetary magnetism – although I cannot see any hope of establishing a general scaling law. Let me list just 3 problems that have, and still need more attention paid to them:

(1) Lunar magnetism – I need say no more on this, since Professor Runcorn has long championed the idea of a small lunar core which acted as a dynamo in the moon's early history.

(2) The absence of a magnetic field on Venus. One would expect that a body similar in many respects to the Earth would have a magnetic field. It's absence cannot be put down to its much slower rotation – the Coriolis force would still be dominant.

(3) The small tilt ($< 1°$) of the dipole axis of Saturn's magnetic field.

6. Aeronomy and Space Research

It is remarkable what progress has been made in geomagnetism and aeronomy during the last 2 decades in a field where a controlled experiment is rare. Progress has come about in a variety of ways, some of which are not to be found in conventional 'laboratory' physics. The sheer vastness of the phenomena to be observed has demanded, and continues to demand, numerous observations and often the accumulation of lengthy time series of data. Whilst Gauss was able, with relatively few observations, to determine some of the harmonics of the geomagnetic field, we need, for some purposes, (e.g. the investigation of the interaction of the solar wind with the magnetic field), many orders of magnitude more information than was available to Gauss. Outside of geophysics, the exploration of the natural plasma in the environment of the Earth in space has provided, and will continue to provide, tests of physical theories and the finding of new phenomena in plasmas unattainable so far in Earth-bound laboratories. There are, of course, still many unresolved problems of geophysical interest such as: What are the conditions for creating an aurora and what is the true nature of the mechanism of the interaction of the solar wind with the geomagnetic field? We also need to know far more about the chemistry of the upper atmosphere. There is still much argument and conflicting reports of the effect of the release of chlorofluoro-carbons into the atmosphere – whether it presents a hazard to the Earth's ozone layer and entails the threat of drastic climatic changes. The same can be said for the effect of increased carbon dioxide in the atmosphere.

7. Data analysis

Geophysics is essentially an observational science. Data acquisition, analysis and storage have increased in complexity and related demands exceed the capacity of existing information-handling systems, and demand new models of collective, multi-group or even international data analysis and interpretation. The complexity and nonlinearity of many of the physical processes involved require massive studies with computer simulation and numerical modelling techniques. The demands on rapid information exchange on who is doing what, where and when are beginning to exceed the capability of present-day scientific information services, especially in their currently reduced operational state. These more general problems will affect all branches of science in the future.

Appendix

Following the discovery (Jacobs, 1973, 1981) that Lewis Carroll had made significant contributions in the field of Geophysics, further research has brought to light his early interest in plate tectonics. The following is a preliminary draft of one of his verses in Through the Looking-Glass,

The ocean floor was moving fast
Moving with all its might
It did its very best to reach
The subduction zone that night
And this was odd, because you know
No zone was yet in sight

Ray Lyttleton and Keith Runcorn
Were walking close at hand
They wept like anything to see
Such quantities of sand
Said Ray "Where did it all come from?
I cannot understand"

"The time has come", Keith Runcorn said,
"To talk of plate tectonics
It's simple if you will expand
In spherical harmonics
The motions of the continents
At speeds that are subsonic."

"Expand, expand," Lyttleton cried
"You really are exacting
You cannot build your mountains
Unless the Earth's contracting
You need no sea-floor spreading
The oceans are compacting."

"As Ramsey showed us long ago
But few Earth scientists heeded
The mantle and the core are one
No iron at all is needed
A phase change will just come about
When some pressure's exceeded."

"I weep for you," Keith Runcorn said
"I deeply sympathize
That the power of plate tectonics
You cannot realize
That you cannot see whence comes this sand
Is really no surprize."

References

Alvarez, L. W., Alvarez, W., Asaro, F., and Michel, H. V.: 1980, 'Extra Terrestrial Cause for the Cretaceous-Tertiary Extinction', *Science* **208**, 1095–1108.

Anderle, R. J. and Malyevac, C. A.: 1983, 'Current Plate Motions Based on Doppler Satellite Observations', *Geophys. Res. Letters* **10**, 67–70.

Canuto, V. M.: 1981, 'The Earth's Radius and the G Variation', *Nature* **290**, 739–744.

Jacobs, J. A.: 1973, 'A Magneto-tale', *Quart. J. Roy. Astron. Soc.* **14**, 459–460.

Jacobs, J. A.: 1981, 'A Seismo-tale', *Quart. J. Roy. Astron. Soc.* **22**, 421–422.

Okubo, S.: 1982, 'Is the Chandler Wobble Period Variable?' *Geophys. J. Roy. Astron. Soc.* **71**, 629–646.

Rampino, M. R. and Reynolds, R. C.: 1983, 'Clay Mineralogy of the Cretaceous-Tertiary Boundary Clay', *Science* **219**, 495–498.

S. K. RUNCORN'S COMMENTARY

It is most valuable to have as the final paper this futuristic contribution by Professor Jacobs one of the first geophysicists of the Canadian school I got to know.

Long a friend of the department, he spent a year at Newcastle upon Tyne as the first Faculty of Science Visiting professor – and as it turned out the last: not I hasten to add due to him, as his visit was judged a great success, not least by our research students. Sadly this enlightened innovation, which I had suggested and fought for in the University, was naturally the first casualty of the first economic cuts imposed by the Government on British universities in recent years (as our Registrar E. M. Bettenson pointed out, not wholly in jest, that he would have to have an increase in staff to administer the cuts). All will gain from pondering this contribution, but both Jack Jacobs and I know from experience how the progress of a science takes quite unexpected turns in directions, and probably one is on safest ground in predicting that now entirely separate specialisms will be finding they have common interests. Who would have supposed a few years ago that the isotope geochemists and the fluid dynamicists would be eagerly debating the relevance of the latter's observations on mid-ocean ridge basalts and 'hot spots' to the pattern of mantle convection! And who would have supposed a few years ago that lunar palaeomagnetism would have yielded evidence that the primeval Moon had a system of satellites? But Professor Jacobs is absolutely right to direct our view to the future of this great adventure in understanding the physics of the Earth and planets, to the great expansion of which all the contributors to this symposium have been privileged to make great and most varied contributions.

EPILOGUE

To those old friends who attended the meeting and have contributed these interesting papers on very fundamental questions of the Earth and planets, I now give my thanks. We have all been privileged to work in geophysics when it burst into its Spring. I am conscious that in my preface I told how our work was greatly helped by the interest of distinguished British scientists: that it is an important part of how science advances – we were fortunate in that backing. I soon began to experience that geophysics is international when I was sent by Blackett to the first International Union of Geodesy and Geophysics assembly after the war to discuss the problem of the main field. The Scandinavians had developed new geomagnetic instruments and helped our mine experiment a great deal. I remember the impressive opening by King Haakon of the assembly in the Aula of Oslo University and Edward Munch's great mural – perhaps the first great painting that struck me. Hannes Alfvén was present and took a great interest in our work.

In 1952 I was invited by the Director, Louis B. Slichter, to spend the summer at the then newly established Institute of Geophysics at UCLA, and then, in the following summer, I worked in David Grigg's laboratory on some optical measurements at high pressure in connection with the semiconducting properties of olivine. Thus began a very happy association with UCLA which I have kept up. I had already met Harold Urey in Cambridge when he came to give a colloquium and I was invited many times to lecture at what is now the Fermi Institute. In 1953 I was introduced by Professor Bucher to the work at Lamont and met Maurice Ewing and Frank Press. I thus began to be stimulated by the developments in geophysics in the United States. I have, of course, experienced criticism and jokes because of my travels: still I think as a "wandering scholar" I have contributed to the science – not least by telling friends of "the interesting work going on in the next building which they would find relevant".

Above all I gained from friendship with Harold Urey and his strong advocacy of studies of the Moon. I have written about his work elsewhere,[1] but he expanded our horizons to the solar system as a whole and, as a result of his influence, our group embarked on applying palaeomagnetism to the Apollo rocks. I believe that this work – although much criticised and widely ignored – has now yielded evidence of the internal dynamics of the Moon as striking as did the application of these methods to the Earth 25 years ago. Thus our work has come full cycle.

It was indeed my great good fortune in 1946 to go to Manchester University to work with Professor Patrick Blackett. I have written of the debt which modern geoscience owes him.[2] He was a most inspiring leader. Above all he was a man of great generosity of spirit and I think he would have liked this book and would have shared my pride in this tribute to his pupil.

S. K. RUNCORN

[1] H. C. Urey. *Biographical Memoirs Roy. Soc. London* **29**, 623–659, 1983.
[2] P. M. S. Blackett, memorial meeting. *Notes and Records of the Royal Soc.* **29**, 156–158, 1975.

ANNOUNCEMENT

Magnetism, Planetary Rotation, and Convection in the Solar System:
Retrospect and Prospect

Editor W. O'Reilly

Please note that a hardbound edition of these special issues of *Geophysical Surveys*, Vol. 7, Nos. 1, 2, and 3 (December 1984, March and June 1985), is available from the the publishers.

ISBN-13: 978-94-010-8886-2 Prices: Dfl. 140,– $49.00 £38.95.